Drug Delivery Systems

METHODS IN MOLECULAR BIOLOGY™

John M. Walker, Series Editor

459. **Prion Protein Protocols**, edited by *Andrew F. Hill, 2008*
458. **Artificial Neural Networks: Methods and Applications**, edited by *David S. Livingstone, 2008*
457. **Membrane Trafficking**, edited by *Ales Vancura, 2008*
456. **Adipose Tissue Protocols**, *Second Edition*, edited by *Kaiping Yang, 2008*
455. **Osteoporosis**, edited by *Jennifer J. Westendorf, 2008*
454. **SARS- and Other Coronaviruses: Laboratory Protocols**, edited by *Dave Cavanagh, 2008*
453. **Bioinformatics, Volume 2:** *Structure, Function, and Applications*, edited by *Jonathan M. Keith, 2008*
452. **Bioinformatics, Volume 1:** *Data, Sequence Analysis, and Evolution*, edited by *Jonathan M. Keith, 2008*
451. **Plant Virology Protocols: From Viral Sequence to Protein Function**, edited by *Gary Foster, Elisabeth Johansen, Yiguo Hong, and Peter Nagy, 2008*
450. **Germline Stem Cells**, edited by *Steven X. Hou and Shree Ram Singh, 2008*
449. **Mesenchymal Stem Cells: Methods and Protocols**, edited by *Darwin J. Prockop, Douglas G. Phinney, and Bruce A. Brunnell, 2008*
448. **Pharmacogenomics in Drug Discovery and Development**, edited by *Qing Yan, 2008*
447. **Alcohol: Methods and Protocols**, edited by *Laura E. Nagy, 2008*
446. **Post-translational Modification of Proteins: Tools for Functional Proteomics**, *Second Edition*, edited by *Christoph Kannicht, 2008*
445. **Autophagosome and Phagosome**, edited by *Vojo Deretic, 2008*
444. **Prenatal Diagnosis**, edited by *Sinhue Hahn and Laird G. Jackson, 2008*
443. **Molecular Modeling of Proteins**, edited by *Andreas Kukol, 2008*
442. **RNAi: Design and Application**, edited by *Sailen Barik, 2008*
441. **Tissue Proteomics: Pathways, Biomarkers, and Drug Discovery**, edited by *Brian Liu, 2008*
440. **Exocytosis and Endocytosis**, edited by *Andrei I. Ivanov, 2008*
439. **Genomics Protocols:** *Second Edition*, edited by *Mike Starkey and Ramnanth Elaswarapu, 2008*
438. **Neural Stem Cells: Methods and Protocols**, *Second Edition*, edited by *Leslie P. Weiner, 2008*
437. **Drug Delivery Systems**, edited by *Kewal K. Jain, 2008*
436. **Avian Influenza Virus**, edited by *Erica Spackman, 2008*
435. **Chromosomal Mutagenesis**, edited by *Greg Davis and Kevin J. Kayser, 2008*
434. **Gene Therapy Protocols: Volume 2:** *Design and Characterization of Gene Transfer Vectors*, edited by *Joseph M. LeDoux, 2008*
433. **Gene Therapy Protocols: Volume 1:** *Production and In Vivo Applications of Gene Transfer Vectors*, edited by *Joseph M. LeDoux, 2008*
432. **Organelle Proteomics**, edited by *Delphine Pflieger and Jean Rossier, 2008*
431. **Bacterial Pathogenesis: Methods and Protocols**, edited by *Frank DeLeo and Michael Otto, 2008*
430. **Hematopoietic Stem Cell Protocols**, edited by *Kevin D. Bunting, 2008*
429. **Molecular Beacons: Signalling Nucleic Acid Probes, Methods and Protocols**, edited by *Andreas Marx and Oliver Seitz, 2008*
428. **Clinical Proteomics: Methods and Protocols**, edited by *Antonio Vlahou, 2008*
427. **Plant Embryogenesis**, edited by *Maria Fernanda Suarez and Peter Bozhkov, 2008*
426. **Structural Proteomics: High-Throughput Methods**, edited by *Bostjan Kobe, Mitchell Guss, and Huber Thomas, 2008*
425. **2D PAGE: Sample Preparation and Fractionation**, *Volume 2*, edited by *Anton Posch, 2008*
424. **2D PAGE: Sample Preparation and Fractionation**, *Volume 1*, edited by *Anton Posch, 2008*
423. **Electroporation Protocols: Preclinical and Clinical Gene Medicine**, edited by *Shulin Li, 2008*
422. **Phylogenomics**, edited by *William J. Murphy, 2008*
421. **Affinity Chromatography: Methods and Protocols**, *Second Edition*, edited by *Michael Zachariou, 2008*
420. **Drosophila: Methods and Protocols**, edited by *Christian Dahmann, 2008*
419. **Post-Transcriptional Gene Regulation**, edited by *Jeffrey Wilusz, 2008*
418. **Avidin–Biotin Interactions: Methods and Applications**, edited by *Robert J. McMahon, 2008*

Drug Delivery Systems

Edited by

Kewal K. Jain, MD
Jain PharmaBiotech, Basel, Switzerland

Editor
Kewal K. Jain
Jain PharmaBiotech
Basel, Switzerland

Series Editor
John M. Walker
School of Life Sciences
University of Hertfordshire
Hatfield, Herts., UK

ISBN: 978-1-58829-891-1 e-ISBN: 978-1-59745-210-6
ISSN: 1064-3745

Library of Congress Control Number: 2007935100

Printed on acid-free paper

9 8 7 6 5 4 3 2 1

springer.com

Preface

Drug delivery systems (DDS) are an important component of drug development and therapeutics. The field is quite extensive and requires an encyclopedia to describe all the technologies. The aim of this book is to put together descriptions of important selective technologies used in DDS. Important drugs, new technologies such as nanoparticles, as well as important therapeutic applications, are taken into consideration in this selection. This book will be an important source of information for pharmaceutical scientists and pharmacologists working in the academia as well as in the industry. It has useful information for pharmaceutical physicians and scientists in many disciplines involved in developing DDS such as chemical engineering, protein engineering, gene therapy, and so on. This will be an important reference for executives in charge of research and development at several hundred companies that are developing drug delivery technologies.

Kewal K. Jain, MD

Contents

Contributors

Mavis Agbandje-McKenna, PhD
Department of Biochemistry and Molecular Biology, University of Florida, Gainesville, FL

Franklin K. Akomeah, PhD
Department of Pharmacy, King's College London, London, UK

Veronique Blouin, Diplôme EPHE
Department of Molecular Genetics and Microbiology, University of Florida, Gainesville, FL, Laboratoire de Therapie Genique, Nantes Cedex 1, France

Albertus G. de Boer, PhD
Blood-Brain Barrier Research Group, Division of Pharmacology, Leiden-Amsterdam Center for Drug Research, University of Leiden, Leiden, The Netherlands

Marc B. Brown, PhD
School of Pharmacy, University of Hertfordshire, College Lane Campus, Hatfield, Herts., UK

Nicole Brument, Diplôme EPHE
Department of Molecular Genetics and Microbiology, University of Florida, Gainesville, FL, Laboratoire de Therapie Genique, Nantes Cedex 1, France

Benjamin S. Carson, Sr., MD
Johns Hopkins Neurological Surgery, Baltimore, MD

Crispin R. Dass, PhD
Department of Orthopaedics, St. Vincent's Hospital Melbourne, Fitzroy, Vic., Australia

Pieter J. Gaillard, PhD
to-BBB technologies BV, Bio-Science Park Leiden, Leiden, The Netherlands

Michael Guarnieri, PhD
Johns Hopkins Neurological Surgery, Baltimore, MD

Kewal K. Jain, MD
Jain PharmaBiotech, Basel, Switzerland

George I. Jallo, MD
Johns Hopkins Neurological Surgery, Baltimore, MD

Patrick Y. Lu, PhD
Sirnaomics, Inc. (Advancing RNAi Technology), Rockville, MD

Gary P. Martin, PhD
Department of Pharmacy, King's College London, London, UK

William J. Murdoch, PhD
Department of Animal Science, University of Wyoming, Laramie, WY

Maciej Radosz, PhD
Soft Materials Laboratory, Department of Chemical and Petroleum Engineering, University of Wyoming, Laramie, WY

Ali R. Rajabi-Siahboomi, PhD
Colorcon, West Point, PA

Youqing Shen, PhD
Soft Materials Laboratory, Department of Chemical and Petroleum Engineering, University of Wyoming, Laramie, WY

Sunday A. Shoyele, PhD
University of Bradford, Bradford, UK

Richard O. Snyder, PhD
Department of Molecular Genetics and Microbiology, University of Florida, Gainesville, FL

Huadong Tang, PhD
Soft Materials Laboratory, Department of Chemical and Petroleum Engineering, University of Wyoming, Laramie, WY

Sandip B. Tiwari, PhD
Modified Release Technologies, Colorcon, West Point, PA

Matthew J. Traynor, PhD
School of Pharmacy, University of Hertfordshire, College Lane Campus, Hatfield, Herts., UK

Edward Van Kirk, MS
Department of Animal Science, University of Wyoming, Laramie, WY

Kim M. Van Vliet, MS
Department of Molecular Genetics and Microbiology, University of Florida, Gainesville, FL

Martin C. Woodle, PhD
Aparna Biosciences Corp., Rockville, MD

Chapter 1
Drug Delivery Systems – An Overview

Kewal K. Jain

Abstract This is an overview of drug delivery systems (DDS), starting with various routes of drug administration. Various drug formulations, as well as devices used for drug delivery and targeted drug delivery, are then described. Delivery of proteins and peptides presents special challenges. Nanoparticles are considered to be important in refining drug delivery; they can be pharmaceuticals as well as diagnostics. Refinements in drug delivery will facilitate the development of personalized medicine, which includes pharmacogenomics, pharmacogenetics, and pharmacoproteomics. The ideal DDS, commercial aspects, current achievements, challenges, and future prospects are also discussed.

Keywords Drug delivery systems; Targeted drug delivery; Nanoparticles; Nanobiotechnology; Personalized medicine; Routes of drug administration; Drug delivery devices; Controlled release; Protein/peptide delivery; Drug formulations

1 Introduction

A drug delivery system (DDS) is defined as a formulation or a device that enables the introduction of a therapeutic substance in the body and improves its efficacy and safety by controlling the rate, time, and place of release of drugs in the body. This process includes the administration of the therapeutic product, the release of the active ingredients by the product, and the subsequent transport of the active ingredients across the biological membranes to the site of action. The term therapeutic substance also applies to an agent such as gene therapy that will induce in vivo production of the active therapeutic agent. Gene therapy can fit in the basic and broad definition of a drug delivery system. Gene vectors may need to be introduced into the human body by novel delivery methods. However, gene therapy has its own special regulatory control.

Drug delivery system is an interface between the patient and the drug. It may be a formulation of the drug to administer it for a therapeutic purpose or a device used to deliver the drug. This distinction between the drug and the device is important, as it is the criterion for regulatory control of the delivery system by the drug or

From: *Methods in Molecular Biology, Vol. 437: Drug Delivery Systems*
Edited by: Kewal K. Jain © Humana Press, Totowa, NJ

medicine control agency. If a device is introduced into the human body for purposes other than drug administration, such as therapeutic effect by a physical modality or a drug may be incorporated into the device for preventing complications resulting from the device, it is regulated strictly as a device. There is a wide spectrum between drugs and devices, and the allocation to one or the other category is decided on a case by case basis.

2 Drug Delivery Routes

Drugs may be introduced into the human body by various anatomical routes. They may be intended for systemic effects or targeted to various organs and diseases. The choice of the route of administration depends on the disease, the effect desired, and the product available. Drugs may be administered directly to the organ affected by disease or given systemically and targeted to the diseased organ. A classification of various methods of systemic drug delivery by anatomical routes is shown in Table 1.1.

2.1 Oral Drug Delivery

Historically, the oral route of drug administration has been the one used most for both conventional as well as novel drug delivery. The reasons for this preference are obvious because of the ease of administration and widespread acceptance by patients. Major limitations of oral route of drug administration are as follows:

1. Drugs taken orally for systemic effects have variable absorption rates and variable serum concentrations which may be unpredictable. This has led to the development of sustained release and controlled-release systems.

Table 1.1 A classification of various anatomical routes for systemic drug delivery

Gastrointestinal system
Oral
Rectal
Parenteral
Subcutaneous injection
Intramuscular injection
Intravenous injection
Intra-arterial injection
Transmucosal: buccal and through mucosa lining the rest of gastrointestinal tract
Transnasal
Pulmonary: drug delivery by inhalation
Transdermal drug delivery
Intra-osseous infusion

2. The high acid content and ubiquitous digestive enzymes of the digestive tract can degrade some drugs well before they reach the site of absorption into the bloodstream. This is a particular problem for ingested proteins. Therefore, this route has limitations for administration of biotechnology products.
3. Many macromolecules and polar compounds cannot effectively traverse the cells of the epithelial membrane in the small intestines to reach the bloodstream. Their use is limited to local effect in the gastrointestinal tract.
4. Many drugs become insoluble at the low pH levels encountered in the digestive tract. Since only the soluble form of the drug can be absorbed into the bloodstream, the transition of the drug to the insoluble form can significantly reduce bioavailability.
5. The drug may be inactivated in the liver on its way to the systemic circulation. An example of this is the inactivation of glyceryl trinitrate by hepatic mon-oxygenase enzymes during the first pass metabolism.
6. Some drugs irritate the gastrointestinal tract and this is partially counteracted by coating.
7. Oral route may not be suitable for drugs targeted to specific organs.
8. Despite disadvantages, the oral route remains the preferred route of drug delivery. Several improvements have taken place in the formulation of drugs for oral delivery for improving their action.

2.2 *Parenteral Drug Delivery*

Parenteral literally means introduction of substances into the body by routes other than the gastrointestinal tract but practically the term is applied to injection of substances by subcutaneous, intramuscular, intravenous, and intra-arterial routes. Injections made into specific organs of the body for targeted drug delivery will be described under various therapeutic areas.

Parenteral administration of the drugs is now an established part of medical practice and is the most commonly used invasive method of drug delivery. Many important drugs are available only in parenteral form. Conventional syringes with needles are either glass or plastic (disposable). Non-reusable syringe and needle come either with autodestruct syringes, which lock after injection, or with retractable needles. Advantages of parenteral administration are as follows:

1. Rapid onset of action.
2. Predictable and almost complete bioavailability.
3. Avoidance of the gastrointestinal tract with problems of oral drug administration.
4. Provides a reliable route for drug administration in very ill and comatose patients, who are not able to ingest anything orally.

Major drawbacks of parenteral administration are as follows:

1. Injection is not an ideal method of delivery because of pain involved and patient compliance becomes a major problem.

2. Injections have limitations for the delivery of protein products, particularly those that require sustained levels.

Comments on various types of injections are given in the following text.

Subcutaneous. This involves the introduction of the drug to a layer of subcutaneous fatty tissue by the use of a hypodermic needle. Large portions of the body are available for subcutaneous injection, which can be given by the patients themselves as in the case of insulin for diabetes. Various factors that influence drug delivery by subcutaneous route are as follows:

1. Size of the molecules, as larger molecules have slower penetration rates than do smaller ones.
2. Viscosity may impede the diffusion of drugs into body fluids.
3. The anatomical characteristics of the site of injection, such as vascularity and amount of fatty tissue, influence the rate of absorption of the drug.

Subcutaneous injections usually have a lower rate of absorption and slower onset of action than intramuscular or intravenous injections. The rate of absorption may be enhanced by infiltration with the enzyme hyaluronidase. Disadvantages of subcutaneous injection are as follows:

4. The rate of absorption is difficult to control from the subcutaneous deposit.
5. Local complications, which include irritation and pain at site of injection.
6. Injection sites have to be changed frequently to avoid accumulation of the unabsorbed drug, which may cause tissue damage.

Several self-administration subcutaneous injection systems are available and include conventional syringes, prefilled glass syringes, autoinjectors, pen pumps, and needleless injectors. Subcutaneous still remains a predictable and controllable route of delivery for peptides and macromolecules.

Intramuscular injections. These are given deep into skeletal muscles, usually the deltoids or the gluteal muscles. The onset of action after intramuscular injection is faster than with subcutaneous injection but slower than with intravenous injection. The absorption of the drug is diffusion controlled but it is faster because of high vascularity of the muscle tissue. Rate of absorption varies according to physicochemical properties of the solution injected and physiological variables such as blood circulation of the muscle and the state of muscular activity. Disadvantages of intramuscular route for drug delivery are as follows:

1. Pain at the injection site.
2. Limitation of the amount injected according to the mass of the muscle available.
3. Degradation of peptides at the site of injection.
4. Complications include peripheral nerve injury and formation of hematoma and abscess at the site of injection.
5. Inadvertent puncture of a blood vessel during injection may introduce the drug directly into the blood circulation.

Most injectable products can be given intramuscularly. Numerous dosage forms are available for this route: oil in water emulsions, colloidal suspensions, and reconstituted powders. The product form in which the drug is not fully dissolved generally results in slower, more gradual absorption and slower onset of action with longer lasting effects. Intramuscularly administered drugs typically form a depot in the muscle mass from which the drug is slowly absorbed. Peak drug concentrations are usually seen from 1 to 2 h. Factors that affect the rate of release of a drug from such a depot include the following:

1. Compactness of the depot, as the release is faster from a less compact and more diffuse depot
2. Concentration and particle size of drug in the vehicle
3. Nature of solvent in the injection
4. Physical form of the product
5. The flow characteristics of the product
6. Volume of the injection

Intravenous administration. This involves injection in the aqueous form into a superficial vein or continuous infusion via a needle or a catheter placed in a superficial or deep vein. This is the only method of administration available for some drugs and is chosen in emergency situations because the onset of action is rapid following the injection. Theoretically, none of the drug is lost, and smaller doses are required than with other routes of administration. The rate of infusion can be controlled for prolonged and continuous administration. Devices are available for timed administration of intermittent doses via an intravenous catheter. The particles in the intravenous solution are distributed to various organs depending on the particle size. Particles larger than 7 μm are trapped in the lungs and those smaller than 0.1 μm accumulate in the bone marrow. Those with diameter between 0.1 and 7 μm are taken up by the liver and the spleen. This information is useful in targeting of a drug to various organs. Disadvantages of the intravenous route are as follows:

1. Immune reactions may occur following injections of proteins and peptides.
2. Trauma to veins can lead to thrombophlebitis.
3. Extravasation of the drug solution into the extravascular space may lead to irritation and tissue necrosis.
4. Infections may occur at the site of catheter introduction.
5. Air embolism may occur because of air sucked in via the intravenous line.

It is now possible to modify the kinetics of disposition and sometimes the metabolic profile of a drug given by intravenous route. This can be achieved by incorporating the drug into nanovesicles such as liposomes.

Intra-arterial. Direct injection into the arteries is not a usual route for therapeutic drug administration. Arterial puncture and injection of contrast material has been carried out for angiography. Most of the intra-arterial injections or arterial perfusions via catheters placed in arteries are for regional chemotherapy of some organs and limbs. Intra-arterial chemotherapy has been used for malignant tumors of the brain.

2.3 Transdermal Drug Delivery

Transdermal drug delivery is an approach used to deliver drugs through the skin for therapeutic use as an alternative to oral, intravascular, subcutaneous, and transmucosal routes. It includes the following categories of drug administration:

1. Local application formulations, e.g., transdermal gels
2. Penetration enhancers
3. Drug carriers, e.g., liposomes and nanoparticles
4. Transdermal patches
5. Transdermal electrotransport
6. Use of physical modalities to facilitate transdermal drug transport
7. Minimally invasive methods of transdermal drug delivery, e.g., needle-free injections

Chapter 5 of this book deals with transdermal drug delivery. A detailed description of technologies and commercial aspects of development are described in a special report on this topic [1].

2.4 Transmucosal Drug Delivery

Mucous membrane covers all the internal passages and orifices of the body, and drugs can be introduced at various anatomical sites. Only some general statements applicable to all mucous membranes will be made here and the details will be described according to the locations such as buccal, nasal, rectal.

Movement of penetrants across the mucous membranes is by diffusion. At steady state, the amount of a substance crossing the tissue per unit of time is constant and the permeability coefficients are not influenced by the concentration of the solutions or the direction of nonelectrolyte transfer. As in the epidermis of the skin, the pathways of permeation through the epithelial barriers are intercellular rather than intracellular. The permeability can be enhanced by the use surfactants such as sodium lauryl sulfate (a cationic surfactant). An unsaturated fatty acid, oleic acid, in a propylene glycol vehicle can act as a penetration enhancer for diffusion of propranolol through the porcine buccal mucosa in vitro. Delivery of biopharmaceuticals across mucosal surfaces may offer several advantages over injection techniques, which include the following:

1. Avoidance of an injection
2. Increase of therapeutic efficiency
3. Possibility of administering peptides
4. Rapid absorption when compared with oral administration
5. Bypassing first pass metabolism by the liver
6. Higher patient acceptance when compared with injectables
7. Lower cost when compared with injectables

Mucoadhesive controlled-release devices can improve the effectiveness of transmucosal delivery of a drug by maintaining the drug concentration between the effective and toxic levels, inhibiting the dilution of the drug in the body fluids, and allowing targeting and localization of a drug at a specific site. Acrylic-based hydrogels have been used extensively as mucoadhesive systems. They are well suited for bioadhesion because of their flexibility and nonabrasive characteristics in the partially swollen state, which reduce damage-causing attrition to the tissues in contact. Cross-linked polymeric devices may be rendered adhesive to the mucosa. For example, adhesive capabilities of these hydrogels can be improved by tethering of long flexible poly(ethylene glycol) chains. The ensuing hydrogels exhibit mucoadhesive properties due to enhanced anchoring of the chains with the mucosa.

Buccal and sublingual routes. Buccal absorption is dependent on lipid solubility of the nonionized drug, the salivary pH, and the partition coefficient, which is an index of the relative affinity of the drug for the vehicle than for the epithelial barrier. A large partition coefficient value indicates a poor affinity of vehicle for the drug. A small partition coefficient value means a strong interaction between the drug and the vehicle, which reduces the release of the drug from the vehicle. The ideal vehicle is the one in which the drug is minimally soluble. Buccal drug administration has the following attractive features:

1. Quick absorption into the systemic circulation with rapid onset of effect due to absorption from the rich mucosal network of systemic veins and lymphatics.
2. The tablet can be removed in case of an undesirable effect.
3. Oral mucosal absorption avoids the first pass hepatic metabolism.
4. A tablet can remain for a prolonged period in the buccal cavity, which enables development of formulations with sustained-release effect.
5. This route can be used in patients with swallowing difficulties.

Limitations to the use of buccal route are as follows:

1. The tablet must be kept in place and not chewed or swallowed.
2. Excessive salivary flow may cause a very rapid dissolution and absorption of the tablet or wash it away.
3. A bad-tasting tablet will have a low patient acceptability.
4. Some of these disadvantages have been overcome by the use of a patch containing the drug that is applied to the buccal mucosa or by using the drug as a spray.

2.5 Nasal Drug Delivery

Drugs have been administered nasally for several years both for topical and systemic effect. Topical administration includes agents for the treatment of nasal congestion, rhinitis, sinusitis, and related allergic and other chronic conditions. Various medications include corticosteroids, antihistaminics, anticholinergics, and vasoconstrictors. The focus in recent years has been on the use of nasal route for systemic drug delivery.

Surface epithelium of the nasal cavity. The anterior one third of the nasal cavity is covered by a squamous and transitional epithelium, the upper part of the cavity by an olfactory epithelium, and the remaining portion by a typical airway epithelium, which is ciliated, pseudostratified, and columnar. The columnar cells are related to neighboring cells by tight junctions at the apices as well as by interdigitations of the cell membrane. The cilia have an important function of propelling the mucous into the throat. Toxic effect of the drug on the cilia impairs the mucous clearance. Safety of drugs for nasal delivery has been studied by in vitro effect on ciliary beating and its reversibility as well as on physical properties of the mucous layer.

Another cell type characteristic of the airway epithelium is the goblet cell. The contribution of the goblet cells to the nasal secretion is less than that of the submucosal glands, which are the main source of mucous. The tight junctions of the columnar cells have gaps around filled goblet cells and this may be relevant to the absorption of aerosolized drugs which are deposited on the airway epithelium. Airway mucous is composed mostly of water but contains some proteins, inorganic salts, and lipids. The mucous layer is about 5 μm in thickness and has an aqueous phase in which the cilia beat and a superficial blanket of gel which is moved forward by the tips of cilia. Mucociliary clearance depends on the beating of cilia, which in turn is influenced by the thickness and composition of the mucous layer. This is not the only mechanism for clearing nasal mucous, and sniffing, sneezing and blowing the nose helps in moving airway secretions.

Intranasal drug delivery. Intranasal route is considered for drugs that are ineffective orally, are used chronically, require small doses, and where rapid entry into the circulation is desired. The rate of diffusion of the compounds through the nasal mucous membranes, like other biological membranes, is influenced by the physicochemical properties of the compound. However, in vivo nasal absorption of compounds of molecular weight less than 300 is not significantly influenced by the physicochemical properties of the drug. Factors such as the size of the molecule and the ability of the compound to hydrogen bond with the component of the membrane are more important than lipophilicity and ionization state. The absorption of drugs from the nasal mucosa most probably takes place via the aqueous channels of the membrane. Therefore, as long as the drug is in solution and the molecular size is small, the drug will be absorbed rapidly via the aqueous path of the membrane. The absorption from the nasal cavity decreases as the molecular size increases. Factors that affect the rate and extent of absorption of drugs via the nasal route are as follows:

1. The rate of nasal secretion. The greater the rate of secretion the lesser the bioavailability of the drug.
2. Ciliary movement. The faster the ciliary movement, the lesser the bioavailability of the drug.
3. Vascularity of the nose. Increase of blood flow leads to faster drug absorption and vice versa.
4. Metabolism of drugs in the nasal cavity. Although enzymes are found in the nasal tissues, they do not significantly affect the absorption of most compounds

except peptides which can be degraded by aminopeptidases. This may be due to low levels of enzymes and short exposure time of the drug to the enzyme.
5. Diseases affecting nasal mucous membrane. Effect of the common cold on nasal drug absorption is also an important consideration.

Enhancement of nasal drug delivery. Complete mechanism of drug absorption enhancement through the nasal mucosa is not known. Nasal drug delivery can be enhanced by reducing drug metabolism, prolonging the drug residence time in the nasal cavity, and by increasing absorption. The last is the most important strategy and will be discussed here.

Nasal drug absorption can be accomplished by use of prodrugs, chemical modification of the parent molecule, and use of physical methods of increasing permeability. Special excipient used in the nasal preparations comes into contact with the nasal mucosa and may exert some effect to facilitate the drug transport. The mucosal pores are easier to open than those in the epidermis. The following characteristics should be considered in choosing an absorption enhancer:

1. The enhancer should be pharmacologically inert.
2. It should be nonirritating, nontoxic, and nonallergic
3. Its effect on the nasal mucosa should be reversible.
4. It should be compatible with the drug.
5. It should be able to remain in contact with the nasal mucosa long enough to achieve maximal effects.
6. It should not have any offensive odor or taste.
7. It should be relatively inexpensive and readily available.

The effect of nasal absorption enhancers on ciliary beating needs to be tested, as any adverse effect on mucociliary clearance will limit the patient's acceptance of the nasal formulation. Chitosan, a naturally occurring polysaccharide that is extracted from the shells of crustaceans, is an absorption enhancer. It is bioadhesive and binds to the mucosal membrane, prolonging retention time of the formulation on the nasal mucosa. Chitosan may also facilitate absorption through promoting paracellular transport or through other mechanisms. The chitosan nasal technology can be exploited as solution, dry powders, or microsphere formulations to further optimize the delivery system for individual compounds. Impressive improvements in bioavailability have been achieved with a range of compounds. For compounds requiring rapid onset of action, the nasal chitosan technology can provide a fast peak concentration, compared with oral or subcutaneous administration.

Advantages of nasal drug delivery:

1. High permeability of the nasal mucosa, compared with the epidermis or the gastrointestinal mucosa
2. Highly vascularized subepithelial tissue
3. Rapid absorption, usually within half an hour
4. Avoidance of first pass effect that occurs after absorption of drugs from the gastrointestinal tract

5. Avoidance of the effects of gastric stasis and vomiting, for example, in migraine patients
6. Ease of administration by the patients, who are usually familiar with nasal drops and sprays
7. Higher bioavailability of the drugs than in the case of gastrointestinal route or pulmonary route
8. Most feasible route for the delivery of peptides

Disadvantages of nasal drug delivery:

1. Diseases conditions of the nose may result in impaired absorption.
2. Dose is limited because of relatively small area available for absorption.
3. Time available for absorption is limited.
4. Little is known of the effect of common cold on transnasal drug delivery, and it is likely that instilling a drug into a blocked nose or a nose with surplus of watery rhinorrhea may expel the medication from the nose.
5. The nasal route of delivery is not applicable to all drugs. Polar drugs and some macromolecules are not absorbed in sufficient concentration because of poor membrane permeability, rapid clearance, and enzymatic degradation into the nasal cavity.

Alternative means that help overcome these nasal barriers are currently in development. Absorption enhancers such as phospholipids and surfactants are constantly used, but care must be taken in relation to their concentration. Drug delivery systems, including liposomes, cyclodextrins, and micro- and nanoparticles are being investigated to increase the bioavailability of drugs delivered intranasally [2].

After a consideration of advantages as well as disadvantages, nasal drug delivery turns out to be a promising route of delivery and competes with pulmonary drug, which is also showing great potential. One of the important points is the almost complete bioavailability and precision of dosage.

2.6 Colorectal Drug Delivery

Although drug administration to the rectum in human beings dates back to 1,500 B.C., majority of pharmaceutical consumers are reluctant to administer drugs directly by this route. However, the colon is a suitable site for the safe and slow absorption of drugs which are targeted at the large intestine or designed to act systematically. Although the colon has a lower absorption capacity than the small intestine, ingested materials remain in the colon for a much longer time. Food passes through the small intestine within a few hours but it remains in the colon for 2–3 days. Basic requirements of drug delivery to the colorectal area are as follows:

1. The drug should be delivered to the colon either in a slow release or targeted form ingested orally or introduced directly by an enema or rectal suppository.

2. The drug must overcome the physical barrier of the colonic mucous.
3. Drugs must survive metabolic transformation by numerous bacterial species resident in the colon, which are mainly anaerobes and possess a wide range of enzymatic activities.

Factors that influence drug delivery to colorectal area:

1. The rate of absorption of drugs from the colon is influenced by the rate of blood flow to and from the absorptive epithelium.
2. Dietary components such as complex carbohydrates trap molecules within polysaccharide chains.
3. Lipid-soluble molecules are readily absorbed by passive diffusion.
4. The rate of gastric emptying and small bowel transit time.
5. Motility patterns of the colon determine the rate of transit through the colon and hence the residence time of a drug and its absorption.
6. Drug absorption varies according to whether the drug is targeted to the upper colon, lower colon. or the rectum.

Drugs administered by rectal route. Advantages of the rectal route for drug administration are as follows:

1. A relatively large amount of the drug can be administered.
2. Oral delivery of drugs that are destroyed by the stomach acid and/or metabolized by pancreatic enzymes.
3. This route is safe and convenient particularly for the infants and the elderly.
4. This route is useful in the treatment of emergencies such as seizures in infants when the intravenous route is not available.
5. The rate of drug absorption from the rectum is not influenced by ingestion of food or rate of gastric emptying.
6. The effect of various adjuvants is generally more effective in the rectum than in the upper part of the gastrointestinal tract.
7. Drugs absorbed from the lower part of the rectum bypass the liver.
8. Degradation of the drugs is much less in the rectal lumen than in the upper gastrointestinal tract.

Disadvantages of the rectal route for drug administration are as follows:

1. Some hydrophilic drugs such as antibiotics and peptide drugs are not easily absorbed from the rectum and absorption enhancers are required.
2. Drugs may cause rectal irritation and sometimes proctitis with ulceration and bleeding.

Drugs targeted for action in the colon can also be administered orally. Oral drug delivery to the colon has attracted significant attention during the past 20 years. Colon targeting is recognized to have several therapeutic advantages, such as the oral delivery of drugs that are destroyed by the stomach acid and/or metabolized by pancreatic enzymes. Sustained colonic release of drugs can be useful in the treatment of nocturnal asthma, angina, and arthritis. Local treatment of colonic pathologies, such as ulcerative colitis, colorectal cancer, and Crohn's disease, is more effective with the delivery of drugs to the affected area [3]. Likewise, colonic delivery of vermicides and colonic diagnostic agents requires smaller doses.

2.7 *Pulmonary Drug Delivery*

Although aerosols of various forms for treatment of respiratory disorders have been in use since the middle of the twentieth century, the interest in the use of pulmonary route for systemic drug delivery is recent. Interest in this approach has been further stimulated by the demonstration of potential utility of lung as a portal for entry of peptides and the feasibility of gene therapy for cystic fibrosis. It is important to understand the mechanism of macromolecule absorption by the lungs for an effective use of this route.

2.7.1 Mechanisms of Macromolecule Absorption by the Lungs

The lung takes inhaled breaths of air and distributes them deep into the tissue to a very large surface, known as the alveolar epithelium, which is ~100 m^2 in adults. This very large surface has approximately a half billion tiny air sacs known as alveoli, which are enveloped by an equally large capillary network. The delivery of inhaled air to the alveoli is facilitated by the airways, which start with the single trachea and branch several times to reach the grape-like clusters of tiny alveoli. The alveolar volume is 4,000–6,000 ml when compared to the airway volume of 400 ml, thus providing a greater area for absorption for the inhaled substances. Large molecule drugs, such as peptides and proteins, do not easily pass through the airway surface because it is lined with a thick, ciliated mucus-covered cell layer making it nearly impermeable. The alveoli, on the other hand, have a thin single cellular layer enabling absorption into the bloodstream. Some barriers to the absorption of substances in the alveoli are as follows:

1. Surfactant, a thin layer at the air/water interface, may trap the large molecules.
2. A molecule must traverse the surface lining fluid which is a reservoir for the surfactant and contains many components of the plasma as well as mucous.
3. The single layer of epithelial cells is the most significant barrier.
4. The extracellular space inside the tissues and the basement membrane to which the epithelial cells are attached.
5. The vascular endothelium, which is the final barrier to systemic absorption, is more permeable to macromolecules than is the pulmonary epithelium.

Although the mechanism of absorption of macromolecules by the lungs is still poorly understood, the following mechanisms are considered to play a part:

1. Transcytosis (passage through the cells). This may occur and may be receptor-mediated but it is not very significant for macromolecules >40 kDa.
2. Paracellular absorption. This is usually thought to occur through the junctional complex between two cells. The evidence for this route of absorption is not very convincing in case of the lungs. Molecules smaller than 40 kDa may enter via the junctional pores.

Once past the epithelial barrier, the entry of macromolecules into the blood is easier to predict. Venules and lymph vessels provide the major pathway for absorption. Direct absorption may also occur across the tight junctions of capillary endothelium.

2.7.2 Pharmacokinetics of Inhaled Therapeutics for Systemic Delivery

An accurate estimation of pharmacokinetics of inhaled therapeutics for systemic delivery is a challenging experimental task. Various models for in vivo, in vitro, and ex vivo study of lung absorption and disposition for inhaled therapeutic molecules have been described [4]. In vivo methods in small rodents continue to be the mainstay of assessment, as it allows direct acquisition of pharmacokinetic data by reproducible dosing and control of regional distribution in the lungs through use of different methods of administration. In vitro lung epithelial cell lines provide an opportunity to study the kinetics and mechanisms of transepithelial drug transport in more detail. The ex vivo model of the isolated perfused lung resolves some of the limitations of in vivo and in vitro models. While controlling lung-regional distributions, the preparation alongside a novel kinetic modeling analysis enables separate determinations of kinetic descriptors for lung absorption and nonabsorptive clearances, i.e., mucociliary clearance, phagocytosis, and/or metabolism. There are advantages and disadvantages of each model, and scientists must make appropriate selection of the best model at each stage of the research and development program, before proceeding to clinical trials for future inhaled therapeutic entities for systemic delivery.

2.7.3 Advantages of Pulmonary Drug Delivery

Advantages of lungs for drug delivery are as follows:

1. Large surface area available for absorption.
2. Close proximity to blood flow.
3. Avoidance of first pass hepatic metabolism.
4. Smaller doses are required than by the oral route to achieve equivalent therapeutic effects.

2.7.4 Disadvantages of Pulmonary Drug Delivery

Disadvantages of pulmonary drug delivery are as follows:

1. The lungs have an efficient aerodynamic filter, which must be overcome for effective drug deposition to occur.
2. The mucous lining the pulmonary airways clears the deposited particles toward the throat.
3. Only 10–40% of the drug leaving an inhalation device is usually deposited in the lungs by using conventional devices.

2.7.5 Techniques of Systemic Drug Delivery via the Lungs

Drugs may be delivered to the lungs for local treatment of pulmonary conditions, but here the emphasis is on the use of lungs for systemic drug delivery. Simple inhalation devices have been used for inhalation anesthesia, and aerosols containing various drugs have been used in the past. The current interest in delivery of peptides and proteins by this route has led to the use of dry powder formulations for deposition in the deep lung, which requires placement within the tracheal bronchial tree rather than simple aerosol inhalation. Various technologies that are in development for systemic delivery of drugs by pulmonary route are as follows:

Dry powders. For many drugs, more active ingredients can be contained in dry powders than in liquid forms. In contrast to aqueous aerosols, where only 1–2% of the aerosol particle is drug (the rest is water), dry powder aerosol particles can contain up to 50–95% of pure drug. This means that therapeutic doses of most drugs can be delivered as a dry powder aerosol in one to three puffs. Dry powder aerosols can carry ~5 times more drug in a single breath than can metered dose inhaler (MDI) systems and many more times than can currently marketed liquid or nebulizer systems. It is possible that a dry powder system for drugs requiring higher doses, such as insulin or α1-antitrypsin, could decrease dosing time when compared with nebulizers. For example, delivery of insulin by nebulizer requires many more puffs per dose, e.g., up to 50–80 per dose in one study of diabetics. A final reason for focusing on dry powders concerns the microbial growth in the formulation. The risk of microbial growth, which can cause serious lung infections, is greater in liquids than in solids.

Inhalers. Various aerosols can deliver liquid drug formulations. The liquid units are inserted into the device which generates the aerosol and delivers it directly to the patient. This avoids any problems associated with converting proteins into powders. This method has applications in delivery of morphine and insulin.

Controlled-release pulmonary drug delivery. This is suitable for drug agents that are intended to be inhaled, for either local action in lungs or for systemic absorption. Potential applications for controlled release of drugs delivered through the lungs are as follows:

1. It enables reduction of dosing frequency for drugs given several times per day.
2. It increases the half-life of drugs which are absorbed very rapidly into the blood circulation and are rapidly cleared from blood.
3. An inhaled formulation may lead to the development of products that might otherwise be abandoned because of unfavorable pharmacokinetics.
4. Pulmonary controlled release could decrease development cycles for drug molecules by obviating the need for chemical modification.

2.7.6 Conclusions and Future Prospects of Pulmonary Drug Delivery

The pulmonary route for drug administration is now established for systemic delivery of drugs. A wide range of drugs can be administered by this route, but the special attraction is for the delivery of peptides and proteins. Considering the growing number of peptide and protein therapeutic products, several biotechnology companies will get involved in this area. Advances in the production of dry powder formulation will be as important as design of devices for delivery of drugs to the lungs. Effervescent carrier particles can be synthesized with an adequate particle size for deep lung deposition. This opens the door for future research to explore this technology for delivery of a large range of substances to the lungs with possible improved release compared to conventional carrier particles [5].

Issues of microparticle formation for lung delivery will become more critical in the move from chemically and physically robust small particles to more sensitive and potent large molecules. In spite of these limitations, pulmonary delivery of biopharmaceuticals is an achievable and worthwhile goal. Nanoparticles have been investigated for pulmonary drug delivery, but there is some concern about the adverse effects of nanoparticle inhalation and these issues are under investigation. Available evidence suggests that biodegradable polymeric nanoparticles designed for pulmonary drug delivery may not induce the same inflammatory response as does nonbiodegradable polystyrene particles of comparable size [6].

Drugs other than biotherapeutics are being developed for inhalation and include treatments either on the market or under development to reduce the symptoms of influenza, to minimize nausea and vomiting following cancer chemotherapy, and to provide vaccinations. Future applications could find inhalable forms of antibiotics to treat directly lung diseases such as tuberculosis with large, local doses. Or medications known to cause stomach upsets could be packaged for inhalation, including migraine pain medications, erythromycin, or antidepressants. Inhalable drugs hold the possibility of eliminating common side effects of oral dosages, including low solubility, interactions with food, and low bioavailability. Because inhalables reach the blood stream faster than pills and some injections, many medical conditions, including pain, spasms, anaphylaxis, and seizures, could benefit from fast-acting therapies.

The medicine cabinet of the future may hold various types of inhalable drugs that will replace not only dreaded injections, but also drugs with numerous side effects when taken orally. New approaches will lend support to the broad challenge of delivering biotherapeutics and other medications to the lungs.

2.8 Cardiovascular Drug Delivery

Drug delivery to the cardiovascular system is different from delivery to other systems because of the anatomy and physiology of the vascular system; it supplies blood and nutrients to all organs of the body. Drugs can be introduced into the vascular system for systemic effects or targeted to an organ via the regional blood supply. In addition to the usual formulations of drugs such as controlled release, devices are used as well. A considerable amount of cardiovascular therapeutics, particularly for major and serious disorders, involves the use of devices. Some of these may be implanted by surgery whereas others are inserted via minimally invasive procedures involving catheterization. Use of sophisticated cardiovascular imaging systems is important for the placement of devices. Drug delivery to the cardiovascular system is not simply formulation of drugs into controlled release preparation but it includes delivery of innovative therapeutics to the heart. Details of cardiovascular drug delivery are described elsewhere [7].

Methods for local administration of drugs to the cardiovascular system include the following:

1. Drug delivery into the myocardium: direct intramyocardial injection, drug-eluting implanted devices
2. Drug delivery via coronary venous system
3. Injection into coronary arteries via cardiac catheter
4. Intrapericardial drug delivery
5. Release of drugs into arterial lumen from drug-eluting stents

2.9 Drug Delivery to the Central Nervous System

The delivery of drugs to the brain is a challenge in the treatment of central nervous system (CNS) disorders. The major obstruction to CNS drug delivery is the blood-brain barrier, which limits the access of drugs to the brain substance. In the past, treatment of CNS disease was mostly by systemically administered drugs. This trend continues. Most CNS-disorder research is directed toward the discovery of drugs and formulations for controlled release; little attention has been paid to the method of delivery of these drugs to the brain. Various methods of delivering drugs to the CNS are shown in Table 1.2 and are described in detail elsewhere [8].

2.10 Intra-osseous Infusion

The use of this route was initially limited to young children because of the replacement of the red marrow by the less vascular yellow marrow at the age of five years. Intra-osseous (IO) infusion provides an alternative route for the administration of

Table 1.2 Various methods of drug delivery to the central nervous system (CNS)

Systemic administration of therapeutic substances for CNS action
Intravenous injection for targeted action in the CNS
Direct administration of therapeutic substances to the CNS
Introduction into cerebrospinal fluid pathways: intraventricular, subarachnoid pathways
Introduction into the cerebral arterial circulation
Introduction into the brain substance
Direct positive pressure infusion
Drug delivery by manipulation of the blood-brain barrier
Drug delivery using novel formulations
Conjugates
Gels
Liposomes
Microspheres
Nanoparticles
Chemical delivery systems
Drug delivery devices
Pumps
Catheters
Implants releasing drugs
Use of microorganisms for drug delivery to the brain
Bacteriophages for brain penetration
Bacterial vectors
Cell therapy
CNS implants of live cells secreting therapeutic substances
CNS implants of encapsulated genetically engineered cells producing therapeutic substances
Cells for facilitating crossing of the blood-brain barrier
Gene transfer
Direct injection into the brain substance
Intranasal instillation for introduction into the brain along the olfactory tract
Targeting of CNS by retrograde axonal transport

fluids and medications when difficulty with peripheral or central lines is encountered during resuscitation of critically ill and injured patients. The anatomical basis of this approach is that the sinusoids of the marrow of long bones drain into the systemic venous system via medullary venous channels. Substances injected into the bone marrow are absorbed almost immediately into the systemic circulation. The technique involves the use of a bone marrow aspiration needle in the tibia bone of leg or the sternum. The advantage of this route is that the marrow cavity functions as a rigid vein that does not collapse like the peripheral veins in case of shock and vascular collapse.

Now IO infusion can be given into the sternum in adults. Indications for use included adult patient, urgent need for fluids or medications, and unacceptable delay or inability to achieve standard vascular access. The overall success rate for achieving vascular access with the system is high and no complications or complaints have been reported. Sternal IO infusion may provide rapid, safe vascular access and may be a useful technique for reducing unacceptable delays in the provision of emergency treatment. This route has not been developed for drug delivery.

2.11 Concluding Remarks on Routes of Drug Delivery

A comparison of common routes of drug delivery is shown in Table 1.3. Owing to various modifications of techniques, the characteristics can be changed from those depicted in this table. For example, injections can be needle-less and do not have the discomfort, leading to better compliance.

3 Drug Formulations

There is constant evolution of the methods of delivery, which involves modifications of conventional methods and discovery of new devices. Some of the modifications of drugs and the methods of administration will be discussed in this section. A classification of technologies that affect the release and availability of drugs is shown in Table 1.4.

3.1 Sustained Release

Sustained release (SR) preparations are not new but several new modifications are being introduced. They are also referred to as "long acting" or "delayed release" when compared to "rapid" or "conventional" release preparations. The term sometimes overlaps with "controlled release," which implies more sophisticated control of release and not just confined to the time dimension. Controlled release implies consistency, but release of drug in SR preparations may not be consistent. The following are the rationale of developing SR:

1. To extend the duration of action of the drug
2. To reduce the frequency of dosing
3. To minimize the fluctuations in plasma level
4. Improved drug utilization
5. Less adverse effects

Limitations of SR products are as follows:

1. Increase of drug cost.
2. Variation in the drug level profile with food intake and from one subject to another.

Table 1.3 Comparison of major routes of drug delivery for systemic absorption

Issue	Oral	Intravenous	Intramuscular/ subcutaneous	Transnasal	Transdermal	Pulmonary
Delivery to blood circulation	Indirect through GI tract	Direct	Indirect absorption from tissues	Indirect	Indirect	Indirect
Onset of action	Slow	Rapid	Moderate to rapid	Rapid	Moderate to rapid	Rapid
Bioavailability	Low to high	High	High	Moderate	Low	Moderate to high
Dose control	Moderate	Good	Moderate	Moderate to good	Poor	Moderate to good
Administration	Self	Health professional	Self or health professional	Self	Self	Self
Patient convenience	High	Low	Low	High	Moderate	High
Adverse effects	GI upset	Acute reactions	Acute reactions	Insignificant	Skin irritation	Insignificant
Use for proteins and peptides	No	Yes	Yes	Yes	No	Yes

GI gastrointestinal

Table 1.4 Classification of DDS that affect the release and availability of drugs

Systemic versus localized drug delivery
General nontargeted delivery to all tissues
Targeted delivery to a system or organ
Controlled release delivery systems (systemic delivery)
Release on timescale
Immediate release
Programmed release at a defined time/pulsatile release
Delayed, sustained, or prolonged release, long acting
Targeted release (see also drug delivery devices)
Site-specific controlled release following delivery to a target organ
Release in response to requirements or feedback
Receptor-mediated targeted drug delivery
Type of drug delivery device

3. The optimal release form is not always defined, and multiplicity of SR forms may confuse the physician as well as the patient.
4. SR is achieved by either chemical modification of the drug or modifying the delivery system, e.g., use of a special coating to delay diffusion of the drug from the system. Chemical modification of drugs may alter such properties as distribution, pharmacokinetics, solubility, or antigenicity. One example of this is attachment of polymers to the drugs to lengthen their lifetime by preventing cells and enzymes from attacking the drug.

3.2 Controlled Release

Controlled release implies regulation of the delivery of a drug usually by a device. The control is aimed at delivering the drug at a specific rate for a definite period of time independent of the local environments. The periods of delivery are usually much longer than in case of SR and vary from days to years. Controlled release may also incorporate methods to promote localization of drug at an active site. Site-specific and targeted delivery systems are the descriptive terms used to denote this type of control.

3.3 Programming the Release at a Defined Time

Approaches used for achieving programmed or pulsatile release may be physical mechanisms such as swelling with bursting or chemical actions such as enzymatic degradation. Capsules have been designed that burst after a predetermined exposure to an aqueous environment. Physical factors that can be controlled are the radius of the sphere, osmotic pressure of the contents, and wall thickness as well as elasticity.

Various pulsatile release methods for oral drug delivery include the Port system (a semipermeable capsule containing an osmotic charge and an insoluble plug) and Chronset system (an osmotically active compartment in a semipermeable cap).

3.4 Prodrugs

A prodrug is a pharmacologically inert form of an active drug that must undergo transformation to the parent compound in vivo either by a chemical or an enzymatic reaction to exert its therapeutic effect. The following are required for a prodrug to be useful for site-specific delivery:

1. Prodrug must have adequate access to its pharmacological receptors.
2. The enzyme or chemical responsible for activating the drug should be active only at the target site.
3. The enzyme should be in adequate supply to produce the required level of the drug to manifest its pharmacological effects.
4. The active drug produced at the target site should be retained there and not diffuse into the systemic circulation.

An example of prodrugs is L-dopa, the precursor of dopamine, which when administered orally, is distributed systemically. Its conversion to dopamine in the corpus striatum of the brain produces the desired therapeutic effects.

3.5 Novel Carriers and Formulations for Drug Delivery

Various novel methods of delivery have evolved since the simple administration of pills and capsules as well as injections. These involve formulations shown in Table 1.5 and carriers shown in Table 1.6. Biodegradable implants are shown in Table 1.7.

Table 1.5 Novel preparations for improving bioavailability of drugs

Oral drug delivery
Fast-dissolving tablets
Technologies to increase gastrointestinal retention time
Technologies to improve drug release mechanisms of oral preparations
Adjuvants to enhance absorption
Methods of increasing bioavailability of drugs
Penetration enhancement
Improved dissolution rate
Inhibition of degradation prior to reaching site of action
Production of therapeutic substances inside the body
Gene therapy
Cell therapy

Table 1.6 Novel carriers for drug delivery

Novel carriers for drug delivery
Polymeric carriers for drug delivery
Collagen
Particulate drug delivery systems: microspheres
Nanobiotechnology-based methods, including nanoparticles such as liposomes
Glass-like sugar matrices
Resealed red blood cells
Antibody-targeted systems

Table 1.7 Biodegradable implants for controlled sustained drug delivery

Biodegradable implants for controlled sustained drug delivery
Injectable implants
Gels
Microspheres
Surgical implants
Sheets/films
Foams
Scaffolds

3.6 Ideal Properties of Material for Drug Delivery

Properties of an ideal macromolecular drug delivery or biomedical vector are as follows:

1. Structural control over size and shape of drug or imaging-agent cargo-space.
2. Biocompatible, nontoxic polymer/pendant functionality.
3. Precise, nanoscale-container and/or scaffolding properties with high drug or imaging-agent capacity features.
4. Well-defined scaffolding and/or surface modifiable functionality for cell-specific targeting moieties.
5. Lack of immunogenicity.
6. Appropriate cellular adhesion, endocytosis, and intracellular trafficking to allow therapeutic delivery or imaging in the cytoplasm or nucleus.
7. Acceptable bioelimination or biodegradation.
8. Controlled or triggerable drug release.
9. Molecular level isolation and protection of the drug against inactivation during transit to target cells.
10. Minimal nonspecific cellular and blood-protein binding properties.
11. Ease of consistent, reproducible, clinical-grade synthesis.

3.7 Innovations for Improving Oral Drug Delivery

3.7.1 Fast-Dissolving Tablets

Fast-disintegration technology is used for manufacturing these tablets. The advantages of fast-dissolving tablets are as follows:

1. Convenient to take without use of water
2. Easier to take by patients who cannot swallow
3. Rapid onset of action due to faster absorption
4. Less gastric upset because the drug is dissolved before it reaches the stomach
5. Improved patient compliance

3.7.2 Softgel Formulations

Capsules and other protective coatings have been used to protect the drugs in their passage through the upper gastrointestinal tract for delayed absorption. The coatings also serve to reduce stomach irritation. The softgel delivers drugs in solution and yet offers advantages of solid dosage form. Softgel capsules are particularly suited for hydrophobic drugs which have poor bioavailability because these drugs do not dissolve readily in water and gastrointestinal juices. If hydrophobic drugs are compounded in solid dosage forms, the dissolution rate may be slow, absorption is variable, and the bioavailability is incomplete. Bioavailability is improved in the presence of fatty acids, e.g., mono- or diglycerides. Fatty acids can solubilize hydrophobic drugs such as hydrochlorothiazide, isotretinoin, and griseofulvin in the gut and facilitate rapid absorption. Hydrophobic drugs are dissolved in hydrophilic solvent and encapsulated. When softgels are crushed or chewed, the drug is released immediately in the gastric juice and is absorbed from the gastrointestinal tract into the blood stream. This results in rapid onset of desired therapeutic effects. Advantages of softgels over tablets are as follows:

1. The development time for softgel is shorter because of lower bioavailability concerns, and such solutions can be marketed at a fraction of cost.
2. Softgel formulations, e.g., that of ibuprofen, have a shorter time to peak plasma concentration and greater peak plasma concentration when compared with a marketed tablet formulation. Cyclosporin in softgel form can produce therapeutic levels in blood that are not achievable from tablet form. Similarly, oral hypoglycemic glipizide in softgel form is known to have better bioavailability results when compared with tablet form.
3. Softgel delivery systems can also incorporate phospholipids or polymers or natural gums to entrap the drug active in the gelatin layer with an outer coating to give desired delayed/controlled-release effects.

Advantages of softgel capsule over other hardshell capsules are as follows:

1. Sealed tightly in automatic manner
2. Easy to swallow
3. Allow product identification, using colors and several shapes
4. Better stability than other oral delivery systems
5. Good availability and rapid absorption
6. Offer protection against contamination, light, and oxidation
7. Unpleasant flavors are avoided because of content encapsulation

3.7.3 Improving Drug Release Mechanisms of Oral Preparations

Drug release rates of orally administered products tend to decrease from the matrix system as a function of time based on the nature and method of preparation. Various approaches to address the problems associated with drug release mechanisms and release rates use geometric configurations, including the cylindrical rod method and the cylindrical donut method. The three-dimensional printing (3DP) provides the following advantages:

1. Zero-order drug delivery
2. Patterned diffusion gradient by microstructure diffusion barrier technique
3. Cyclic drug release

3DP method utilizes ink-jet printing technology to create a solid object by printing a binder into selected areas of sequentially deposited layers of powder. The active agent can be embedded into the device either as dispersion along the polymeric matrix or as discrete units in the matrix structure. The drug release mechanism can be tailored for a variety of requirements such as controlled release by a proper selection of polymer material and binder material.

3.8 *Drug Delivery Devices*

One of the most obvious ways to provide sustained-release medication is to place the drug in a delivery device and implant the system into body tissue. A classification of drug delivery devices is shown in Table 1.8.

The concept of drug delivery devices is old, but new technologies are being applied. Surgical techniques and special injection devices are sometimes required for implantation. The materials used for these implants must be biocompatible,

Table 1.8 Classification of drug delivery devices

Surgically implanted devices for prolonged sustained drug release
Drug reservoirs
Surgically implanted devices for controlled/intermittent drug delivery
Pumps and conduits
Implants for controlled release of drugs (nonbiodegradable)
Implantable biosensor-drug delivery system
Microfluidics device for drug delivery
Controlled-release microchip
Implants that could benefit from local drug release
Vascular stents: coronary, carotid, and peripheral vascular
Ocular implants
Dental implants
Orthopedic implants

i.e., the polymers used should not cause any irritation at the site of implantation or promote an abscess formation. Subcutaneous implantation is currently one of the routes utilized to investigate the potential of sustained delivery systems. Favorable absorption sites are available and the device can be removed at any time. Most notable implantable product is Norplant (Wyeth), a contraceptive device releasing levonorgestrel for up to 5 years. However, acceptance of Norplant has been less than optimal after its initial success, and some of the reasons for this are as follows:

1. Because of cultural differences in populations around the world, the use of a preparation approved in a developed country may not be appropriate in a developing country.
2. Women with implants are less likely to have annual Papanicolaou smears because they do not revisit their doctors as often as they do when using another form of contraceptive.
3. Serious adverse events have been reported in some implant recipients.

A variety of other drugs have been implanted subcutaneously, including thyroid hormones, cardiovascular agents, insulin, and nerve growth factor. Some implantable devices extend beyond simple sources of drug diffusion. Some devices can be triggered by changes in osmotic pressure to release insulin, and pellets can be activated by magnetism to release their encapsulated drug load. Such external control of an embedded device would eliminate many of the disadvantages of most implantable drug delivery systems.

3.8.1 Implantable Biosensor-Drug Delivery System

Implantable biosensor-drug delivery system (ChipRx Inc.) integrates genetically engineered reagents with drug storage into an implantable biosensor-drug reservoir system constructed from electroactive polymers (EAPs) that could be implanted into the body for controlled release of medication. The device is the size of a small matchstick and comes equipped with a sensor and a battery and is covered with a series of EAP valves. When the sensor detects a certain chemical change, it signals the battery, which emits an electrical charge. This charge activates the polymer valves, causing them to flap open and expose tiny perforations on the capsule surface. Medication stored in the capsule then seeps through the perforations until the sensor determines that a sufficient amount has been released. The sensor signals the battery again, which triggers the polymer flaps to close; the perforations are covered and the flow of medication stops. Telemetry enables physician/patient regulation of drug release. The flagship product of ChipRx is a fully integrated, self-regulated therapeutic system that eliminates the need for telemetry and human intervention. This system is a true "responsive therapeutic device"; biosensors, electronic feedback, and drug/countermeasure release are fully integrated.

3.8.2 Drug Delivery Device Based on Microfluidics

Computer-logic-like circuits, which control the flow of fluid through a chamber rather than the flow of electricity through a solid, have been constructed and have potential application in drug delivery [9]. The microfluidic circuits could eventually be used to deliver constant flows of medicine to specific points in the human body and to control other microfluidic devices. The key to the circuit-like behavior is an elastic polymer fluid that has nonlinear properties similar to those of electronics components. In a linear system the output is proportional to the input; nonlinear output, however, increases or decreases at a different rate than the input. The fluid circuits have different-shaped channels that cause the molecules of the elastic fluid to align or scramble, changing the fluid's viscosity and therefore its flow rate. Such miniaturized fluidic circuits are insensitive to electromagnetic interference and may also find medical applications for implanted drug delivery devices. No commercial development has been reported so far. As miniaturization continues to nanoscale, a microelectromechanical systems micropump with circular bossed membrane designed for nanoliter drug delivery has been characterized [10].

3.8.3 Controlled-Release Microchip

The conventional controlled drug release from polymeric materials is in response to specific stimuli such as electric and magnetic fields, ultrasound, light, enzymes. Microchip technology has been applied to achieve pulsatile release of liquid solutions. A solid-state silicon microchip was invented at the Massachusetts Institute of Technology (Cambridge, MA), which incorporates micrometer-scale pumps and flow channels to provide controlled release of single or multiple chemical substances on demand. The release mechanism is based on the electrochemical dissolution of thin anode membranes covering microreservoirs filled with chemicals in various forms. Various amounts of chemical substances in solid, liquid, or gel form can be released either in a pulsatile or in a continuous manner or a combination of both. The entire device can be mounted on the tip of a small probe or implanted in the body. In future, proper selection of a biocompatible material may enable the development of an autonomous controlled-release implant that has been dubbed as "pharmacy-on-a-chip" or a highly controlled tablet (smart tablet) for drug delivery. The researchers hope to engineer the chips so that they can change the drug release schedule or medication type in response to commands beamed through the skin. Commercial development is being done by MicroChips Inc. Products currently in development include external and implantable microchips for the delivery of proteins, hormones, pain medications, and other pharmaceutical compounds. Controlled pulsatile release of the polypeptide leuprolide has been demonstrated from microchip implants over 6 months in dogs [11]. Each microchip contains an array of discrete reservoirs from which dose delivery can be controlled by telemetry.

3.8.4 Pumps and Conduits for Drug Delivery

Mechanical pumps are usually miniature devices such as implantable infusion pumps and percutaneous infusion catheters which deliver drugs into appropriate vessels or other sites in the body. Several pumps, implantable catheters, and infusion devices are available commercially. Examples of applications of these devices are as follows:

1. Intrathecal morphine infusion for pain control
2. Intraventricular drug administration for disorders of the brain
3. Hepatic arterial chemotherapy
4. Intravenous infusion of heparin in thrombotic disorders
5. Intravenous infusion of insulin in diabetes

The advantages of these devices are as follows:

1. The rate of drug diffusion can be controlled.
2. Relatively large amounts of drugs can be delivered.
3. The drug administration can be changed or stopped when required.

3.9 Targeted Delivery Systems

For targeted and controlled delivery, a number of carrier systems and homing devices are under development: glass-like matrices, monoclonal antibodies, resealed erythrocytes, microspheres, and liposomes. There are more sophisticated systems based on molecular mechanisms, nanotechnology, and gene delivery. These will be discussed in the following pages.

3.9.1 Polymeric Carriers for Drug Delivery

The limitation of currently available drug therapies, particularly for the treatment of diseases localized to specific organs, has led to efforts to develop alternative methods of drug administration to increase their specificity. One approach for this purpose is the use of degradable polymeric carriers for drugs which are delivered to and deposited at the site of the disease for extended periods with minimal systemic distribution of the drug. The polymeric carrier is degraded and eliminated from the body shortly after the drug has been released. The polymers are divided into three groups:

1. *Nondegradable polymers*. These are stable in biological systems. They are mostly used as components of implantable devices for drug delivery.
2. *Drug-conjugated polymers*. In these the drug is attached to a water-soluble polymer carrier by a cleavable bond. These polymers are less accessible to healthy tissues when compared with the diseased tissues. These conjugates can be used

for drug targeting via systemic administration or by implanting them directly at the desired site of action where the drug is released by cleavage of the drug-polymer bond. Examples of such polymers are dextrans, polyacrylamides, and albumin.
3. *Biodegradable polymers*. These degrade under biological conditions to nontoxic products that are eliminated from the body.

Macromolecular complexes of various polymers can be divided into the following categories according to the nature of molecular interactions:

1. Complexes formed by interaction of oppositely charged polyelectrolytes
2. Charge transfer complexes
3. Hydrogen-bonding complexes
4. Stereocomplexes

Polyelectrolyte complexes can be used as implants for medical use, as microcapsules, or for binding of pharmaceutical products, including proteins. In recent years, a new class of organometallic polymers, polyphosphazenes, has become available. Synthetic flexibility of polyphosphazenes makes them a suitable material for controlled-release technologies. Desirable characteristics of a polymeric system used for drug delivery are as follows:

1. Minimal tissue reaction after implantation
2. High polymeric purity and reproducibility
3. A reliable drug-release profile

In Vivo Degradation at a Well-Defined Rate in Case of Biodegradable Implants

Polymeric delivery systems for implanting at specific sites are either a reservoir type where the drug is encapsulated into a polymeric envelope that serves as a diffusional rate-controlled membrane or a matrix type where the drug is evenly dispersed in a polymer matrix. Most of the biodegradable systems are of the matrix type, where drug is released by a combination of diffusion, erosion, and dissolution. Disadvantages of the implants are that once they are in place, the dose cannot be adjusted and the discontinuation of therapy requires a surgical procedure to remove the implant. For chronic long-term release repeated implantations are required.

The development of injectable biodegradable drug-delivery systems has provided new opportunities for controlled drug delivery as they have advantages over traditional ones such as ease of application, and prolonged localized drug delivery [12]. Both natural as well as synthetic polymers have been used for this purpose. Following injection in fluid state, they solidify at the desired site. These systems have been explored widely for the delivery of various therapeutic agents ranging from antineoplastic agents to proteins and peptides such as insulin. Polymers are also being used as nanoparticles for drug delivery, as described later in the section on nanobiotechnology-based drug delivery.

3.9.2 Evaluation of Polymers In Vivo

Biodegradable polymers have attracted much attention as implantable drug delivery systems. Uncertainty in extrapolating in vitro results to in vivo systems due to the difficulties of appropriate characterization in vivo, however, is a significant issue in the development of these systems. To circumvent this limitation, nonelectron paramagnetic resonance (EPR) and magnetic resonance imaging (MRI) were applied to characterize drug release and polymer degradation in vitro and in vivo. MRI makes it possible to monitor water content, tablet shape, and response of the biological system such as edema and encapsulation. The results of the MRI experiments give the first direct proof in vivo of postulated mechanisms of polymer erosion. Using nitroxide radicals as model drug-releasing compounds, information on the mechanism of drug release and microviscosity inside the implant can be obtained by means of EPR spectroscopy. The use of both these noninvasive methods to monitor processes in vivo leads to new insights into the understanding of the mechanisms of drug release and polymer degradation.

3.9.3 Collagen

Collagen, being a major protein of connective tissues in animals, is widely distributed in skin, bones, teeth, tendons, eyes, and most other tissues in the body and accounts for about one-third of the total protein content in mammals. It also plays an important role in the formation of tissues and organs and is involved in various cells in terms of their functional expression. Collagen as a biomaterial has been used for repair and reconstruction of tissues and as an agent for wound dressing.

Several studies have already been conducted on the role of collagen as a carrier in drug delivery. In vivo absorption of collagen is controlled by the use of a cross-linking agent such as glutaraldehyde or by induction of cross-linking through ultraviolet or gamma ray irradiation in order to enhance the sustained-release effects. Release rate of drugs can be controlled by (1) collagen gel concentration during preparation of the drug delivery system, (2) the form of drug delivery system, and (3) the degree of cross-linking of the collagen.

3.10 Particulate Drug Delivery Systems

The concept of using particles to deliver drugs to selected sites of the body originated from their use as radiodiagnostic agents in medicine in the investigation of the reticuloendothelial system (liver, spleen, bone marrow, and lymph nodes). Particles ranging from 20 to 300 μm have been proposed for drug targeting. Because of the small size of the particles, they can be injected directly into the systemic circulation or a certain compartment of the body. Particulate drug delivery systems may contain an intimate mixture of the drug and the core material or the drug may be dispersed as an emulsion in the carrier material, or the drug may be encapsulated by the carrier material. Factors that influence the release of drugs from particulate carriers are as follows:

1. The drug: its physicochemical properties, position in the particle, and drug-carrier interaction
2. Particles: type, size, and density of the particle
3. Environment: temperature, light, presence of enzymes, ionic strength, and hydrogen ion concentration

Various particulate drug carrier systems can be grouped into the following classes:

1. Microspheres are particles larger than 1 μm but small enough not to sediment when suspended in water (usually 1–100 μm).
2. Nanoparticles are colloidal particles ranging in size between 10 and 1000 nm.
3. Glass-like sugar matrices.
4. Liposomes.
5. Cellular particles such as resealed erythrocytes, leukocytes, and platelets.

3.10.1 Microspheres

Microspheres prepared from cross-linked proteins have been used as biodegradable drug carriers. The rate of release of small drug molecules from protein microspheres is relatively rapid, although various strategies, such as complexing the drug with macromolecules, can be adopted to overcome this problem. Polysaccharides (e.g., starch) and a wide range of synthetic polymers have been used to manufacture microspheres. Microcapsules differ from microspheres in having a barrier membrane surrounding a solid or liquid core, which is an advantage in case of peptides and proteins. Special applications of microspheres and microcapsules are as follows:

1. Poly-DL-lactide-*co*-glycolide-agarose microspheres can encapsulate protein and stabilize them for drug delivery.
2. Multicomponent, environmentally responsive, hydrogel microspheres, coated with a lipid bilayer, can be used to mimic the natural secretory granules for drug delivery.
3. Microencapsulation of therapeutic agents provides local controlled drug release in the central nervous system across the blood-brain barrier.
4. Microspheres can be used for chemoembolization of tumors in which the vasculature is blocked while anticancer agent is released from the trapped microparticles.
5. Microcapsules, produced at ideal size for inhalation (1–5 μm), can be used in formulating drugs for pulmonary delivery, both for local delivery and for systemic absorption.
6. Microspheres can be used as nasal drug delivery systems for systemic absorption of peptides and proteins.

7. Poly-DL-lactide-*co*-glycolide microspheres can be used as a controlled-release antigen delivery system – parenteral or oral.
8. Delivery of antisense oligonucleotides.
9. Nanoencapsulation of DNA in bioadhesive particles can be used for gene therapy by oral administration.

3.10.2 Glass-like Sugar Matrices

These are microparticles made of glass-like sugar matrix. The solution of sugar and insulin is sprayed as a mist into a stream of hot, dry air, which quickly dries the mist to a powder, a process known as spray drying. The transformation from liquid to a glassy powder is rapid and prevents denaturation of the insulin. Sugar microspheres can also be used for preserving drugs and vaccines which normally require refrigeration for travel to remote parts of the world. Sugar molecules protect the drug molecules by "propping up" the active structure, preventing it from denaturing when the water molecules are removed.

3.10.3 Resealed Red Blood Cells

Red blood cells (RBCs) have been studied the most of all the cellular drug carriers. When RBCs are placed in a hypotonic medium, they swell, leading to rupture of the membrane and formation of pores. This allows encapsulation of 25% of the drug or enzyme in solution. The membrane is resealed by restoring the tonicity of the solution. The following are the potential uses of loaded RBCs as drug delivery systems:

1. They are biodegradable and nonimmunogenic.
2. They can be modified to change their resident circulation time; depending on their surface, cells with little surface damage can circulate for a longer time.
3. Entrapped drug is shielded from immunological detection and external enzymatic degradation.
4. The system is relatively independent of the physicochemical properties of the drug.

The drawbacks of using RBCs are that the damaged RBCs are sequestered in the spleen and the storage life is limited to about 2 weeks.

3.11 Nanotechnology-Based Drug Delivery

Nanotechnology is the creation and utilization of materials, devices, and systems through the control of matter on the nanometer-length scale, i.e., at the level of atoms, molecules, and supramolecular structures. It is the popular term for the construction and utilization of functional structures with at least one characteristic dimension measured in nanometer – a nanometer is one billionth of a meter (10^{-9} m). Nanotechnologies are described in detail in a special report on this topic [13].

Trend toward miniaturization of carrier particles had already started prior to the introduction of nanotechnology in drug delivery. The suitability of nanoparticles for use in drug delivery depends on a variety of characteristics, including size and porosity. Nanoparticles can be used to deliver drugs to patients through various routes of delivery. Nanoparticles are important for delivering drugs intravenously so that they can pass safely through the body's smallest blood vessels, for increasing the surface area of a drug so that it will dissolve more rapidly, and for delivering drugs via inhalation. Porosity is important for entrapping gases in nanoparticles, for controlling the release rate of the drug, and for targeting drugs to specific regions.

It is difficult to create sustained-release formulations for many hydrophobic drugs because they release too slowly from the nanoparticles used to deliver the drug, diminishing the efficacy of the delivery system. Modifying water uptake into the nanoparticles can speed the release, while retaining the desired sustained-release profile of these drugs. Water uptake into nanoparticles can be modified by adjusting the porosity of the nanoparticles during manufacturing and by choosing from a wide variety of materials to include in the shell.

Nanobiotechnology provides the following solutions to the problems of drug delivery:

1. Improving solubilization of the drug.
2. Using noninvasive routes of administration eliminates the need for administration of drugs by injection.
3. Development of novel nanoparticle formulations with improved stabilities and shelf-lives.
4. Development of nanoparticle formulations for improved absorption of insoluble compounds and macromolecules enables improved bioavailability and release rates, potentially reducing the amount of dose required and increasing safety through reduced side effects.
5. Manufacture of nanoparticle formulations with controlled particle sizes, morphology, and surface properties would be more effective and less expensive than other technologies.
6. Nanoparticle formulations that can provide sustained-release profiles up to 24 h can improve patient compliance with drug regimens.
7. Direct coupling of drugs to targeting ligand restricts the coupling capacity to a few drug molecules, but coupling of drug carrier nanosystems to ligands allows import of thousands of drug molecules by means of one receptor targeted ligand. Nanosystems offer opportunities to achieve drug targeting with newly discovered disease-specific targets.

3.11.1 Nanomaterials and Nanobiotechnologies Used for Drug Delivery

Various nanomaterials and nanobiotechnologies used for drug delivery are shown in Table 1.9.

3.11.2 Liposomes

Liposomes are stable microscopic vesicles formed by phospholipids and similar amphipathic lipids. Liposome properties vary substantially with lipid composition, size, surface charge, and the method of preparation. They are therefore divided into three classes based on their size and number of bilayers.

1. Small unilamellar vesicles are surrounded by a single lipid layer and are 25–50 nm in diameter.
2. Large unilamellar vesicles are a heterogeneous group of vesicles similar to small unilamellar vesicles and are surrounded by a single lipid layer.
3. Multilamellar vesicles consist of several lipid layers separated from each other by a layer of aqueous solution.

Lipid bilayers of liposomes are similar in structure to those found in living cell membranes and can carry lipophilic substances such as drugs within these layers in the same way as cell membranes. The pharmaceutical properties of the liposomes depend on the composition of the lipid bilayer and its permeability and fluidity. Cholesterol, an important constituent of many cell membranes, is frequently included in liposome formulations because it reduces the permeability and increases the stability of the phospholipid bilayers.

Until recently, the use of liposomes as therapeutic vectors was hampered by their toxicity and lack of knowledge about their biochemical behavior. The simplest use of liposomes is as vehicles for drugs and antibodies targeted for the targeted delivery of anticancer agents. The use of liposomes may be limited because of problems related to stability, the inability to deliver to the right site, and the inability to release the drug when it gets to the right site. However, liposome surfaces can be readily modified by attaching polyethylene glycol (PEG) units to the bilayer (producing what is known as stealth liposomes) to enhance their circulation time in the bloodstream. Furthermore, liposomes can be conjugated to antibodies or ligands to enhance target-specific drug therapy.

Polymer Nanoparticles

Biodegradable polymer nanoparticles are PEG-coated poly(lactic acid) (PLA) nanoparticles, chitosan (CS)-coated poly(lactic acid–glycolic acid) (PLGA) nanoparticles, and chitosan (CS) nanoparticles. These nanoparticles can carry and

Table 1.9 Nanomaterials and nanobiotechnologies used for drug delivery

Structure	Size	Role in drug delivery
Bacteriophage NK97 (a virus that attacks bacteria)		Emptied of its own genetic material, HK97, which is covered by 72 interlocking protein rings, can act as a nanocontainer to carry drugs and chemicals to targeted locations
Canine parvovirus (CPV) particles	26 nm	Tumor-targeted drug delivery: CPV binds to transferrin receptors, which are overexpressed by a variety of tumor cells
Carbon magnetic nanoparticles	40–50 nm	For drug delivery and targeted cell destruction
Dendrimers	1–20 nm	Holding therapeutic substances such as DNA in their cavities
Ceramic nanoparticles	~35 nm	Accumulate exclusively in the tumor tissue and allow the drug to act as sensitizer for PDT without being released
HTCC nanoparticles	110–180 nm	Encapsulation efficiency is up to 90%. In vitro release studies show a burst, effect followed by a slow and continuous release
Liposomes	25–50 nm	Incorporate fullerenes to deliver drugs that are not water-soluble, that tend to have large molecules
Micelle/Nanopill	25–200 nm	Made from 2 polymer molecules – one water-repellant and the other hydrophobic – that self-assemble into a sphere called a micelle, which can deliver drugs to specific structures in the cell
Low-density lipoproteins	20–25 nm	Drugs solubilized in the lipid core or attached to the surface
Nanocochleates		Nanocochleates facilitate delivery of biologicals such as DNA and genes
Nanocrystals	<1,000 nm	NanoCrystal technology (Elan) has the potential to rescue a significant number of poorly soluble chemical compounds by increasing solubility
Nanoemulsions	20–25 nm	Drugs in oil and/or liquid phases to improve absorption

Nanolipispheres	25–50 nm	Carrier incorporation of lipophilic and hydrophilic drugs
Nanoparticle composites	~40 nm	Attached to guiding molecules such as MAbs for targeted drug delivery
Nanoparticles	25–200 nm	Continuous matrices containing dispersed or dissolved drug
Nanospheres	50–500 nm	Hollow ceramic nanospheres created by ultrasound
Nanostructured organogels	50 nm	Made by mixing olive oil, liquid solvents, and a simple enzyme to chemically activate a sugar and used to encapsulate drugs
Nanotubes	20–60 nm	Offer some advantages over spherical nanoparticles
Nanovalve	500 nm	Externally controlled release of drug into a cell
Nanovesicles	25–3,000 nm	Multilamellar bilayer spheres containing the drugs in lipids
Polymer nanocapsules	50–200 nm	Enclosing drugs
PEG-coated PLA nanoparticles		PEG coating improves the stability of PLA nanoparticles in the gastrointestinal fluids and helps the transport of encapsulated protein across the intestinal and nasal mucus membranes
Superparamagnetic iron oxide nanoparticles	10–100 nm	As drug carriers for intravenous injection to evade RES of the body as well as penetrate the very small capillaries within the body tissues and therefore offer the most effective distribution

PDT photodynamic therapy, *MAbs* monoclonal antibodies, *PEG* poly(ethylene glycol), *PLA* poly(lactic acid), *HTCC* *N*-(2-hydroxyl) propyl-3-trimethyl ammonium chitosan chloride, *RES* reticuloendothelial system deliver proteins in an active form, and transport them across the nasal and intestinal mucosa. Additionally, PEG-coating improves the stability of PLA nanoparticles in the gastrointestinal fluids and helps the transport of the encapsulated protein, tetanus toxoid, across the intestinal and nasal mucous membranes [14]. Furthermore, intranasal administration of these nanoparticles provided high and long-lasting immune responses.

N-(2-Hydroxyl) propyl-3-trimethyl ammonium chitosan chloride (HTCC) is a water-soluble derivative of chitosan (CS), synthesized by the reaction between glycidyl-trimethyl-ammonium chloride and CS. HTCC nanoparticles have been formed based on ionic gelation process of HTCC and sodium tripolyphosphate (TPP). Bovine serum albumin (BSA), as a model protein drug, was incorporated into the HTCC nanoparticles. HTCC nanoparticles were 110–180 nm in size, and their encapsulation efficiency was up to 90%. In vitro release studies showed a burst effect, followed by a slow and continuous release. Encapsulation efficiency was obviously increased with increase in initial BSA concentration [15].

Coating of PLGA nanoparticles with the mucoadhesive CS improves the stability of the particles in the presence of lysozyme and enhanced the nasal transport of the encapsulated tetanus toxoid. Nanoparticles made solely of CS are stable upon incubation with lysozyme. Moreover, these particles are very efficient in improving the nasal absorption of insulin as well as the local and systemic immune responses to tetanus toxoid, following intranasal administration.

Polymeric Micelles

Micelles are biocompatible nanoparticles varying in size from 50 to 200 nm in which poorly soluble drugs can be encapsulated. They represent a possible solution to the delivery problems associated with such compounds and could be exploited to target the drugs to particular sites in the body, potentially alleviating toxicity problems. pH-sensitive drug delivery systems can be engineered to release their contents or change their physicochemical properties in response to variations in the acidity of the surroundings. One example of this is the preparation and characterization of novel polymeric micelles (PM) composed of amphiphilic pH-responsive poly(*N*-isopropylacrylamide) (PNIPAM) or poly(alkyl(meth)acrylate) derivatives [16]. On one hand, acidification of the PNIPAM copolymers induces a coil-to-globule transition that can be exploited to destabilize the intracellular vesicle membranes. PNIPAM-based PMs, loaded with either doxorubicin or aluminum chloride phthalocyanine, are cytotoxic in murine tumor models. On the other hand, poly(alkyl(meth)acrylate) copolymers can be designed to interact with either hydrophobic drugs or polyions and release their cargo upon an increase in pH. Micelle-forming polymeric drugs such as NK911 (doxorubicin-incorporating micelle) and NK105 (taxol-incorporating micelle) are in clinical trials sponsored by Nippon Kayaku Co. and conducted at the National Cancer Center Hospital, Tokyo, Japan [17].

3.11.3 Future Prospects of Nanotechnology-Based Drug Delivery

A desirable situation in drug delivery is to have smart drug delivery systems that can integrate with the human body. This is an area where nanotechnology will play an extremely important role. Even time-release tablets, which have a relatively simple coating that dissolves in specific locations, involve the use of nanoparticles. Pharmaceutical companies are already involved in using nanotechnology to create intelligent drug release devices. For example, control of the interface between the drug/particle and the human body can be programmed so that when the drug reaches its target, it can become active. The use of nanotechnology for drug release devices requires autonomous device operation. For example, in contrast to converting a biochemical signal into a mechanical signal and being able to control and communicate with the device, autonomous device operation would require biochemical recognition to generate forces to stimulate various valves and channels in the drug delivery systems, so that it does not require any external control.

3.12 Antibody-Targeted Systems

Drug delivery systems can make use of macromolecular attachment for delivery using immunoglobulins as the macromolecule. The obvious advantage of this system is that it can be targeted to the site of antibody specificity. The advantages are that less amount of drug is required and side effects are reduced considerably.

Drugs are linked, covalently or noncovalently to the antibody, or placed in vesicles such as liposomes or microspheres and the antibody is used to target the liposome. Covalent attachments are generally not very efficient and diminish the antigen-binding capacity. If conjugation is done through an intermediate carrier molecule, it is possible to increase the drug/antibody ratio. Such intermediates include dextran or poly-L-glutamic acid.

Examples of drugs that have been conjugated to antibodies or their fragments are anticancer drugs. Numerous antibody-liposome combinations have been investigated for delivering drugs and genes. The term immunoliposomes is used for liposomes loaded with drug cargo that have been surface-conjugated to antibodies. The main advantage of antibody-targeted system is that the adverse effects of anticancer drugs can be reduced by use of monoclonal antibodies that recognize only tumor antigens.

3.13 Receptor-Mediated Targeted Drug Delivery

Receptor-mediated endocytosis is a process whereby extracellular macromolecules and particles gain entry to the intracellular environments. Cell surface receptors are complex transmembrane proteins that mediate highly specific interactions between cells and their extracellular environment. The cells use receptor-mediated endocytosis for nutrition, defense, transport, and processing. Cellular uptake of drugs bound to a targeting carrier or to a targetable polymeric carrier is mostly restricted to receptor-mediated endocytosis. Because receptors are differentially expressed in various cell types and tissues, using receptors as markers may be an advantageous strategy for drug delivery. Receptor-mediated uptake can achieve the specific transport of the drug to the receptor-bearing target cells. Many receptors such as receptors for transferrin, low-density lipoprotein, and asialoglycoprotein have been used to deliver drugs to specific types of cells or tissues.

Many recent advances in targeted drug delivery have focused on regulation of the endogenous membrane trafficking machinery in order to facilitate uptake of drugs via receptor-mediated endocytosis into target tissues. Vesicle motor proteins (kinesin, cytoplasmic dynein, and myosin) play an important role in membrane trafficking events and are referred to as molecular motors. It is important to understand the events involved in the movement of surface-bound and extracellular components by endocytosis into the cell. Knowledge of sorting events within the endocytic pathways that govern the intracellular destination and the ultimate fate of the drug is also important. If internalization of the drug is followed by recycling or degradation, no accumulation can occur within the cell. Strategies for regulation of such events can enhance drug delivery. One example is delivery of genes via receptor-mediated endocytosis in hepatocytes. Factors that influence the efficiency of the receptor-mediated uptake of targeted drug conjugate are as follows:

1. Affinity of the targeting moieties
2. The affinity and nature of target antigen
3. Density of the target antigen
4. The type of cell targeted
5. The rate of endocytosis
6. The route of internalization of the receptor-ligand complex
7. The ability of the drug to escape from the vesicular compartment into the cytosol
8. The affinity of the carrier to the drug

Receptor-mediated drug delivery is particularly applicable to cytotoxic therapy for cancer and gene therapy.

3.14 Methods of Administration of Proteins and Peptides

Various possible routes for administration of proteins and peptides are as follows:

1. Parenteral
2. Transdermal
3. Inhalation
4. Transnasal
5. Oral

6. Rectal
7. Implants
8. Cell and gene therapies
9. Use of special formulations

Injection still remains the most common method for administration of proteins and peptides. Efforts are being made to use needle-free or painless injections and also to improve the controlled delivery by parenteral route.

3.14.1 Delivery of Peptides by Subcutaneous Injection

Subcutaneous still remains the predictable and controllable route of delivery for peptides and macromolecules. However, there is need for greater convenience and lower cost for prolonged and repeated delivery. An example of refinement of subcutaneous delivery is MEDIPAD (Elan Pharmaceutical Technologies), which is a combination of "patch" concept and a sophisticated miniaturized pump operated by gas generation. It was described in the report on transdermal drug delivery.

3.14.2 Depot Formulations and Implants

These are usually administered by injection and must ensure protein/peptide stability. One of the formulations used is poly(lactide-*co*-glycide) sustained release. Example of an approved product in the market is leuprolide. Implants involve invasive administration and also must ensure protein/peptide stability. Implantable titanium systems provide drug release driven by osmotic pumps. This technology has been extended to other proteins such as growth hormone. Nutropin Depot (Genentech/Alkermes) is the first long-acting form of growth hormone that encapsulates the drug in biodegradable microspheres that release the hormone slowly after injection. It reduces the frequency of injection in children with growth hormone deficiency from once daily to once a month.

3.14.3 Poly(ethylene glycol) Technology

Poly(ethylene glycol) or PEG, a water-soluble polymer, is a well recognized treatment for constipation. When covalently linked to proteins, PEG alters their properties in ways that extend their potential uses. Chemical modification of proteins and other bioactive molecules with PEG – a process referred to as PEGylation – can be used to tailor molecular properties to particular applications, eliminating disadvantageous properties or conferring new molecular functions. This approach can be used to improve delivery of proteins and peptides. Advantages of PEG technology are as follows:

1. Increase of drug solubility
2. Increase of drug stability

3. Reduction of immunogenicity
4. Increase in circulation life-time
5. Improvement of release profile

Enzyme deficiencies for which therapy with the native enzyme is inefficient (due to rapid clearance and/or immunological reactions) can now be treated with equivalent PEG-enzymes.

4 New Concepts in Pharmacology That Influence Design of DDS

Pharmacology, particularly pharmacokinetics and pharmacodynamics, have traditionally influenced drug delivery formulations. Some of the newer developments in pharmacology and therapeutics that influence the development of DDSs are as follows:

1. Pharmacogenetics
2. Pharmacogenomics
3. Pharmacoproteomics
4. Pharmacometabolomics
5. Chronopharmacology

The first four items are linked together and form the basis of personalized medicine, which will be discussed later in this chapter.

4.1 Pharmacogenetics

Pharmacogenetics, a term recognized in pharmacology in the pregenomic era, is the study of influence of genetic factors on action of drugs as opposed to genetic causes of disease. Now it is the study of the linkage between the individual's genotype and the individual's ability to metabolize a foreign compound. The pharmacological effect of a drug depends on pharmacodynamics (interaction with the target or the site of action) and pharmacokinetics (absorption, distribution, and metabolism). It also covers the influence of various factors on these processes. Drug metabolism is one of the major determinants of drug clearance and the factor that is most often responsible for interindividual differences in pharmacokinetics.

The differences in response to medications are often greater among members of a population than they are within the same person or between monozygotic twins at different times. The existence of large population differences with small intrapatient variability is consistent with inheritance as a determinant of drug response. It is estimated that genetics can account for 20–95% of variability in drug disposition and effects. Genetic polymorphisms in drug-metabolizing enzymes, transporters, receptors, and other drug targets have been linked. From this initial definition, the scope has broadened so that it overlaps with pharmacogenomics. Genes influence pharmacodynamics and pharmacokinetics. Pharmacogenetics has a threefold role in the pharmaceutical industry, which is relevant to the development of personalized medicines. The three roles are as follows:

1. For study of the drug metabolism and pharmacological effects
2. For predicting genetically determined adverse reactions
3. Drug discovery and development and as an aid to planning clinical trials

4.2 Pharmacogenomics

Pharmacogenomics, a distinct discipline within genomics, carries on that tradition by applying the large-scale systemic approaches of genomics to understand the basic mechanisms and apply them to drug discovery and development. Pharmacogenomics now seeks to examine the way drugs act on the cells as revealed by the gene expression patterns and thus bridges the fields of medicinal chemistry and genomics. Some of the drug response markers are examples of interplay between pharmacogenomics and pharmacogenetics; both are playing an important role in the development of personalized medicines [18]. The two terms – pharmacogenetics and pharmacogenomics – are sometimes used synonymously but one must recognize the differences between the two.

Various technologies enable the analysis of these complex multifactorial situations to obtain individual genotypic and gene expression information. These same tools are applicable to study the diversity of drug effects in different populations. Pharmacogenomics promises to enable the development of safer and more effective drugs by helping to design clinical trials such that nonresponders would be eliminated from the patient population and take the guesswork out of prescribing medications. It will also ensure that the right drug is given to the right person from the start. In clinical practice, doctors could test patients for specific single nucleotide polymorphisms (SNPs) known to be associated with nontherapeutic drug effects before prescribing in order to determine which drug regimen best fits their genetic make-up. Pharmacogenomic studies are rapidly elucidating the inherited nature of these differences in drug disposition and effects, thereby enhancing drug discovery and providing a stronger scientific basis for optimizing drug therapy on the basis of each patient's genetic constitution.

Pharmacogenomics provides a new way of looking at the old problems, i.e., how to identify and target the essential component of disease pathway(s). These changes will increase the importance of drug delivery systems, which need to be adapted to our changing concept of the disease. Drug delivery problems have to be considered parallel to all stages of drug development from discovery to clinical use.

4.3 Pharmacoproteomics

The term *proteomics* indicates PROTEins expressed by a genOME and is the systematic analysis of protein profiles of tissues. There is an increasing interest in proteomics technologies now because deoxyribonucleic acid (DNA) sequence information provides only a static snapshot of the various ways in which the cell might use its proteins, whereas the life of the cell is a dynamic process. Role of proteomics in drug development can be termed *pharmacoproteomics*. Proteomics-based characterization of multifactorial diseases may help to match a particular target-based therapy to a particular marker in a subgroup of patients. The industrial sector is taking a lead in developing this area. Individualized therapy may be based on differential protein expression rather than a genetic polymorphism.

4.4 Chronopharmacology

The term *chronopharmacology* is applied to variations in the effect of drugs according to the time of their administration during the day. Mammalian biological functions are organized according to circadian rhythms (lasting about 24 h). They are coordinated by a biological clock situated in the suprachiasmatic nuclei (SCN) of the hypothalamus. These rhythms persist under constant environmental conditions, demonstrating their endogenous nature. Some rhythms can be altered by disease. The rhythms of disease and pharmacology can be taken into account to modulate treatment over the 24-h period.

The knowledge of such rhythms appears particularly relevant for the understanding and/or treatment of hypertension and ischemic coronary artery disease. In rats and in man, the circadian rhythm of systolic or diastolic blood pressure can be dissociated from the rest-activity cycle, suggesting that it is controlled by an oscillator which can function independently of the SCN, which could justify modification of treatment according to the anomalies of the blood pressure rhythm. The morning peak of myocardial infarction in man is due to the convergence of several risk factors, each of which has a 24-h cycle: blood coagulability, BP, oxygen requirements, and myocardial susceptibility to ischemia. The existence of these rhythms, and the chronopharmacology of cardiovascular drugs such as nitrate derivatives, constitute clinical prerequisites for the chronopharmacotherapy of heart disease.

It is known that the sensitivity of tumor cells to chemotherapeutic agents can depend on circadian phase. There are possible differences in rhythmicity of cells within tissues. If cells within a tumor are not identically phased, this may allow some cells to escape from the drug's effect. Perhaps synchronizing the cells prior to drug treatment would improve tumor eradication. Wild-type and circadian mutant mice demonstrate striking differences in their response to the anticancer drug cyclophosphamide. The sensitivity of wild-type mice varies greatly, depending on the time of drug administration, while Clock mutant and Bmal1 knockout mice are highly sensitive to treatment at all times tested. Both time-of-day and allelic-dependent variations in response to chemotherapy correlate with the functional status of the circadian Clock/Bmal1 transactivation complex, which affects the lethality of chemotherapeutic agents by modulating the survival of the target cells necessary for the viability of the organism [19]. These findings will provide a rationale not only for adjusting the

Table 1.10 Current trends in pharmaceutical product development

Use of recombinant DNA technology
Expansion of use of protein and peptide drugs in current therapeutics
Introduction of antisense, RNA interference, and gene therapy
Advances in cell therapy: introduction of stem cells
Miniaturization of drug delivery: microparticles and nanoparticles
Increasing use of bioinformatics and computer drug design
A trend toward development of target-organ-oriented dosage forms
Increasing emphasis on controlled-release drug delivery
Use of routes of administration other than injections
Increasing alliances between pharmaceutical companies and DDS companies

DDS drug delivery systems

timing of chemotherapeutic treatment to be less toxic but also for providing a basis for a search for pharmacological modulators of drug toxicity acting through circadian system regulators. This result may significantly increase the therapeutic index and reduce morbidity associated with anticancer treatment.

Chronopharmacological drug formulations can provide the optimal serum levels of the drug at the appropriate time of the day or night. For example, if the time of action desired is early morning, drug release is optimized for that time, whereas with conventional methods of drug administration, the peak will be reached in the earlier part of the night and with controlled release the patient will have a constant high level throughout the night. Effective chronopharmacotherapeutics will not only improve the efficacy of treatment but will open up new markets. This approach to treatment requires suitable drug delivery systems.

4.5 Impact of Current Trends in Pharmaceutical Product Development on DDS

Considerable advances have taken place in pharmaceutical industry during the past two decades. Contemporary trends in pharmaceutical product development that are relevant to DDS are listed in Table 1.10. Drug delivery technologies have become an important part of the biopharmaceutical industry. Drug delivery systems, pharmaceutical industry, and biotechnology interact with each other as shown in Fig. 1.1.

4.6 Impact of New Biotechnologies on Design of DDS

New biotechnologies have a great impact on the design of DDS during the past decade. The most significant of these technologies is nanobiotechnology.

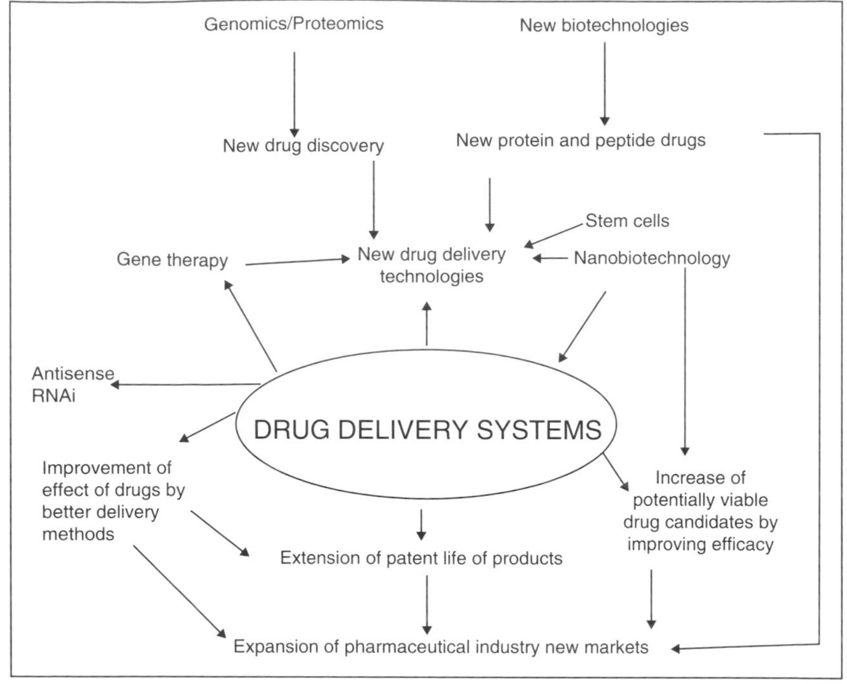

Fig. 1.1 Interrelationship of DDS, pharmaceutical industry, and biotechnology

5 Aims of DDS Development

Drug delivery technologies are aimed at improving efficacy and safety of medicines as well as commercial pharmaceutical development. The following are the important points:

1. Improvement of drug safety and efficacy
2. Improved compliance
3. Chronopharmacological benefits
4. Reduction of cost of drug development
5. Life extension of the products
6. Reduction of risk of failure in new product development

5.1 Improvement of Safety and Efficacy

Improvement of the safety and efficacy of existing medications is a common objective for all those involved in healthcare. There is no doubt that improved delivery of medications with longer duration of action leads to increased efficacy. Lesser quantities of the active ingredients are required, and targeted application can spare the rest of the body from side effects.

5.2 Improved Compliance

Compliance is a big problem in medical care. Most patients do not like to take medications or fail to take them as instructed. Drugs with sustained release can remedy some of these problems. Once daily dosage with sustained action is likely to improve the compliance rate to 80% when compared with 40% for three or four times a day. Some of the novel delivery methods such as transdermal or buccal are preferred by most patients to oral intake or injections. Lack of compliance is responsible for a significant number of hospital admissions in the USA (as high as 10% in some estimates). Overall cost of noncompliance in the USA was more than $100 billion in 2006. Improvement of compliance can lead to significant reductions in healthcare costs.

5.3 Life Extension of the Products

Patents of several proprietary drugs are about to expire. The introduction of an improved dosage delivery form prior to the expiry of the patent allows the manufacturer to maintain the product with some advantages over generic copies. Several aspects of this are as follows:

1. Alternative dosage form may offer advantages over the old product such as improved compliance, increased safety, and enhanced efficacy.
2. Development of a previously unknown delivery formulation of an old product would enable a new patent to be obtained even though the active ingredient is the same.
3. Sustained released versions of an older drug are easy to copy in the generic form but high tech drug delivery forms are not easy to copy.
4. Drugs with expiring patents can be converted into proprietary over-the-counter products to maintain brand franchise.

5.4 Economic Factors

Economics is the most important driver for the development of drug delivery technology. Benefits of new formulations are perceptible at various levels of drug development and patient care as follows:

Continued revenues after expiry of patent. A new drug delivery method can continue to generate revenues for the manufacturer years after expiry of the patent for the original active ingredient. This may sometimes exceed the earnings from the original product.

Market extension. New formulations based on novel drug delivery systems can open up new indications and new markets for the old product. Calcium antagonists were originally launched for angina but achieved more success in the management of hypertension following development of immediate-release formulations.

Drug rescue. Several drugs are discontinued at various stages of development process because of lack of suitable delivery technology. Some promising products get into clinical trials only to be dismissed because of unacceptable toxicity or lack of efficacy. This means considerable financial loss for the companies. An appropriate drug delivery system may rescue some of these products by overcoming these difficulties and also increase the number of potential drugs for clinical trials from the preclinical pipeline of a company.

Reduction of cost of drug development. Compared to the high cost and long development time of a pharmaceutical product, a drug delivery system takes a much shorter time to develop – usually 2–4 years – and costs much less. Development for a new formulation of a generic preparation has more potential of profit in relation to investment when compared with developing a new chemical entity.

Reduction of financial risk. With large investments in R&D of biotechnology products, there is a considerable element of risk involved, and a number of biotechnology companies have failed. In contrast, development of drug delivery products requires only a fraction of capital investment and less time for approval. The use of already approved drugs takes away the element of risk involved in approval of the main ingredient of the DDS.

Reduction of healthcare costs. Lack of compliance is a considerable burden on the healthcare systems because of increased hospitalizations. Improvement of compliance by appropriate drug delivery systems reduces the cost of healthcare. An appropriate delivery system also reduces the amount of drug used and thus reduces the costs.

Competitive advantage. In today's pharmaceutical market place, with several products for the same pharmaceutical category, a suitable drug delivery system may help in providing an advantage in competition. A product with a better and more appropriate drug delivery system may move ahead of its competitors if the economic advantages based on improved efficacy, safety, and compliance can be demonstrated.

6 Impact of Current Trends in Healthcare on DDS

Medicine is constantly evolving from the impact of new technologies. In the past, medicine was more of an art than a science, but the effect of new discoveries in life sciences is having its impact on the practice of medicine. Revolutionary discoveries in molecular biology did not have an immediate impact on medicine and there is a lag period before changes are noticeable in the practice of medicine. Many of these changes come from better understanding of the disease, whereas others come from improvements in pharmaceuticals and their delivery. One of the most important trends in healthcare is the concept of personalized medicine.

6.1 Personalized Medicine

Personalized medicine simply means the prescription of specific treatments and therapeutics best suited for an individual. It is also referred to as individualized or individual-based therapy. Personalized medicine is based on the idea of using a patient's genotype as a factor in deciding on treatment options, but other factors are also taken into consideration. Genomic/proteomic technologies have facilitated the development of personalized medicines but other technologies are also contributing to this effort. This process of personalization starts at the development stage of a medicine and is based on pharmacogenomics and pharmacogenetics. Selection of a DDS most appropriate for a patient would be a part of personalized medicine.

Because all major diseases have a genetic component, knowledge of genetic basis helps in distinguishing between clinically similar diseases. Classifying diseases based on genetic differences in affected individuals rather than by clinical symptoms alone makes diagnosis and treatment more effective. Identifying human genetic variations will eventually allow clinicians to subclassify diseases and adapt therapies to the individual patients.

Several diseases can now be described in molecular terms. Some defects can give rise to several disorders, and diseases will be reclassified on molecular basis rather than according to symptoms and gross pathology. The implication of this is that the same drug can be used to treat a number of diseases with the same molecular basis. Another way of reclassification of human diseases will be subdivision of patient populations within the same disease group according to genetic markers and response to medications.

Along with other technologies, refinements in drug delivery will play an important role in the development of personalized medicine. One well-known example is glucose sensors regulating the release of insulin in diabetic patients. Gene therapy, as a sophisticated drug delivery method, can be regulated according to the needs of individual patients. ChipRx Inc. is developing a true "responsive therapeutic device" in which biosensors, electronic feedback, and drug/countermeasure release are fully integrated.

7 Characteristics of an Ideal DDS

Characteristics of an ideal drug delivery system are as follows:

1. It should increase the bioavailability of the drug.
2. It should provide for controlled drug delivery.

3. It should transport the drug intact to the site of action while avoiding the nondiseased host tissues.
4. The product should be stable and delivery should be maintained under various physiological variables.
5. A high degree of drug dispersion.
6. The same method should be applicable to a wide range of drugs.
7. It should be easy to administer to the patients.
8. It should be safe and reliable.
9. It should be cost-effective.

8 Integration of Diagnostics, Therapeutics, and DDS

Combination of diagnostics with therapeutics is an important component of personalized medicine. A DDS can be integrated into this combination to control the delivery of therapeutics in response to variations in the patient's condition as monitored by diagnostics. Nanobiotechnology has helped in this integration. One example is use of quantum dots in cancer, where diagnosis, therapeutics, and drug delivery can be combined using quantum dots (QDs) as the common denominator [20].

The pharmaceutical industry is taking an active part in the integration of diagnostics and therapeutics. During drug development, there is an opportunity to guide the selection, dosage, route of administration, and multidrug combinations to increase the efficacy and reduce toxicity of pharmaceutical products.

9 Current Achievements, Challenges and Future of Drug Delivery

Considerable advances have occurred in DDS within the past decade. Extended release, controlled release, and once-a-day medications are available for several commonly used drugs. Global vaccine programs are close to becoming a reality with the use of oral, transmucosal, transcutaneous, and needle-less vaccination. Considerable advances have been made in gene therapy and delivery of protein therapeutics. Many improvements in cancer treatment can be attributed to novel drug delivery technologies.

New drug delivery systems will develop during the next decade by interdisciplinary collaboration of material scientists, engineers, biologists, and pharmaceutical scientists. Progress in microelectronics and nanotechnology is revolutionalizing drug delivery. However, DDS industry is still facing challenges and some of these are as follows:

1. Drug delivery technologies require constant redesigning to keep up with new methods of drug design and manufacture, particularly of biotechnology products.

2. As the costs of drugs are rising, drug delivery aims to reduce the costs by improving the bioavailability of drugs so that lesser quantities need to be taken in by the patient.
3. More fundamental research needs to be done to characterize the physiological barriers to therapy such as the blood-brain barrier.
4. New materials that are being discovered, such as nanoparticles, need to have safety studies and regulatory approval.

Future of DDS can be predicted to some extent for the next ten years. To go beyond that, e.g., to 20 years from now, would be very speculative. Some of the drugs currently used would disappear from the market and no one knows for sure what drugs would be discovered 20 years from now. Some of the diseases may be partially eliminated and new variants may appear, particularly in infections. With this scenario, it would be difficult to say what methods will evolve to deliver the drugs that we do not know as yet. Some of the drug treatments may be replaced by devices that do not involve the use of drugs.

References

1. Jain KK. Transdermal drug delivery: technologies, markets and companies. Basel: Jain PharmaBiotech Publications, 2008:1–230.
2. Mainardes RM, Urban MC, Cinto PO, et al. Liposomes and micro/nanoparticles as colloidal carriers for nasal drug delivery. Curr Drug Deliv 2006;3:275–85.
3. Van den Mooter G. Colon drug delivery. Expert Opin Drug Deliv 2006;3:111–25.
4. Sakagami M. In vivo, in vitro and ex vivo models to assess pulmonary absorption and disposition of inhaled therapeutics for systemic delivery. Adv Drug Deliv Rev 2006;58:1030–60.
5. Ely L, Roa W, Finlay WH, Lobenberg R. Effervescent dry powder for respiratory drug delivery. Eur J Pharm Biopharm 2007;65:346–53.
6. Dailey LA, Jekel N, Fink L, et al. Investigation of the proinflammatory potential of biodegradable nanoparticle drug delivery systems in the lung. Toxicol Appl Pharmacol 2006;215:100–8.
7. Jain KK. Cardiovascular drug delivery: technologies, markets and companies. Basel: Jain PharmaBiotech Publications, 2008:1–230.
8. Jain KK. Drug delivery in CNS disorders: technologies, markets and companies. Basel: Jain PharmaBiotech Publications, 2008:1–310.
9. Groisman A, Enzelberger M, Quake SR. Microfluidic memory and control devices. Science 2003;300:955–8.
10. Yih TC, Wei C, Hammad B. Modeling and characterization of a nanoliter drug-delivery MEMS micropump with circular bossed membrane. Nanomedicine 2005;1:164–75.
11. Prescott JH, Lipka S, Baldwin S, et al. Chronic, programmed polypeptide delivery from an implanted, multireservoir microchip device. Nat Biotechnol 2006;24:437–8.
12. Chitkara D, Shikanov A, Kumar N, Domb AJ. Biodegradable injectable in situ depot-forming drug delivery systems. Macromol Biosci 2006;6:977–90.
13. Jain KK. Nanobiotechnology: applications, markets and companies. Basel: Jain PharmaBiotech Publications, 2008:1–690.
14. Vila A, Gill H, McCallion O, Alonso MJ. Transport of PLA-PEG particles across the nasal mucosa: effect of particle size and PEG coating density. J Control Release 2004;98:231–44.
15. Xu Y, Du Y, Huang R, Gao L. Preparation and modification of *N*-(2-hydroxyl) propyl-3-trimethyl ammonium chitosan chloride nanoparticle as a protein carrier. Biomaterials 2003;24:5015–22.

16. Dufresne MH, Le Garrec D, Sant V, et al. Preparation and characterization of water-soluble pH-sensitive nanocarriers for drug delivery. Int J Pharm 2004;277:81–90.
17. Matsumura Y. Micelle carrier system in clinical trial. Nippon Rinsho 2006;64:316–21.
18. Jain KK. Personalized medicine: scientific and commercial aspects. Basel: Jain PharmaBiotech Publications, 2007:1–570.
19. Gorbacheva VY, Kondratov RV, Zhang R, et al. Circadian sensitivity to the chemotherapeutic agent cyclophosphamide depends on the functional status of the CLOCK/BMAL1 transactivation complex. Proc Natl Acad Sci USA 2005;102:3407–12.
20. Jain KK. Drug delivery in cancer: technologies, markets and companies. Basel: Jain PharmaBiotech Publications, 2008:1–550.

Chapter 2
The Role of the Adeno-Associated Virus Capsid in Gene Transfer

Kim M. Van Vliet, Veronique Blouin, Nicole Brument, Mavis Agbandje-McKenna, and Richard O. Snyder

Abstract Adeno-associated virus (AAV) is one of the most promising viral gene transfer vectors that has been shown to effect long-term gene expression and disease correction with low toxicity in animal models, and is well tolerated in human clinical trials. The surface of the AAV capsid is an essential component that is involved in cell binding, internalization, and trafficking within the targeted cell. Prior to developing a gene therapy strategy that utilizes AAV, the serotype should be carefully considered since each capsid exhibits a unique tissue tropism and transduction efficiency. Several approaches have been undertaken in an effort to target AAV vectors to specific cell types, including utilizing natural serotypes that target a desired cellular receptor, producing pseudotyped vectors, and engineering chimeric and mosaic AAV capsids. These capsid modifications are being incorporated into vector production and purification methods that provide for the ability to scale-up the manufacturing process to support human clinical trials. Protocols for small-scale and large-scale production of AAV, as well as assays to characterize the final vector product, are presented here.

The structures of AAV2, AAV4, and AAV5 have been solved by X-ray crystallography or cryo-electron microscopy (cryo-EM), and provide a basis for rational vector design in developing customized capsids for specific targeting of AAV vectors. The capsid of AAV has been shown to be remarkably stable, which is a desirable characteristic for a gene therapy vector; however, recently it has been shown that the AAV serotypes exhibit differential susceptibility to proteases. The capsid fragmentation pattern when exposed to various proteases, as well as the susceptibility of the serotypes to a series of proteases, provides a unique fingerprint for each serotype that can be used for capsid identity validation. In addition to serotype identification, protease susceptibility can also be utilized to study dynamic structural changes that must occur for the AAV capsid to perform its various functions during the virus life cycle. The use of proteases for structural studies in solution complements the crystal structural studies of the virus. A generic protocol based on proteolysis for AAV serotype identification is provided here.

Keywords AAV; capsid; Proteolysis; capsid structure; Serotype; Vector production; Column chromatography; Gene therapy

From: *Methods in Molecular Biology, Vol. 437: Drug Delivery Systems*
Edited by: Kewal K. Jain © Humana Press, Totowa, NJ

1 Introduction

Adeno-associated virus (AAV) is a single-stranded DNA virus that is currently being utilized for gene therapy applications. AAV is a member of the family Parvoviridae, genus *Dependovirus*. Several features of AAVs that make them promising candidates as gene transfer vectors are as follows: (1) Good safety profile: the members of this genus are not associated with disease in humans. The virus is replication-deficient, and, except under special circumstances, it will not replicate or spread [1], and AAV has been well tolerated in human clinical trials. (2) Stable long-term expression of the transgene: AAV has successfully been used to express transgenes in several tissues, including brain, muscle, liver, lung, vascular endothelium, and hematopoietic cells [2–9]. (3) Ability to transduce dividing and nondividing cells [10]. (4) Episomal maintenance: wild-type AAV exhibits site-specific integration of its genome into chromosome 19; however, the majority of vector genomes appear to be maintained episomally [11–13]. Therefore, the risk of integration is minimal, compared with retroviral vectors that require integration into the host genome and have the potential to activate proto-oncogenes. (5) Low immunogenicity: unlike adenoviral vectors, a robust T cell response to the vector is not generated by AAV [14], although humoral immune responses are generated, which may result in viral neutralization [15]. The use of other AAV serotypes may circumvent this response and allow for repeated treatments [16]. Alternatively, immunosuppressive therapy may be used during treatment with AAV to avoid the immune response [17]. (6) Physiochemical stability: AAV virions are highly stable over a wide range of pH and temperature, a feature that is important for production and purification methods for clinical-grade AAV vectors, as well as for stability of the final vector product [18].

The genome of AAV is ~4,700 bases of linear, single-stranded DNA. There are two genes, *rep* and *cap*, which are flanked by the inverted terminal repeats (ITRs). The ITRs consist of 145 nucleotides, which form a characteristic T-shaped hairpin. These are the only sequences required in *cis* for viral DNA replication and packaging. The *rep* gene encodes four nonstructural proteins, Rep78, Rep68, Rep52, and Rep40, which play a role in viral genome replication and transcription, as well as packaging. Rep78 and Rep68 are translated from mRNAs transcribed from the p5 promoter, while Rep52 and Rep 40 are derived from mRNAs transcribed from the p19 promoter. Alternative splicing replaces a 92 amino acid C-terminal element in Rep78 and Rep52 with a 9 amino acid element in Rep68 and Rep40 [19]. The *cap* gene encodes the three structural proteins of the AAV capsid, VP1 (87 kDa), VP2 (72 kDa), and VP3 (63 kDa), translated from mRNA transcribed from the p40 promoter. Differential splicing yields major and minor spliced products. VP1 is translated from the minor spliced mRNA, yielding less VP1 protein. VP2 and VP3 are both translated from the more abundant major spliced mRNA; however, VP2 is translated less efficiently because it initiates at an ACG codon, while VP3 is translated very efficiently because of a favorable Kozak context [20]. As a result, the AAV capsid proteins, which differ only in their N-terminal region, are present in the mature virion in a ratio of 1:1:10 (VP1:VP2:VP3). The single-stranded DNA genome of AAV and its products are depicted in Fig. 2.1.

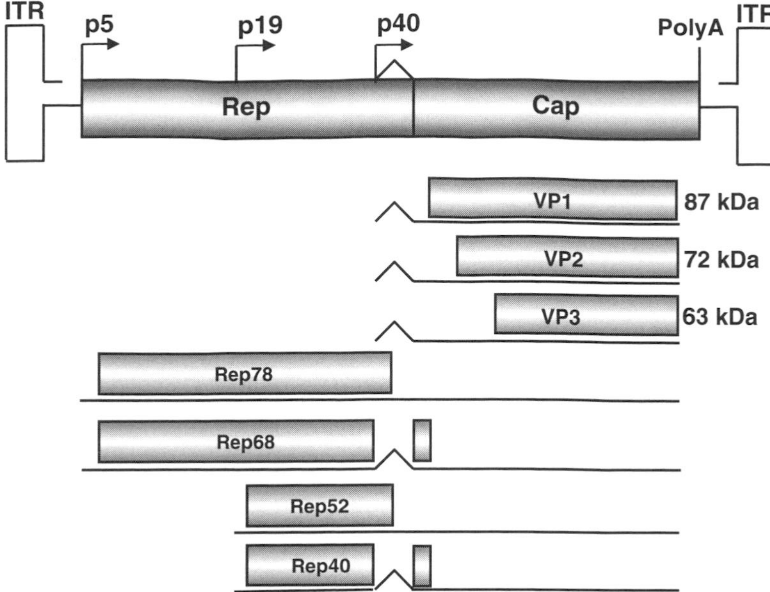

Fig. 2.1 The single-stranded DNA genome of AAV. The inverted terminal repeats (ITRs) flank the two open reading frames *rep* and *cap*. The *rep* gene encodes four nonstructural proteins – Rep78, Rep68, Rep52, and Rep40. The *cap* gene encodes three structural proteins – VP1, VP2, and VP3. The location of the promoters, p5, p19, and p40 are depicted by arrows

2 The AAV Capsid

2.1 Capsid Assembly and Packaging

In 1980, the work of Myers and Carter provided evidence that the structural proteins for AAV assemble into empty capsids, and then the genome is packaged into these preformed capsids [21]. Pulse-chase experiments showed that empty particles rapidly accumulate (10–20 min), but that mature "full" virions accumulated more slowly (4–8 h). They also showed that the number of empty viral particles decreases at the same rate as the number of DNA-containing mature virions increases during the course of infection. Additionally, de la Maza and Carter showed that DI particles, particles with deletions in the AAV genome, are packaged into apparently normal capsids, indicating that full-length viral genomic DNA is not required for the assembly or structural integrity of the AAV capsid [22]. In the absence of capsid assembly, ssDNA does not accumulate, further suggesting that empty capsids form first. Interactions between the preformed AAV capsid and Rep52 provides a mechanism where the nonstructural protein's helicase activity inserts the viral DNA [23]. Within the cell, capsid assembly occurs at centers within the nucleus where there is colocalization of Rep proteins, capsid proteins, and DNA [24]. Empty AAV capsids can be produced by expressing the AAV *cap* gene in insect cells using a baculovirus expression system [25] or in mammalian cells utilizing a recombinant adenovirus expressing the AAV capsid proteins. These systems have advanced the structural studies of the AAV capsid as a result of the large amount of empty capsids or virus-like particles (VLPs) that can be produced. Studies of AAV assembly have demonstrated that VP3 alone is sufficient to form VLPs [26], but VP1 is required for infectivity [27]. Subsequently, it was shown that the unique N-terminus of VP1 has phospholipase A2 activity and contains a nuclear localization signal (NLS) [28]. The N-terminus of VP2 also has a NLS and may play a role in transporting VP3 into the nucleus; however, it has been shown that the N-terminus of VP2 is nonessential and that infectious virus can be produced that lack VP2 entirely [29]. Additionally, the N-terminus of VP2 has been replaced with green fluorescent protein and these capsids still assemble and maintain infectivity. This demonstrates that VP2 can tolerate peptide insertions and may be useful for incorporating peptides into the capsid for cell-specific targeting of AAV.

2.2 Capsid Structure

The adeno-associated virus capsid is ~25 nm in diameter and is composed of 60 subunits arranged in T = 1 icosahedral symmetry. Because AAV has a number of features that make it attractive as a gene transfer vector, many studies have focused on the basic biology of the virus, including studies that address the structural characteristics. Cryo-EM or crystal structures for AAV2, AAV4, and AAV5 have been determined [30–34], and the crystal structure for AAV8 is currently in progress [35]. Dependoviruses share the same subunit fold as the other members of the family Parvoviridae, including the insect densoviruses, and the autonomously replicating parvoviruses such as canine parvovirus (CPV) and minute virus of mice (MVM), even though AAV shares low capsid primary sequence identity (7–22%) [36–38]. The monomeric subunit of AAV has a conserved β-barrel core that is common in viral capsid proteins. Figure 2.2 depicts the structure of a monomeric subunit of AAV2 as determined by Xie et al. [31].

The motif is an eight-stranded antiparallel β-barrel motif (jelly-roll β-barrel), with the β strands labeled B-I [36]. The β-strand labeled A is present in some parvoviruses, including AAV. Not surprisingly, genetic capsid mutants of AAV2 in the conserved core β-barrel, such as mut19 (βA mutation), mut20 (βB mutation), mut25 (βD mutation), and mut46 (βI mutation), are unable to assemble into capsids [40]. AAV has long loop insertions between the strands of the core β-barrel that are labeled according to the β strands that they flank. These long interstrand loops contain β ribbons and elements of secondary structure that form much of the outer surface features of the AAV capsid. The GH loop is the longest interstrand loop, and three VPs interact extensively at each three-fold axis of symmetry, forming a prominent spike. Five DE loops each form an antiparallel β ribbon at the five-fold axis of symmetry that results in a cylindrical structure surrounding a canyon-like

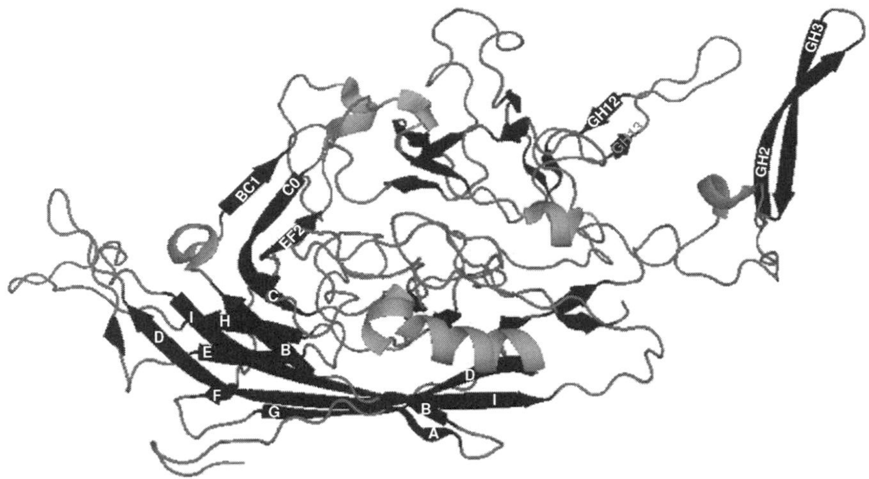

Fig. 2.2 The structure of a monomeric subunit of AAV2 as determined by Xie et al. [31]. This image was produced using the AAV2 coordinates from the Protein Databank, (PDB Accession no. 1lp3), with the molecular modeling software PyMOL (www.pymol.org) provided by DeLano Scientific, Palo Alto, CA [39]

depression. At the two-fold axis of symmetry there is a small depression, often referred to as the two-fold dimple [41, 42]. Analysis of newly discovered AAV genotypes identified a total of 12 hypervariable regions on the AAV capsid [43]. Overlaying these regions onto the X-ray crystallographic model of AAV2 showed that these regions are exposed on the capsid surface. Most of the variability is located between the G and H β strands, which are implicated in the formation of the valley and the peaks of the protruding three-fold axis of symmetry. These surface features of the virus are responsible for the interactions of the capsid with cellular receptors, as well as antibodies.

2.3 Cellular Receptors

Motifs on the capsid surface are critical for attachment to the host cell, which is the first step required for infection. Different serotypes of AAV utilize unique cellular receptors. The primary receptor for AAV2 is heparin sulfate proteoglycan (HSPG). After binding the cell surface, AAV2 can utilize secondary receptors, such as αVβ5 integrin, human fibroblast growth factor receptor-1 (FGFR-1), or hepatocyte growth factor (c-met), which mediate entry [44–47]. Recently, it was demonstrated that AAV2 utilizes α5β1 as an alternative co-receptor in HEK293 cells which lack αVβ5 integrin. The integrin recognition sequence, NGR, is at amino acid 511–513 within VP3, which, in assembled capsids, is located at the three-fold axis of symmetry, adjacent to R585 and R588 of AAV2, which have previously been implicated in heparin binding [48]. The NGR motif is conserved among AAV serotypes 1–11, with the exception of AAV4, AAV5, and AAV11. Similar to AAV2, AAV3H has also been reported to bind heparin, heparan sulfate, as well as FGFR-1 [49]. The receptor for AAV5 has been shown to be PDGFR [50], and AAV5 binds to α-2-3-*N*-linked sialic acids [51], while AAV4 binds to α2-3-*O*-linked sialic acids [52]. AAV1 and AAV6 utilize α-2-3-*N*-linked sialic acids, as well as α-2-6-*N*-linked sialic acids; however, unlike AAV4, they are unable to utilize *O*-linked sialic acids [53]. Domains on the surface of the AAV capsid interact with the cellular receptors found on specific cells, resulting in the range of tissue tropism seen for the various AAV serotypes.

2.4 Serotypes

There are 11 known serotypes of AAV with different cellular targets and antigenic properties. Recently, about 100 genomic variants of these primary AAV serotypes have been discovered [54–56]. These new variants may provide expanded tropism and unique cellular targets for AAV-mediated gene delivery. Utilizing the natural differences in tropism of the AAV serotypes provides one strategy to efficiently deliver AAV vectors to specific target tissues, and selecting the appropriate capsid serotype for the target tissue is an important consideration. The rAAV genome can be packaged into capsids of its own serotype, "isotype," or alternatively the rAAV genome can be "cross-packaged" into capsids derived from another serotype, a process called pseudotyping [57–59]. For pseudotyped vectors, the capsid gene of each serotype can be cloned with AAV2 *rep,* and the transgene is flanked by ITRs of AAV2. The resulting vector has a capsid of some serotype other than AAV2 and the packaged transgene flanked by the ITRs of AAV2. This provides a method to compare the transduction efficiency of various serotypes and the ability to choose the most efficient one for the specific application [58]. Table 2.1 lists several target tissues and the comparative transduction efficiency of selected AAV serotypes studied in animal models. Efficient transduction could involve cell entry associated with receptor binding and internalization, or post-entry events such as cellular processing pathways, intracellular transport, nuclear entry, or processing of virions and vector genomes. Thomas et al. showed that AAV6 and AAV8 vectors uncoat faster and, as a result, transgene expression is seen sooner than AAV2 [90, 91].

Repeated vector delivery might be needed in order to increase transgene expression or to deliver two genes in a coordinated fashion. For these applications, the same transgene can be packaged into capsids from different AAV serotypes for readministration and to circumvent an antibody neutralization response [92]. Differences in the amino acid composition of the viral capsid present unique epitopes to the host's immune system. Studies with rAAV2 and rAAV8 vectors

Table 2.1 Transduction efficiencies of AAV serotypes in different tissues

Tissue type	Serotype	Reference
Muscle	AAV1 > AAV2	[60]
	AAV1-AAV6-AAV7 > AAV5 > AAV3 > AAV2 > AAV4	[58]
	AAV6 > AAV5 or AAV2	[61]
	AAV5 > AAV2	[62]
	AAV7	[63]
	AAV8 highest systemic; AAV1 and AAV6 highest local injection	[64]
	Transducing remote sites when injected locally:	
	AAV7 = AAV8 > AAV1, AAV5 and AAV6. AAV8 > AAV2; AAV1 > AAV2 and AAV8	
Liver (hepatocytes)	AAV8 > AAV7 > AAV5 > AAV2; AAV1 > AAV5 > AAV3 > AAV2 > AAV4	[58]
	AAV5 > AAV2	[65]
	AAV1 > AAV8 > AAV6 > AAV2	[66]
	AAV8	[63]
	AAV9 = AAV8	[67]
	AAV8-AAV9 > AAV2	[54]
Pancreas	AAV8 > AAV2	[91]
	AAV1 > AAV2	[68]
Kidney	AAV1 = AAV2 = AAV5	[69]
	AAV2 transduces tubular epithelium but not glomerular, blood vessel or interstitial cells	
Lung	AAV5 and AAV6	[70]
	AAV9 > AAV5	[71]
Retina	AAV5 > AAV4 > AAV1 = AAV2 = AAV3	[58]
Photoreceptor cells	AAV5 > AAV2	[72]
Hematopoietic stem cells	AAV1 > AAV2-AAV5	[73]
Dendritic cells	AAV6	[74]

(continued)

Table 2.1 (continued)

Tissue type	Serotype	Reference
Cochlear inner ear cells	AAV1 and AAV2 > AAV5	[75]
Solid tumors and melanoma	AAV2 > AAV1 and AAV3	[76]
Glioblastoma	AAV8 = AAV7 > AAV6 > AAV2 > AAV5	[77]
Glioma cells	AAV2 > AAV4 and AAV5	[78]
Brain	AAV5 > AAV1 > AAV2	[58]
	AAV7 > AAV8 > AAV5 > AAV2 = AAV6	[77]
	AAV1 and AAV5 > AAV2; AAV1 and AAV5 transduced pars reticulate; AAV1 transduced entire midbrain; AAV1 and AAV5 transduced pyramidal cell layers; AAV2 transduced the dentate gyrus	[79]
	AAV5 > AAV4 > AAV2	[80]
	AAV1 > AAV2	[81, 82]
	AAV8 > AAV1 or AAV2	[83]
	AAV5 > AAV2	[84]
	AAV2 transduces neurons	[85]
	AAV5 and AAV1 transduce neurons and glial cells	[85]
	AAV4 transduces ependymal cells	[85]
	AAV1 transduces glial cells and ependymal cells	[82]
Cardiac tissue	AAV8 > AAV2	[64]
	AAV1 > AAV2 > AAV5 > AAV4 > AAV3	[86]
	AAV9 > AAV8	[67]
	AAV9 > AAV8 > AAV1	[87]
	AAV1 > AAV2	[88]
	AAV6 > AAV2	[89]

expressing human factor IX indicate that the AAV2 capsid (amino acids 373–381) and the AAV8 capsid (amino acids 50–58) can elicit a cytotoxic T-lymphocyte (CTL) response [93]. In mice, the AAV8 capsid amino acids 126–140 elicited a Th1 response, whereas a Th2 response was elicited by AAV2 amino acids 475–489. The magnitude of the immune response depends on several factors, including vector dose, as well as the route and site of administration. Antibodies formed after transduction with one AAV serotype are likely to show only weak interaction with other serotypes, or may not cross-react. In animals that are transduced with AAV6 and then transduced with AAV2, delivery of the transgene was not hampered; however, animals transduced with AAV2 and then retransduced with AAV2 developed a neutralizing antibody response [92]. The region of the AAV2 capsid that has been shown to be responsible for generating a $CD8^+$ T cell response is the RXXR motif in VP1, VP2, and VP3, which is involved in heparin binding for AAV2. This region was identified as being involved in uptake into dendritic cells, as well as the activation of capsid-specific T cells [94].

Alternative serotypes may also provide for treating cells that have heretofore been refractory to AAV infection, such as stem cells. It has been demonstrated that AAV2 is not efficient at transducing hematopoietic stem cells because of suboptimal levels of expression of the cell surface receptor for the viral vector (HSPG), impaired cellular trafficking of the vector, inefficient vector uncoating, and the lack of viral second-strand DNA synthesis [7]. However, AAV1 is able to transduce hemotopoietic stem/progenitor cells when evaluated in short-term colony assays, as well as in long-term bone marrow transplantation assays *in vivo* [73]. There is also a low transduction efficiency for AAV2 in the airway epithelium of the lung due to a low abundance of HSPG on the apical surface; however, sialic acid is an abundant sugar on the apical surface of airway epithelium and pseudotyped AAV5 and AAV6 vectors have been shown to be efficient at gene transfer to murine airway epithelia *in vivo* [95, 96]. Encapsidated AAV6 vectors achieve transduction rates that should be sufficient for treating lung diseases, such as cystic fibrosis (CF). Recently, it was determined that the transduction efficiency of rAAV2 for the *in vivo* treatment of pancreatic and colon carcinoma is insufficient; however, other AAV serotypes, which have not yet been tested, may result in better transduction of this target tissue [97].

AAV2 and AAV1 bind to different cellular receptors, resulting in better transduction of muscle tissue for AAV1, and since the vectors in these studies contained the same *cis* elements, the AAV capsid is responsible for the differences in transduction efficiency between serotypes [98, 99]. Hauck and Xiao produced a series of capsid mutants to investigate the major regions of the AAV1 capsid responsible for the increased transduction efficiency of AAV1 in muscle tissue [100]; the major tissue tropism determinants were located in the surface region of VP1 (amino acid 350–423). Similar functional studies of the AAV capsid will provide a better understanding of the surface features and the specific requirements for engineering AAV capsids to efficiently target specific cells.

3 Methods

3.1 *Retargeting*

A receptor, like HSPG, which is present on a variety of cell types poses difficulties for targeting AAV2 vectors to a specific tissue. One approach to limiting gene expression in the target tissue is to utilize a viral vector that harbors a tissue-specific promoter [4, 72, 101]. Genetic modifications of the AAV capsid also provide an alternative approach for retargeting AAV2. Regions of the capsid have been identified that will accept insertion of heterologous peptides for retargeting. Girod et al., first demonstrated that the insertion of a 14 amino acid integrin binding peptide at amino acid 587 allowed for retargeting of AAV2 [102]. To obtain a genetic map of the AAV2 capsid, Wu et al. constructed alanine scanning mutants at 59 different positions in the AAV capsid gene by site-directed mutagenesis [103]. Studies of these mutants showed that the capsid can tolerate some modifications and still maintain infectivity. A 34 amino acid IgG binding domain was inserted into rAAV2 capsids at amino acid 587, and this, coupled with antibodies against B1 integrin, CXCR4, or the c-kit receptor, mediated the targeting of these rAAV vectors to specific cell surface receptors [104]. Endothelial cell binding peptides were identified using phage display and inserted into the AAV2 capsid at amino acid 587 to transduce venous endothelial cells and significantly reduce hepatocyte transduction [105]. Recently, it was demonstrated that AAV2 could be retargeted to the heart utilizing a capsid with mutations at amino acids 484 and 585 to eliminate heparin binding, which is required for AAV2 to infect the liver [106]. Perabo et al. also demonstrated that insertion at amino acid 587 with peptides that confer a net negative charge will ablate binding to HSPG, and this correlated with liver and spleen detargeting for AAV2. Insertion of peptides between 585 and 588 can either cause spatial separation or sterically block heparin binding of AAV2 (using bulky amino acids), and the insertion of positively charged peptides reconstitutes the ability to bind HSPG [107]. This provides a strategy to improve AAV targeting in other tissues. For retargeting AAV to tumor cells, an RGD-4C motif can be inserted at amino acids 520 and 584, or 588 to confer a novel tropism and eliminate heparin binding. The RGD motif binds the cellular integrin receptor, which is expressed on tumor cells, including ovarian cancer cells, but these cells express only low levels of HSPG and are nonpermissive for AAV2 transduction [108, 109].

3.2 *Capsid Mosaics*

Introduction of a peptide into the AAV2 capsid is a viable strategy for retargeting AAV2 to additional cell types; however, modifications in the AAV2 capsid often result in a reduction in vector yields, especially if the insertion is large. The unique N-terminal region of VP2 allows for peptide insertions without a loss of titer, when

VP1 and VP3 are supplied in *trans*. AAV2 mosaics that have ligand insertions in a subset of VP1, VP2, and VP3 molecules result in increased vector yields and transducing titers, compared with viruses that carry the insertion in all 60 capsid protein subunits [110]. Insertion of an HA epitope (YPVDVPDYA) at amino acid 522 or 533 of VP1 results in noninfectious particles [103]; however, the inclusion of wild-type capsid proteins can restore viral infectivity. Ried et al. [104] inserted an immunoglobulin-binding fragment of protein A (Z34C) that resulted in at least a tenfold reduction in particle titers, most likely due to its large size (34 amino acids). Compared with wild-type virus, infectious titers were 4 orders of magnitude lower. Gigout et al. generated mosaic viruses that contained between 25 and 75% of Z34C capsid proteins, with the rest of the capsid being composed of wild-type subunits to produce AAV2 mosaics that were infectious. Compared with wild-type virus, the Z34C mosaics showed up to a tenfold increase in titer, and those containing 25% Z34C capsid protein (an average of 15 subunits per capsid) were 4–5 orders of magnitude more infectious than all mutant viruses. By mutating R585 and R588 to alanine, they were able to eliminate HSPG binding and then the insertion of Z34C resulted in retargeting of AAV2 [110]. This work demonstrated for the first time that a combined approach of generating AAV2 mosaics to alter tropism and mutating two residues to reduce HSPG binding could be used to retarget AAV2 to specific cell types. For AAV1, incorporating an RGD4C motif in VP1 at amino acid 590 enables targeting to integrin receptors which are present on vascular endothelial cells [111]. It is also possible to incorporate a small biotin acceptor peptide (BAP) in this position for the purpose of metabolically biotinylating the AAV1 capsid for purification using a commercially available avidin affinity column. Mosaics with both of these modifications have been generated, which enable retargeting of the AAV1 capsid, and simplify purification.

3.3 Capsid Chimeras

Another strategy to broaden tissue tropism is to generate AAV with mixed or chimeric capsids [112]. This is accomplished by supplying the capsid gene from two distinct AAV serotypes during production. By varying the ratio of the two capsid genes, the resulting mixed virus may exhibit altered tropism. Rabinowitz et al. showed that mixed capsids of AAV2/5 at a ratio of 3:1 resulted in a loss of heparin binding. The ability of these virions to transduce HeLa cells, which have high levels of heparin sulfate, decreased, while the ability to transduce the heparin-sulfate-deficient cell line CHO pgsD increased from 2% to > 30% as the composition of the virion changed from AAV2-like to AAV5-like. Chimeras also may show combined tropism of both serotypes, broadening the tropism for these virions. Kohlbrenner et al. utilized chimeric capsids to improve infectivity when expressing AAV5 and AAV8 in insect cells [113]. It was determined that AAV5 and AAV8 vectors generated using the baculovirus system had low infectivity, which was due to insufficient phospholipase A2 activity. Substituting the entire VP1 protein from AAV2 in chimeric

AAV5 or AAV8 capsids resulted in increased phospholipase activity levels and enhanced transduction. Generating chimeras also provides a method to functionally define structural relatedness for newly discovered serotypes. The ability to generate stable chimeric capsids suggests that the subunits from these different serotypes are structurally compatible. Conversely, AAV2 when mixed with AAV4 inefficiently packages genomes. The inability to generate stable virions may reflect a failure of essential structural subunit interactions at one or more axes of symmetry. Alternatively, the nonstructural Rep protein interactions that occur on the surface of the capsid for packaging may not be compatible in the AAV2/AAV4 chimeras, as suggested by the structures for AAV2 and AAV4 [30, 31].

3.4 Vector Utility

The AAV capsid provides a potent gene delivery vehicle and has shown great promise in animal models, as well as in human clinical trials conducted to date. AAV vectors have been developed for a multitude of diseases, including disorders of the central nervous system for which other vectors and methods of treatment have been inadequate. Examples include Parkinson's disease and Alzheimer's disease. Gene therapy approaches utilizing AAV have also addressed classic genetic disorders, such as lysosomal storage disorders, hemophilia, and cystic fibrosis. Many diseases that are amenable to AAV gene therapy approaches are listed in Table 2.2. AAV vectors to treat these diseases have shown significant and persistent gene transfer without toxicity [3–5, 117–124, 138–141, 159–160] in human clinical trials. Vectors are administered to effect intracellular expression of proteins such as dystrophin or expression of secreted therapeutic proteins that result in cross correction of cells. AAV may also be utilized as a vaccine to deliver specific antigens; for example, AAV5 targets dendritic cells, allowing for antigen presentation, and AAV vectors have been produced that express components of papillomavirus, the causative agent of cervical cancer, as well as the receptor for the spike protein of SARS coronavirus [181, 182]. AAV-based gene transfer has the therapeutic potential to arrest or reverse the course of these inherited and acquired diseases.

3.5 Vector Production

An important consideration for the development of a suitable gene delivery vector is the ability to produce the quantities of vector required to treat human patients. Historically, production methods have been a limitation in the development of AAV vectors. Improvements in production and purification methods have resulted in better yields, increased purity, and higher titers through the development of scalable systems [113, 183–187]. Recombinant AAV vectors can be manufactured using a two-plasmid system, as shown in Fig. 2.3. One plasmid harbors the therapeutic

Table 2.2 Disease targets

Disorder	Target tissue	Species	Serotype	Reference
Cystic fibrosis	Nasal and lung epithelium	Mice	AAV2/5, AAV2/9	[71]
	Lung	Mice	AAV1, AAV2, AAV5	[114]
	Airway epithelium	Cell culture	AAV5	[115]
	Lung	Mice	AAV2, AAV5	[116]
	Lung	Mice	AAV2, AAV6	[96]
	Lungs – airway epithelium and maxillary sinus	Human	AAV2	[117–120, 121–124]
Hemophilia B	Liver	Mice	AAV8, AAV9	[67]
	Liver	Mice	AAV1, AAV2, AAV6	[125]
	Liver	Mice	AAV5	[65]
	Liver	Mice, Monkey	AAV2, AAV5, AAV2/8	[126]
	Liver	Mice	AAV2	[127, 128]
	Liver	Dog	AAV2/8	[129]
	Liver	Dog	AAV2	[130, 131]
	Liver	Human	AAV2	[4]
	Muscle	Mice	AAV1-AAV2 hybrid	[60]
	Muscle	Mice	AAV1	[132, 133]
	Muscle	Mice	AAV1, AAV2, AAV3, AAV4, AAV5	[98]
	Muscle	Mice; Dog	AAV1	[134]
	Muscle	Dog	AAV2	[135–137]
	Muscle	Human	AAV2	[138–141]
Anemia	Muscle	Mice	AAV2	[142, 143]
Parkinson's disease	Brain	Rat	AAV2	[144]
	Brain	Monkey	AAV2	[145–149]
	Brain	Human	AAV2	[5]
Lysosomal storage disease	Muscle	Mice	AAV2	[150]
	Muscle	Mice	AAV1 and AAV2	[151]
	Muscle	Mice	AAV2/6	[152]

(continued)

Table 2.2 (continued)

Disorder	Target tissue	Species	Serotype	Reference
	Brain	Mice	AAV2	[153]
	Brain	Mice	AAV2, AAV5	[154]
	Brain	Human	AAV2	[3]
	i.v. – systemic delivery	Mice	AAV8	[155]
	i.v. – systemic delivery	Mice	AAV2/8	[156]
Canavan disease	Brain	Mice	AAV2	[157]
	Brain	Rodents, monkey, human	AAV2	[158]
	Brain	Human	AAV2	[159, 160]
Type I diabetes	Pancreas – acinar cells, beta cells	Mice	AAV6, AAV8	[161]
	Islets	Mice	dsAAV2, dsAAV6, dsAAV8	[162]
	Islet cells	Mice	AAV1, AAV2	[68]
				[68]
Alzheimer's disease	Brain – Abeta vaccine	Mice	AAV2	[163–165]
	Brain	Rat	AAV2/5	[84]
	Brain	Rat	AAV1, AAV2, AAV4, AAV5	[85]
Cardiovascular	Cardiac tissue	Rodent	AAV2	[166]
	Cardiac tissue	Mice	AAV2/1, AAV2/8, AAV2/9	[87]
	Cardiac muscle	Mice	AAV2	[151, 167]
	Cardiac tissue	Rat	AAV6, AAV2	[89]
Cancer	Glioma	Tumor cell lines	AAV2, AAV4, AAV5	[78]
	Ovarian carcinoma	Ovarian carcinoma cell line	RGD modified AAV2	[168]
			AAV2	[169]
		Mice	AAV2	[170]
	Breast carcinoma	Mice	AAV2	[171]
	Glioblastoma	Mice	Pseudotyped AAV7 and AAV8	[77]
		Rat	AAV2	[172]

	Prostate cancer	Mice	AAV2	[173]
	Liver cancer	Mice	AAV2	[174, 175]
	Lung adenocarcinoma	Mice	Hybrid AAV2/5	[176]
Vaccine development	Bone-marrow-derived dendritic cells	Mouse	AAV6	[74]
	Papillomavirus antigen HPV16 L1 protein delivery	Mice – intranasal delivery	AAV5	[182]
	Dendritic cells or muscle, HIV gp160 protein delivery	Mice	AAV1, AAV5, AAV7 and AAV8	[181]
	Dendritic cells, siRNA delivery for dengue virus vaccine	Human cells	AAV2	[177]
Obesity	Vector expressing leptin	Mouse	AAV2	[178]
	Vetor expressing leptin receptor	Rat	AAV2	[179]
	Vector expressing adiponectin	Rat	AAV1, AAV2, AAV3, AAV4, AAV5	[180]

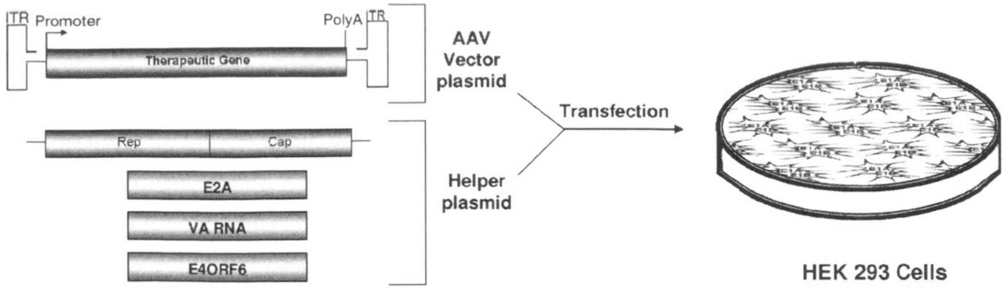

Fig. 2.3 Viral vector production. The rAAV Vector plasmid contains the therapeutic gene flanked by the ITRs, usually of AAV2. The helper plasmid contains the *rep* and *cap* genes, as well as the adenoviral genes needed for replication. Both plasmids are transiently transfected into HEK293 cells that express the adenovirus E1A and E1B gene products

gene flanked by the ITRs and the other one contains the AAV *rep* and *cap* genes, as well as the helper functions for AAV2 replication. The adenovirus helper functions are provided by expression of E2A, E4ORF6, and VA RNAs in the helper plasmid, while the adenovirus E1A and E1B gene products are expressed in the HEK293 cells utilized for production [188]. The two plasmids are co-transfected into 293 cells, and this results in production of an AAV virion that contains the therapeutic gene flanked by the ITRs [189]. The ITRs are the only viral sequences remaining after purification of rAAV vectors.

Protocol 1 (see Appendix) describes the small-scale production and purification of rAAV vectors as described by Zolotukhin et al. [190]. Generally, HEK293 cells are expanded, transfected, and harvested at 72 h post-transfection. The cells are then lysed, and loaded onto either a cesium chloride or an iodixanol gradient to separate infectious virions from empty capsids and other cellular proteins. Virus is then purified using either heparin affinity chromatography or ion-exchange chromatography. The chromatography method used is dependent on the natural receptors for AAV, as well as the charge characteristics of the viral particle, as shown in Fig. 2.4. Following column chromatography, the virus is concentrated and characterized. These protocols yield vector stocks that are relatively pure; however, owing to the density gradient requirement, these methods are not easily scalable.

Protocols 2 and 3 describe the large-scale production and purification of rAAV vectors as described by Snyder et al. [18]. Protocol 3 provides a method suitable for large-scale production of rAAV2 vectors that is scalable for the production of vectors under cGMP conditions for clinical trials [191]. Cell factories of HEK293 cells are transfected, incubated for 72 h, and then harvested. The cell pellet is resuspended in lysis buffer containing 0.5% deoxycholate to reduce viral particle aggregation and 50 U Benzonase/ml to degrade cellular, plasmid, and nonpackaged nucleic acid. A microfluidizer is utilized to lyse the cells in order to form a fine suspension that can be loaded directly onto a Streamline Heparin column. The column is washed and the virus is eluted from the column with phosphate-buffered saline (PBS) containing 0.2 M NaCl (for a total ionic strength of 350 mM). The peak fraction from the Streamline Heparin column is brought to 1 M NaCl and loaded on a Phenyl

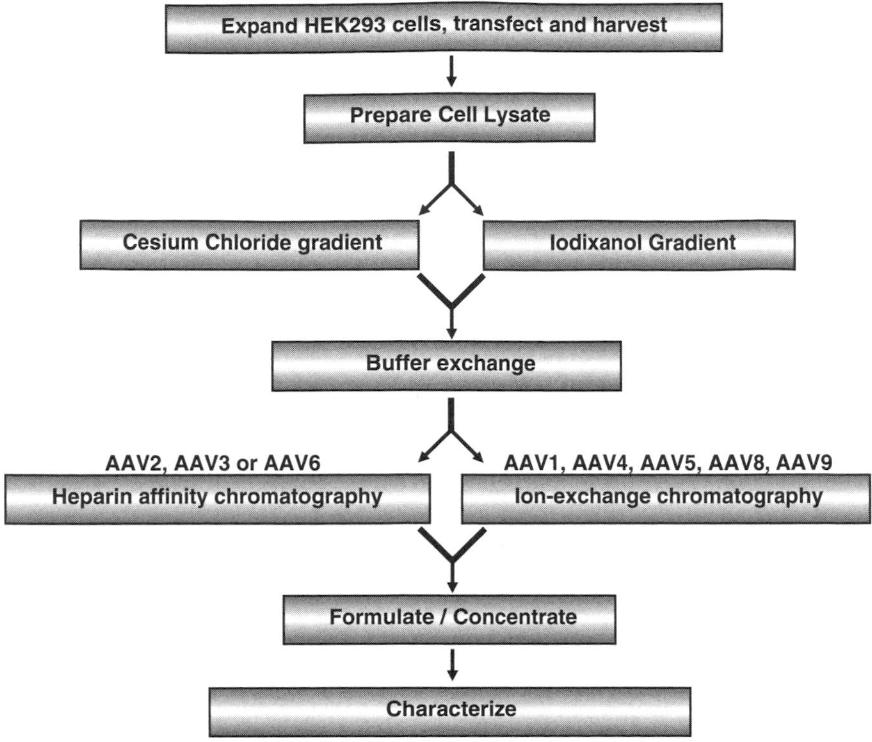

Fig. 2.4 Flowchart of the steps for rAAV production as described in Protocol 1. HEK293 cells are expanded, transfected, and harvested at 60 h posttransfection. The cells are lysed, and loaded onto either a cesium chloride gradient or an iodixanol gradient to separate infectious virions from empty capsids. Virus is purified using either heparin affinity chromatography or ion-exchange chromatography. Virus preparations are formulated and concentrated, and characterized

Sepharose column. This column is a hydrophobic interaction column (HIC) and the virus will remain in the flow through. The flow through from the Phenyl Sepharose column is diluted using sterile water to bring the salt concentration to 0.150 M NaCl. This is loaded onto a SP Sepharose column, washed in column buffer, and eluted in PBS with 0.135 M NaCl (for a total ionic strength of 0.285 M).

3.6 Vector Characterization

Several assays to characterize the final product with respect to titer, purity, identity, potency, particle to infectious ratio, and stability are listed in Table 2.3. In addition, vector product safety testing should be conducted prior to animal or human studies. Assays used for product characterization and safety testing are included in Table 2.3.

Table 2.3 Assays used for vector characterization and safety testing of rAAV vectors produced for clinical trials under cGMP conditions

Assay type	Method and reference	Purpose
Titer		
Infectious	Infectious Center Assay (ICA) [192, 193]	Determine titer of infectious particles produced
	Serial Dilution Replication Assay (dRA) [194]	
Vector genome	Dot-blot hybridization	Determine genome-containing vector concentration
	PCR [195]	
Capsid	ELISA [24]	Determine total capsid protein concentration – enables a determination of empty particles in the prep
	Bradford	
	Western	
	Electron microscopy [196]	
	Optical density (OD) [197]	
Purity		
Protein	SDS-PAGE	Determine the presence or absence of contaminating proteins
Cellular DNA	DNA hybridization or PCR	Determine the presence or absence of cellular DNA
Identity		
Transgene cassette	DNA sequencing or restriction enzyme digestion	Verification of the transgene
Capsid	SDS-PAGE with silver and Coomassie stain	Expected AAV banding pattern
	Limited proteolysis	Serotype identification
Potency		
Transgene expression	Transduction assay in cells or animals	Ensure that the active transgene product is expressed
Safety		
Adventitious agents	qPCR-based assays to detect infectious adventitious viral agents	Detect contaminating infectious viral agents of human or animal origin (serum, trypsin)
Mycoplasma	Growth assays on cells in antibiotic-free conditions, followed by dye or PCR to detect mycoplasma	Determine the presence or absence of mycoplasma
	Growth assay in appropriate agar media	
Endotoxin	Rabbit pyrogen assays	Determine the presence or absence of endotoxin

(continued)

Table 2.3 (continued)

Assay type	Method and reference	Purpose
Sterility	Bacteriostasis/fungistasis	Determine the presence or absence of microbial contaminants
Stability		
Physiochemical	SDS-PAGE	Demonstrate that the product is not degrading over time
Infectious	ICA	Demonstrate that infectivity is maintained over time
Sterility	Bacteriostasis/fungistasis	Demonstrate the integrity of final product container over time

3.7 Proteolytic Structural Mapping

In the near future, customized AAV gene therapy vectors may consist of modified capsids that allow for specific targeting to treat patients with various diseases. The 3D structures of the AAV capsids will provide a basis for rational vector design; however, the 3D structures available for autonomous parvoviruses and dependoviruses provide only a "snapshot" of the capsid topology in a low energy conformation. Our knowledge about the AAV viral capsid structure in solution is limited; however, this structure must be dynamic to carry out the various functions required for viral attachment and entry, as well as trafficking within the cell. Cryo-EM studies have shown that the unique N-terminus of VP1 is internal to the capsid based on additional density at the two-fold axis of symmetry [198]. *In vitro*, upon heat treatment of AAV capsids, it has been shown that this region can be externalized. Mutagenesis experiments have shown that this externalization occurs through the pore at the five-fold axis of symmetry [199]. Previously, antibodies have been produced that detect various regions of the capsid proteins [200]. The B1 antibody epitope is on the C-terminal end of the capsid protein, and is primarily internal at the two-fold axis of symmetry in assembled capsids. This antibody is useful in detecting denatured AAV proteins and the epitope is highly conserved among the AAV serotypes. A1 antibody recognizes the unique N-terminal region of VP1, while the A69 epitope is in the N-terminal region of VP2 for AAV2. Polyclonal antibodies have been produced to AAV2 capsids, as well as to other serotypes. In addition, antibodies have been produced that recognize conformational epitopes that are present only on assembled capsids for AAV2, as well as AAV1. Capsid antibodies can be utilized for serotype identity testing of the final vector product; however, many of these antibodies cross-react with multiple serotypes. Historically,

AAV has been shown to be remarkably stable and is generally resistant to proteases. However, we have demonstrated that when exposed to proteases, specific regions of the capsid are susceptible to proteolysis [201]. Owing to differences in the primary sequence of the capsid proteins of different serotypes, as well as the resulting differences in capsid structure, proteolysis can be utilized for AAV serotype determination. Different serotypes provide unique fragmentation patterns when cleaved with protease, with some serotypes such as AAV5 being relatively resistant, while others such as AAV2 are more susceptible to trypsin and chymotrypsin. Figure 2.5 depicts protease mapping of the AAV capsid for capsid serotype determination. A generic assay is described in Protocol 4 where proteolysis is used together with specific antibodies to provide a powerful mapping technique and a method to identify and confirm the serotype of the AAV capsid.

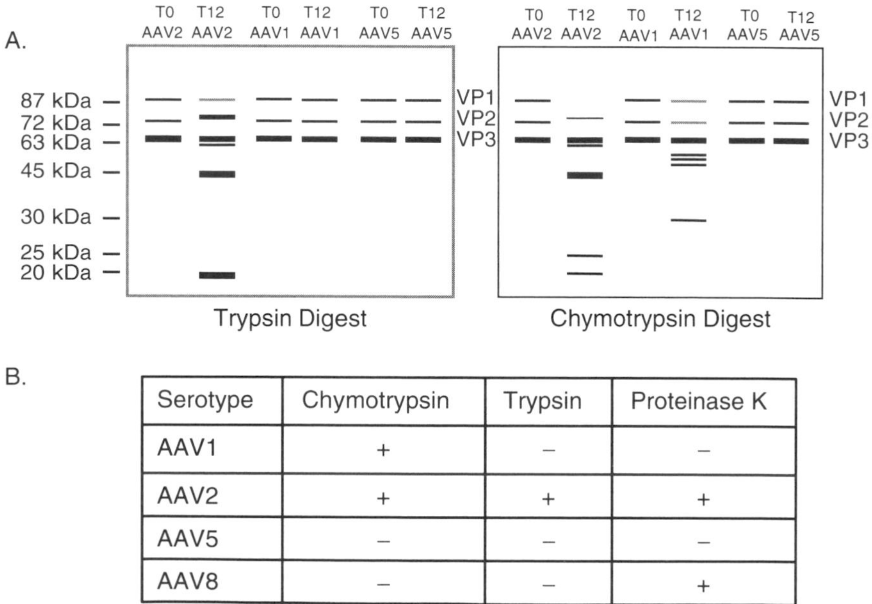

Serotype	Chymotrypsin	Trypsin	Proteinase K
AAV1	+	–	–
AAV2	+	+	+
AAV5	–	–	–
AAV8	–	–	+

Fig. 2.5 Protease mapping of the AAV capsid for capsid serotype determination. **A** Samples are digested with a protease, in this case trypsin for one set of samples and chymotrypsin for the other set. A Western blot is performed using polyclonal antisera to AAV capsids, and based on the fragmentation pattern, a serotype determination can be made. Undigested sample (T_0) and samples digested for 12h (T_{12}) for AAV2, AAV1, and AAV5 are shown. AAV5 is resistant to these proteases, while AAV1 and AAV2 exhibit differences in their fragmentation patterns. T_0 samples represent the undigested capsid proteins VP1, VP2, and VP3. **B** Different AAV serotypes demonstrate different susceptibilities to proteases due to the differences in their primary amino acid sequences. This differential susceptibility provides a unique signature for each serotype and allows for capsid serotype identification

3.8 Concluding Remarks on AAV Capsid

In summary, the AAV capsid is the major determinant in gene transfer efficiency and targeting. Several approaches are available to target AAV viral vectors to specific cell types, including utilizing natural serotypes that target a desired cellular receptor, producing pseudotyped vectors, as well as engineering chimeric and mosaic capsids. Genetic mutations of the AAV capsid have identified regions that will accommodate peptide insertions and modifications for specific targeting. The use of modified capsids may enhance efficiency and target specificity to make a more potent and safe vector. Several genetic diseases are amenable to treatment with AAV vectors, making AAV a valuable gene delivery vehicle, especially for disorders for which current therapies are inadequate or non-existent. Current production methods allow for the generation of quantities of virus needed for clinical trials, and purification methods have been developed for several serotypes. In addition, assays have been developed for final product characterization to ensure patient safety.

Appendix

Protocol 1: AAV Small-Scale Production (~3 × 10^8 cells/prep)

Transfection Protocol

1. Seed [20] 15-cm plates with 293 cells so that they will be 75–80% confluent the following day (12–16 h prior to transfection). 293 cells are cultured in Dulbecco's Modified Eagle's Medium (DMEM) supplemented with 5% fetal bovine serum (FBS), 1% penicillin, and 1% streptomycin.
2. Check plates prior to transfection to ensure optimal confluency, and change media to ensure cells are in a volume of 20 ml. Be careful not to disrupt the monolayer.
3. For transfection, prepare the DNA mixture at room temperature in a 50-ml conical tube as follows: Mix 900 µg of helper plasmid with 300 µg rAAV vector plasmid. This is 60 µg DNA per plate in a 1:1 molar ratio. Bring the final volume of the DNA mixture to 22.5 ml with sterile dH_2O. Add 2.5 ml of 2.5 M $CaCl_2$ to the DNA for a $CaCl_2$ concentration of 0.250 M.
4. Prepare ten 15-ml conical tubes each with 2.5 ml 2× HBS to transfect the 20 plates (25 ml 2× HBS total). Transfect two plates at a time by adding 2.5 ml of DNA mixture to a tube of 2.5 ml 2× HBS, mix, and incubate for 1 min at room temperature to allow precipitate to form. Add 5 ml of media to the tube and mix. (Total volume in the tube is 10 ml.) Pipette 5 ml onto each of the two plates. Repeat for the remainder of the plates.
5. Incubate at 37°C for 72 h.

6. At 72 h, harvest cells. Use media to dislodge the cells. Collect media and cells and centrifuge at 1,000 × *g*, for 10 min at 4°C to pellet the cells. Discard the media. Resuspend pellet in PBS to wash the pellet. Centrifuge at 1,000 × *g*, for 10 min at 4°C. Discard PBS and freeze pellet. Frozen pellets can be stored at −20°C for processing at a later time.

Transfection Reagents and Materials

- 15-cm tissue culture plates.
- 293 cells.
- Culture media: DMEM + 5% FBS + 1% penn/strep.
- Trypsin for splitting cells.
- PBS without magnesium and calcium.
- Vector plasmid, 300 µg.
- Helper plasmid, 900 µg.
- Sterile dH_2O.
- 2.5 M $CaCl_2$ – (stock 147.02 g/mol calcium chloride dihydrate) – for 1 l add 367.55 g $CaCl_2 \cdot 2H_2O$. Prepare 10-ml aliquots and store at −20°C.
- 2× HBS (HEPES-buffered saline; prepare 1 l):

  0.274 M NaCl (stock 58.44 g/mol) – 16.0 g.
  0.010 M KCl (stock 74.56 g/mol) – 0.75 g.
  0.002 M Na_2HPO_4 anhydrous (stock 141.96 g/mol) – 0.28 g.
  0.011 M dextrose (stock 180.16 g/mol) – 2.0 g.
  0.042 M HEPES (stock 238.3 g/mol) – 10.0 g.

  Adjust the pH to 7.05 with 0.5 N NaOH, bring to a final volume of 1 l with dH_2O, and sterile filter. Store in 50-ml aliquots at −20°C. (Note: the optimal pH range for 2× HBS is 7.05–7.12. The pH of the 2× HBS is a key factor that influences precipitate formation.)

Cell Lysate and Cesium Chloride Gradient Protocol

7. Resuspend pellet in 30 ml lysis buffer. Lyse cells by three freeze/thaw cycles. Thaw pellet in 37°C water bath and freeze cells in EtOH/dry ice bath; repeat twice.
8. Transfer the lysate to a 40-ml Dounce homogenizer. Homogenize the lysate with 20 strokes to shear cellular DNA.
9. For each 10 ml of lysate, add 5 g CsCl and homogenize until the CsCl is dissolved completely.
10. Fill six 12.5 ml ultra clear ultracentrifuge tubes (Beckman) with 10 ml of lysate and underlay with 0.5 ml each of CsCl (1.5 g/ml). Balance the tubes using CsCl (1.37 g/ml) prior to ultracentrifugation.

11. Centrifuge for 24 h at 265,000 × *g* (40,000 rpm in SW 41 rotor), 21°C.
12. Collect AAV from the gradient by dripping 1-ml fractions, and verify the density of the fractions by refractometry. The density of infectious AAV virions is 1.40–1.42 g/cm³. Empty AAV particles and DI particles have a density of 1.32–1.35 g/cm³.
13. Dialyze AAV-containing fractions into the buffer that will be used for column chromatography (1× TD buffer for heparin column or 15 mM NaCl, 20 mM Tris (pH 8.5) for Q Sepharose purification).
14. Purify using either heparin affinity chromatography (for AAV2, AAV3, or AAV6) or ion-exchange Q Sepharose chromatography (for AAV1, AAV4, AAV5, AAV8, or AAV9). Use a 1-ml column for 20–40 plates, as described in steps 1–7 under "AAV Purification by Heparin Chromatography" or "AAV Purification by Q Sepharose Chromatography." The purification method is as described by Zolotukhin et al. [190].

Cell Lysate and Cesium Chloride Gradient Reagents

Lysis buffer (prepare 1 l):

- 150 mM NaCl (stock 58.44 g/mol) – 8.766 g.
- 50 mM Tris (stock 121.14 g/mol) – 6.055 g.
- Adjust pH to 8.5 with HCl, bring to a final volume of 1 l with dH_2O, and filter sterilize. Store at room temperature.

1.37 g/ml CsCl in PBS – Dissolve 50 g CsCl in PBS and adjust the final volume to 100 ml. Weigh 1 ml to verify the density is 1.37 g/ml. Filter sterilize. Store at room temperature.

1.5 g/ml CsCl in PBS – Dissolve 67.5 g CsCl in PBS and adjust the final volume to 100 ml. Weigh 1 ml to verify the density is 1.5 g/ml. Filter sterilize. Store at room temperature.

5× TD buffer (prepare 1 l):

- 5× PBS –500 ml of 10× PBS (Invitrogen).
- 5 mM $MgCl_2 \cdot 6H_2O$ (stock 203.3 g/mol) – 1.0165 g.
- 12.5 mM KCl (stock 74.56 g/mol) – 0.932 g.

1× TD buffer – Prepare from 5× TD buffer stock: 200 ml 5× TD stock, 800 ml dH_2O.

Q Sepharose low salt column buffer:

- 15 mM NaCl (stock 58.44 g/mol) – 0.877 g.
- 20 mM Tris (stock 121.14 g/mol) – 2.423 g.
- Adjust pH to 8.5 with HCl and bring to a final volume of 1 l with dH_2O.

Alternative Cell Lysate and Iodixanol Gradient Protocol

7a. Resuspend pellet in 30 ml lysis buffer. Lyse cells by three freeze/thaw cycles. Thaw pellet in 37°C water bath and freeze cells in EtOH/dry ice bath; repeat twice.
8a. To the cell suspension, add 3 µl of 4.82 M $MgCl_2$ and vortex. Add 1 µl of Benzonase (250 U/µl) and vortex. Incubate at 37°C for 30 min. Centrifuge for 20 min at 1,000 × *g*.
9a. Pipette the supernatant into two quick seal tubes for a 70 Ti rotor. Underlay lysate with iodixanol gradient:

- 7.5 ml of 15% iodixanol
- 5.0 ml of 25% iodixanol
- 7.5 ml of 40% iodixanol
- 5.0 ml of 60% iodixanol

10a. Weigh tubes, cap, and seal with heat sealer.
11a. Centrifuge at 350,333 × *g*, 18°C, for 1 h (69,000 rpm in a 70 Ti rotor for 1 h).
12a. Collect the 40% iodixanol band and the interface between the 60% and the 40% bands. This is done by setting up a ring stand and placing the tube in a clamp. Swab the top and bottom of the tube with an alcohol swab. Vent the top of the tube with a needle and collect fractions from bottom of the tube.
13a. Dialyze virus-containing fractions into the buffer that will be used for column chromatography (1× TD buffer for heparin column or 15 mM NaCl, 20 mM Tris (pH 8.5) for Q Sepharose purification).
14a. Purify using either heparin affinity chromatography (for AAV2, AAV3, or AAV6) or ion-exchange Q Sepharose chromatography (for AAV1, AAV4, AAV5, AAV8, or AAV9). Use a 1-ml column for 20–40 plates, as described in steps 1–7 under "AAV Purification by Heparin Chromatography" or "AAV Purification by Q Sepharose Chromatography." The purification method is as described by Zolotukhin et al. [190].

Cell Lysate and Iodixanol Gradient Reagents

Lysis buffer (*150 mM NaCl, 50 mM Tris, pH 8.5*) (prepare 1 l):

- 150 mM NaCl (stock 58.44 g/mol) – 8.766 g.
- 50 mM Tris (stock 121.14 g/mol) – 6.055 g.
- Adjust pH to 8.5 with HCl, bring the final volume to dH_2O with 1 l, and filter sterilize. Store at room temperature.

5 M NaCl (*stock 58.44 g/mol*) – 292.2 g and dH_2O to 1 l.

For iodixanol gradient, mix the following:

Step	Optiprep (ml)	5 M NaCl (ml)	5× TD (ml)	dH_2O (ml)	Phenol red (μl)	Total volume (ml)
15%	45	36	36	63	–	180
25%	50	–	24	46	300	120
40%	68	–	20	12	–	100
60%	100	–	–	–	250	100

AAV Purification by Heparin Chromatography

1. Equilibrate a 1-ml HiTrap heparin HP column (GE Healthcare) by washing with ten column volumes of 1× TD. The flow rate for all chromatography steps is 1 ml/min.
2. Activate the column by washing with five column volumes of 1× TD/1 M NaCl.
3. Re-equilibrate the column by washing with ten column volumes of 1× TD.
4. Apply the dialyzed virus from either the cesium gradient or iodixanol gradient to the column.
5. Wash the column with ten volumes of 1× TD.
6. Elute with five column volumes of elution buffer using either a continuous or step gradient. AAV2 can be eluted in 1× TD/0.5 M NaCl [190] and collect [5] 1-ml fractions.
7. Dialyze the virus preparation into storage buffer or suitable formulation and freeze at −20°C.

Heparin Column Chromatography Reagents

1× TD buffer – prepare from 5× TD buffer stock: 200 ml 5× TD stock, 800 ml dH_2O.
1× TD/1 M NaCl buffer: 200 ml 5× TD buffer, 58.44 g NaCl; bring to 1 l with dH_2O.

Storage buffer:

- 100 mM NaCl (stock 58.44 g/mol) – 5.844 g.
- 50 mM Tris (stock 121.14 g/mol) – 6.055 g.
- Adjust pH to 8.0 and bring to 1 l with dH_2O.

AAV Purification by Q Sepharose Chromatography

1. Equilibrate a 1 ml HiTrap Q HP column (GE Healthcare) by washing with ten column volumes of low salt Q column buffer (0.020 M Tris, 0.015 M NaCl, pH 8.5). The flow rate for all chromatography steps is 1 ml/min.

2. Activate the column by washing with five column volumes of high salt Q column buffer (0.020 M Tris, 1.0 M NaCl, pH 8.5).
3. Re-equilibrate the column by washing in ten column volumes of low salt Q column buffer.
4. Apply the dialyzed virus from either the cesium or iodixanol gradient to the column. (Note: After recovering the virus from the cesium or iodixanol gradient, virus must be dialyzed into low salt buffer or a buffer exchange into low salt buffer must have been performed in order to bind to the Q column.)
5. Wash the column with ten volumes of low salt Q column buffer.
6. Elute with five column volumes of elution buffer (20 mM Tris, 0.5 M NaCl, pH 8.5) [190] and collect [5] 1-ml fractions.
7. Dialyze the virus preparation into storage buffer (50 mM Tris, 100 mM NaCl, pH 8.0) or suitable formulation aliquot, and freeze at −20°C.

Q Sepharose Chromatography Reagents

Q Sepharose low salt column buffer:

- 15 mM NaCl (stock 58.44 g/mol) – 0.877 g.
- 20 mM Tris (stock 121.14 g/mol) – 2.423 g.
- Adjust pH to 8.5 and bring the final volume to 1 l with dH_2O.

Q Sepharose high salt column buffer:

- 1 M NaCl (stock 58.44 g/mol) – 58.44 g.
- 20 mM Tris (stock 121.14 g/mol) – 2.423 g.
- Adjust pH to 8.5 and bring the final volume to 1 l with dH_2O.

Q Sepharose elution buffer:

- 0.5 M NaCl (stock 58.44 g/mol) – 29.22 g.
- 20 mM Tris (stock 121.14 g/mol) – 2.423 g.
- Adjust pH to 8.5 and bring the final volume to 1 l with dH_2O.

Storage buffer:

- 100 mM NaCl (stock 58.44 g/mol) – 5.844 g.
- 50 mM Tris (stock 121.14 g/mol) – 6.055 g.
- Adjust pH to 8.0 with HCl and bring to 1 l with dH_2O.

Protocol 2: AAV Large-Scale Production (10 Cell Factories ~1×10^{10} cells)

Transfection Protocol

1. Seed [10] ten-layer cell factories (Nunc) each with 5×10^8 293 cells so that each will be 75–80% confluent the next day. (Incubate 12–16 h prior to transfection.) 293 cells are cultured in DMEM supplemented with 5% fetal bovine serum, 1% penicillin, and 1% streptomycin.
2. Check each cell factory microscopically prior to transfection to ensure optimal confluency.
3. For transfection, prepare the DNA mixture in a 250-ml conical tube: Each cell factory requires 2,400 µg total DNA: 1,800 µg helper plasmid and 600 µg rAAV vector plasmid. (This is a 1:1 molar ratio.) Calculate the volume of total input DNA required. Bring the final volume of the DNA mixture to 46.8 ml with sterile dH_2O. Add 5.2 ml 2.5 M $CaCl_2$ to the DNA for a final $CaCl_2$ concentration of 0.25 M. The total volume of the DNA/$CaCl_2$/dH_2O is 52 ml.
4. Add the DNA to 52 ml of 2× HBS, mix well, and incubate for 1 min at room temperature to allow precipitate to form. Add the transfection mix to 1,000 ml of prewarmed DMEM-Complete Media. Discard media from cell factory. Pour media with transfection mix into the cell factory. Repeat for all cell factories.
5. Incubate at 37°C for 72 h.
6. At 72 h, harvest cells. Collect media and cells and centrifuge at 1,000 × *g*, for 10 min at 4°C to pellet the cells. Discard the media. Resuspend pellet in 300 ml PBS to wash the pellet. Centrifuge at 1,000 × *g*, for 10 min at 4°C. Discard PBS and freeze pellet. Frozen pellets can be stored at –20°C for processing at a later time.

Protocol 3: Large-Scale Purification Protocol for AAV2

1. Resuspend the cell pellet from one cell factory in 60 ml large-scale lysis buffer (20 mM Tris-Cl, 150 mM NaCl, 0.5% deoxycholate, and Benzonase (50 U/ml)). Deoxycholate is used in the lysis buffer to reduce aggregation.
2. For ten cell factories, lyse the cells in a single-pass by using a microfluidizer (Microfluidics M-110S). This will form a fine suspension that can be loaded onto the column.
3. Load the lysate onto a 150-ml Streamline Heparin column (Pharmacia) at a flow rate of 20 ml/min using an AKTA-FPLC (Pharmacia).
4. Wash the column with four column volumes of lysis buffer, followed by five column volumes of PBS.
5. Elute the virus with PBS containing 0.2 M NaCl (0.350 M total ionic strength) and monitor the absorbance at 280 λ. The peak fraction will be ~90 ml.

6. Bring the NaCl concentration of the peak fraction to 1 M NaCl and load this onto a Phenyl Sepharose column (5 ml column, Pharmacia). This is a hydrophobic interaction column and the flow-through that contains the virus is collected.
7. The flow-through is diluted to 150 mM NaCl with sterile water.
8. The virus is loaded onto a 5 ml SP Sepharose column (Pharmacia) at a flow rate of 5 ml/min.
9. Wash the column with ten column volumes of PBS.
10. Elute the virus with PBS containing 0.135 M NaCl (285 mM total ionic strength) and monitor the absorbance at 280 λ. Aliquot and store at −20°C.
11. Dialyze the virus prep into storage buffer (50 mM Tris-Cl, 100 mM NaCl, pH 8.0), aliquot, and freeze at −20°C.

Protocol 4: Proteolytic Digestion for AAV Capsid Serotype Validation [201]

1. Dialyze AAV vector into protease digestion reaction buffer, 50 mM Tris-Cl, 100 mM NaCl, pH 8.0, if needed.
2. Digest 0.8 μg (~1.2×10^{11}) capsids with either 5 μg (0.02% final concentration) of trypsin, 1 μg of proteinase K, or 80 μg of α-chymotrypsin in a 25 μl reaction at 37°C for up to 24 h. For each serotype, an undigested sample should be included as a control. These proteases are commercially available from Sigma.
3. Add an equal volume of Laemmli sample buffer containing 1% sodium dodecyl sulfate (SDS) and 655 mM β-mercaptoethanol and boil the samples at 100°C for 5 min.
4. Separate the proteolytic fragments on a 10% SDS-PAGE gel for 90 min at 125 V (constant voltage) until the dye front reaches the bottom of the gel.
5. Transfer the proteins to nitrocellulose (Western blot) in transfer buffer (25 mM Tris, 192 mM glycine, 0.1% (w/v) SDS, and 20% methanol) for 2 h at 0.5 A (constant amperes).
6. Probe the membrane with rabbit polyclonal antisera to AAV capsids. Duplicate samples can be run to probe with other AAV antibodies to determine the origin of the fragments that are produced. For example, a signal with monoclonal B1 antibody demonstrates that the fragment is from the common C-terminal end of VP1, VP2, and VP3. Alternatively, by probing with monoclonal A1 antibody, a signal on the Western blot would demonstrate that the origin of thc fragment is the unique N-terminus of VP1. The B1 epitope is conserved among most AAV serotypes and polyclonal antibodies have been developed for a few AAV serotypes. Anti-AAV capsid antibodies are available from Progen or American Research Products, Inc.

Acknowledgments We thank Dr. Philippe Moullier at the Laboratoire de Therapie Genique, INSERM U649, CHU Hotel Dieu, Nantes Cedex 1, France. R.S. owns equity in a gene therapy company that is commercializing AAV for gene therapy applications. This project was funded by

UF College of Medicine start-up funds, The Department of Molecular Genetics and Microbiology, Association Nantaise de Thérapie Génique (ANTG), and Association Francaise contre les Myopathies (AFM) Award 12263 to R.O.S, and by CLINIGENE, an EC-funded Network of Excellence.

References

1. Berns KI, Bohenzky RA. Adeno-associated viruses: an update. Adv Virus Res 1987;32:243–306.
2. Bankiewicz KS, Forsayeth J, Eberling JL, Sanchez-Pernaute R, Pivirotto P, Bringas J, Herscovitch P, Carson RE, Eckelman W, Reutter B, Cunningham J. Long-term clinical improvement in MPTP-lesioned primates after gene therapy with AAV-hAADC. Mol Ther 2006;14:564–570.
3. Crystal RG, Sondhi D, Hackett NR, Kaminsky SM, Worgall S, Stieg P, Souweidane M, Hosain S, Heier L, Ballon D, Dinner M, Wisniewski K, Kaplitt M, Greenwald BM, Howell JD, Strybing K, Dyke J, Voss H. Clinical protocol. Administration of a replication-deficient adeno-associated virus gene transfer vector expressing the human CLN2 cDNA to the brain of children with late infantile neuronal ceroid lipofuscinosis. Hum Gene Ther 2004;15:1131–1154.
4. Manno CS, Pierce GF, Arruda VR, Glader B, Ragni M, Rasko JJ, Ozelo MC, Hoots K, Blatt P, Konkle B, Dake M, Kaye R, Razavi M, Zajko A, Zehnder J, Rustagi PK, Nakai H, Chew A, Leonard D, Wright JF, Lessard RR, Sommer JM, Tigges M, Sabatino D, Luk A, Jiang H, Mingozzi F, Couto L, Ertl HC, High KA, Kay MA. Successful transduction of liver in hemophilia by AAV-factor IX and limitations imposed by the host immune response. Nat Med 2006;12:342–347.
5. During MJ, Kaplitt MG, Stern MB, Eidelberg D. Subthalamic GAD gene transfer in Parkinson disease patients who are candidates for deep brain stimulation. Hum Gene Ther 2001;12:1589–1591.
6. Stedman H, Wilson JM, Finke R, Kleckner AL, Mendell J. Phase I clinical trial utilizing gene therapy for limb girdle muscular dystrophy: alpha-, beta-, gamma-, or delta-sarcoglycan gene delivered with intramuscular instillations of adeno-associated vectors. Hum Gene Ther 2000;11:777–790.
7. Srivastava A. Hematopoietic stem cell transduction by recombinant adeno-associated virus vectors: problems and solutions. Hum Gene Ther 2005;16:792–798.
8. Work LM, Buning H, Hunt E, Nicklin SA, Denby L, Britton N, Leike K, Odenthal M, Drebber U, Hallek M, Baker AH. Vascular bed-targeted in vivo gene delivery using tropism-modified adeno-associated viruses. Mol Ther 2006;13:683–693.
9. De BP, Heguy A, Hackett NR, Ferris B, Leopold PL, Lee J, Pierre L, Gao G, Wilson JM, Crystal RG. High levels of persistent expression of alpha1-antitrypsin mediated by the nonhuman primate serotype rh.10 adeno-associated virus despite preexisting immunity to common human adeno-associated viruses. Mol Ther 2006;13:67–76.
10. Kauffman SL. Cell proliferation in the mammalian lung. Int Rev Exp Pathol 1980; 22:131–191.
11. Duan D, Sharma P, Yang J, Yue Y, Dudus L, Zhang Y, Fisher KJ, Engelhardt JF. Circular intermediates of recombinant adeno-associated virus have defined structural characteristics responsible for long-term episomal persistence in muscle tissue. J Virol 1998;72:8568–8577.
12. McCarty DM, Young SM, Jr, Samulski RJ. Integration of adeno-associated virus (AAV) and recombinant AAV vectors. Annu Rev Genet 2004;38:819–845.
13. Schnepp BC, Jensen RL, Chen CL, Johnson PR, Clark KR. Characterization of adeno-associated virus genomes isolated from human tissues. J Virol 2005;79:14793–14803.

14. Jooss K, Yang Y, Fisher KJ, Wilson JM. Transduction of dendritic cells by DNA viral vectors directs the immune response to transgene products in muscle fibers. J Virol 1998; 72:4212–4223.
15. Zaiss AK, Muruve DA. Immune responses to adeno-associated virus vectors. Curr Gene Ther 2005;5:323–331.
16. Peden CS, Burger C, Muzyczka N, Mandel RJ. Circulating anti-wild-type adeno-associated virus type 2 (AAV2) antibodies inhibit recombinant AAV2 (rAAV2)-mediated, but not rAAV5-mediated, gene transfer in the brain. J Virol 2004;78:6344–6359.
17. Jiang H, Couto LB, Patarroyo-White S, Liu T, Nagy D, Vargas JA, Zhou S, Scallan CD, Sommer J, Vijay S, Mingozzi F, High KA, Pierce GF. Effects of transient immunosuppression on adenoassociated, virus-mediated, liver-directed gene transfer in rhesus macaques and implications for human gene therapy. Blood 2006;108:3321–3328.
18. Snyder RO, Francis J. Adeno-associated viral vectors for clinical gene transfer studies. Curr Gene Ther 2005;5:311–321.
19. Timpe J, Bevington J, Casper J, Dignam JD, Trempe JP. Mechanisms of adeno-associated virus genome encapsidation. Curr Gene Ther 2005;5:273–284.
20. Becerra SP, Koczot F, Fabisch P, Rose JA. Synthesis of adeno-associated virus structural proteins requires both alternative mRNA splicing and alternative initiations from a single transcript. J Virol 1988;62:2745–2754.
21. Myers MW, Carter BJ. Assembly of adeno-associated virus. Virology 1980;102:71–82.
22. de la Maza LM, Carter BJ. Molecular structure of adeno-associated virus variant DNA. J Biol Chem 1980;255:3194–3203.
23. King JA, Dubielzig R, Grimm D, Kleinschmidt JA. DNA helicase-mediated packaging of adeno-associated virus type 2 genomes into preformed capsids. EMBO J 2001; 20:3282–3291.
24. Wistuba A, Kern A, Weger S, Grimm D, Kleinschmidt JA. Subcellular compartmentalization of adeno-associated virus type 2 assembly. J Virol 1997;71:1341–1352.
25. Ruffing M, Zentgraf H, Kleinschmidt JA. Assembly of viruslike particles by recombinant structural proteins of adeno-associated virus type 2 in insect cells. J Virol 1992;66:6922–6930.
26. Steinbach S, Wistuba A, Bock T, Kleinschmidt JA. Assembly of adeno-associated virus type 2 capsids in vitro. J Gen Virol 1997;78 (Pt 6):1453–1462.
27. Hermonat PL, Labow MA, Wright R, Berns KI, Muzyczka N. Genetics of adeno-associated virus: isolation and preliminary characterization of adeno-associated virus type 2 mutants. J Virol 1984;51:329–339.
28. Girod A, Wobus CE, Zadori Z, Ried M, Leike K, Tijssen P, Kleinschmidt JA, Hallek M. The VP1 capsid protein of adeno-associated virus type 2 is carrying a phospholipase A2 domain required for virus infectivity. J Gen Virol 2002;83:973–978.
29. Warrington KH, Jr, Gorbatyuk OS, Harrison JK, Opie SR, Zolotukhin S, Muzyczka N. Adeno-associated virus type 2 VP2 capsid protein is nonessential and can tolerate large peptide insertions at its N terminus. J Virol 2004;78:6595–6609.
30. Govindasamy L, Padron E, McKenna R, Muzyczka N, Kaludov N, Chiorini JA, Agbandje-McKenna M. Structurally mapping the diverse phenotype of adeno-associated virus serotype 4. J Virol 2006;80:11556–11570.
31. Xie Q, Bu W, Bhatia S, Hare J, Somasundaram T, Azzi A, Chapman MS. The atomic structure of adeno-associated virus (AAV-2), a vector for human gene therapy. Proc Natl Acad Sci USA 2002;99:10405–10410.
32. DiMattia M, Govindasamy L, Levy HC, Gurda-Whitaker B, Kalina A, Kohlbrenner E, Chiorini JA, McKenna R, Muzyczka N, Zolotukhin S, Agbandje-McKenna M. Production, purification, crystallization and preliminary X-ray structural studies of adeno-associated virus serotype 5. Acta Crystallogr Sect F Struct Biol Cryst Commun 2005;61:917–921.
33. Padron E, Bowman V, Kaludov N, Govindasamy L, Levy H, Nick P, McKenna R, Muzyczka N, Chiorini JA, Baker TS, Agbandje-McKenna M. Structure of adeno-associated virus type 4. J Virol 2005;79:5047–5058.

34. Walters RW, Agbandje-McKenna M, Bowman VD, Moninger TO, Olson NH, Seiler M, Chiorini JA, Baker TS, Zabner J. Structure of adeno-associated virus serotype 5. J Virol 2004;78:3361–3371.
35. Lane MD, Nam H-J, Padron, E, Gurda-Whitaker B, Kohlbrenner E, Aslanidi G, Byrne B, McKenna R, Muzyczka N, Zolotukhin S, Agbandje-McKenna M. Production, purification, crystallization and preliminary X-ray analysis of adeno-associated virus serotype 8. Acta Crystallogr Sect F 2005;61:558–561.
36. Chapman MS, Rossmann MG. Structure, sequence, and function correlations among parvoviruses. Virology 1993;194:491–508.
37. Xie Q, Chapman MS. Canine parvovirus capsid structure, analyzed at 2.9 A resolution. J Mol Biol 1996;264:497–520.
38. Llamas-Saiz AL, Agbandje-McKenna M, Wikoff WR, Bratton J, Tattersall P, Rossmann MG. Structure determination of minute virus of mice. Acta Crystallogr D Biol Crystallogr 1997;53:93–102.
39 DeLano WL. The PyMOL Molecular Graphics System (2002) on the World Wide Web http://www.pymol.org.
40. Wu P, Xiao W, Conlon T, Hughes J, Agbandje-McKenna M, Ferkol T, Flotte T, Muzyczka N. Mutational analysis of the adeno-associated virus type 2 (AAV2) capsid gene and construction of AAV2 vectors with altered tropism. J Virol 2000;74:8635–8647.
41. Agbandje-McKenna M, Chapman MS. Correlating structure with function in the viral capsid. In: Kerr JR, Cotmore SF, Bloom ME, Linden RM, Parrish CR (eds) Parvoviruses. Edward Arnold, Ltd., New York, 2006:125–139.
42. Chapman MS, Agbandje-McKenna M. Atomic structure of viral particles. In: Kerr JR, Cotmore SF, Bloom ME, Linden RM, Parrish CR (eds) Parvoviruses. Edward Arnold, Ltd., New York, 2006:107–123.
43. Gao G, Alvira MR, Somanathan S, Lu Y, Vandenberghe LH, Rux JJ, Calcedo R, Sanmiguel J, Abbas Z, Wilson JM. Adeno-associated viruses undergo substantial evolution in primates during natural infections. Proc Natl Acad Sci USA 2003;100: 6081–6086.
44. Qing K, Mah C, Hansen J, Zhou S, Dwarki V, Srivastava A. Human fibroblast growth factor receptor 1 is a co-receptor for infection by adeno-associated virus 2. Nat Med 1999; 5:71–77.
45. Summerford C, Samulski RJ. Membrane-associated heparan sulfate proteoglycan is a receptor for adeno-associated virus type 2 virions. J Virol 1998;72:1438–1445.
46. Summerford C, Bartlett JS, Samulski RJ. AlphaVbeta5 integrin: a co-receptor for adeno-associated virus type 2 infection. Nat Med 1999;5:78–82.
47. Kashiwakura Y, Tamayose K, Iwabuchi K, Hirai Y, Shimada T, Matsumoto K, Nakamura T, Watanabe M, Oshimi K, Daida H. Hepatocyte growth factor receptor is a coreceptor for adeno-associated virus type 2 infection. J Virol 2005;79:609–614.
48. Asokan A, Hamra JB, Govindasamy L, Agbandje-McKenna M, Samulski RJ. Adeno-associated virus type 2 contains an integrin alpha5beta1 binding domain essential for viral cell entry. J Virol 2006;80:8961–8969.
49. Blackburn SD, Steadman RA, Johnson FB. Attachment of adeno-associated virus type 3H to fibroblast growth factor receptor 1. Arch Virol 2006;151:617–623.
50. Di Pasquale G, Davidson BL, Stein CS, Martins I, Scudiero D, Monks A, Chiorini JA. Identification of PDGFR as a receptor for AAV-5 transduction. Nat Med 2003;9: 1306–1312.
51. Walters RW, Yi SM, Keshavjee S, Brown KE, Welsh MJ, Chiorini JA, Zabner J. Binding of adeno-associated virus type 5 to 2,3-linked sialic acid is required for gene transfer. J Biol Chem 2001;276:20610–20616.
52. Kaludov N, Brown KE, Walters RW, Zabner J, Chiorini JA. Adeno-associated virus serotype 4 (AAV4) and AAV5 both require sialic acid binding for hemagglutination and efficient transduction but differ in sialic acid linkage specificity. J Virol 2001;75: 6884–6893.

53. Wu Z, Miller E, Agbandje-McKenna M, Samulski RJ. Alpha2,3 and alpha2,6 *N*-linked sialic acids facilitate efficient binding and transduction by adeno-associated virus types 1 and 6. J Virol 2006;80:9093–9103.
54. Gao G, Vandenberghe LH, Wilson JM. New recombinant serotypes of AAV vectors. Curr Gene Ther 2005;5:285–297.
55. Chen CL, Jensen RL, Schnepp BC, Connell MJ, Shell R, Sferra TJ, Bartlett JS, Clark KR, Johnson PR. Molecular characterization of adeno-associated viruses infecting children. J Virol 2005;79:14781–14792.
56. Schmidt M, Grot E, Cervenka P, Wainer S, Buck C, Chiorini JA. Identification and characterization of novel adeno-associated virus isolates in ATCC virus stocks. J Virol 2006;80:5082–5085.
57. Grimm D. Production methods for gene transfer vectors based on adeno-associated virus serotypes. Methods 2002;28:146–157.
58. Rabinowitz JE, Rolling F, Li C, Conrath H, Xiao W, Xiao X, Samulski RJ. Cross-packaging of a single adeno-associated virus (AAV) type 2 vector genome into multiple AAV serotypes enables transduction with broad specificity. J Virol 2002;76:791–801.
59. Grimm D, Kay MA, Kleinschmidt JA. Helper virus-free, optically controllable, and two-plasmid-based production of adeno-associated virus vectors of serotypes 1 to 6. Mol Ther 2003;7:839–850.
60. Hauck B, Xu RR, Xie J, Wu W, Ding Q, Sipler M, Wang H, Chen L, Wright JF, Xiao W. Efficient AAV1-AAV2 hybrid vector for gene therapy of hemophilia. Hum Gene Ther 2006;17:46–54.
61. Ghosh A, Yue Y, Duan D. Viral serotype and the transgene sequence influence overlapping adeno-associated viral (AAV) vector-mediated gene transfer in skeletal muscle. J Gene Med 2006;8:298–305.
62. Duan D, Yan Z, Yue Y, Ding W, Engelhardt JF. Enhancement of muscle gene delivery with pseudotyped adeno-associated virus type 5 correlates with myoblast differentiation. J Virol 2001;75:7662–7671.
63. Denby L, Nicklin SA, Baker AH. Adeno-associated virus (AAV)-7 and -8 poorly transduce vascular endothelial cells and are sensitive to proteasomal degradation. Gene Ther 2005;12:1534–1538.
64. Wang Z, Zhu T, Qiao C, Zhou L, Wang B, Zhang J, Chen C, Li J, Xiao X. Adeno-associated virus serotype 8 efficiently delivers genes to muscle and heart. Nat Biotechnol 2005;23:321–328.
65. Mingozzi F, Schuttrumpf J, Arruda VR, Liu Y, Liu YL, High KA, Xiao W, Herzog RW. Improved hepatic gene transfer by using an adeno-associated virus serotype 5 vector. J Virol 2002;76:10497–10502.
66. Seppen J, Bakker C, de Jong B, Kunne C, van den Oever K, Vandenberghe K, de Waart R, Twisk J, Bosma P. Adeno-associated virus vector serotypes mediate sustained correction of bilirubin UDP glucuronosyltransferase deficiency in rats. Mol Ther 2006;13:1085–1092.
67. Vandendriessche T, Thorrez L, Acosta-Sanchez A, Petrus I, Wang L, Ma L, De Waele L, Iwasaki Y, Gillijns V, Wilson JM, Collen D, Chuah MK. Efficacy and safety of adeno-associated viral vectors based on serotype 8 and 9 versus lentiviral vectors for hemophilia B gene therapy. J Thromb Haemost 2007;5:16–24.
68. Loiler SA, Conlon TJ, Song S, Tang Q, Warrington KH, Agarwal A, Kapturczak M, Li C, Ricordi C, Atkinson MA, Muzyczka N, Flotte TR. Targeting recombinant adeno-associated virus vectors to enhance gene transfer to pancreatic islets and liver. Gene Ther 2003; 10:1551–1558.
69. Takeda S, Takahashi M, Mizukami H, Kobayashi E, Takeuchi K, Hakamata Y, Kaneko T, Yamamoto H, Ito C, Ozawa K, Ishibashi K, Matsuzaki T, Takata K, Asano Y, Kusano E. Successful gene transfer using adeno-associated virus vectors into the kidney: comparison among adeno-associated virus serotype 1–5 vectors in vitro and in vivo. Nephron Exp Nephrol 2004;96:e119–e126.

70. Seiler MP, Miller AD, Zabner J, Halbert CL. Adeno-associated virus types 5 and 6 use distinct receptors for cell entry. Hum Gene Ther 2006;17:10–19.
71. Limberis MP, Wilson JM. Adeno-associated virus serotype 9 vectors transduce murinealveolar and nasal epithelia and can be readministered. Proc Natl Acad Sci USA 2006; 103:12993–12998.
72. Glushakova LG, Timmers AM, Pang J, Teusner JT, Hauswirth WW. Human blue-opsin promoter preferentially targets reporter gene expression to rat s-cone photoreceptors. Invest Ophthalmol Vis Sci 2006;47:3505–3513.
73. Zhong L, Li W, Li Y, Zhao W, Wu J, Li B, Maina N, Bischof D, Qing K, Weigel-Kelley KA, Zolotukhin I, Warrington KH, Jr, Li X, Slayton WB, Yoder MC, Srivastava A. Evaluation of primitive murine hematopoietic stem and progenitor cell transduction in vitro and in vivo by recombinant adeno-associated virus vector serotypes 1 through 5. Hum Gene Ther 2006;17:321–333.
74. Aldrich WA, Ren C, White AF, Zhou SZ, Kumar S, Jenkins CB, Shaw DR, Strong TV, Triozzi PL, Ponnazhagan S. Enhanced transduction of mouse bone marrow-derived dendritic cells by repetitive infection with self-complementary adeno-associated virus 6 combined with immunostimulatory ligands. Gene Ther 2006;13:29–39.
75. Stone IM, Lurie DI, Kelley MW, Poulsen DJ. Adeno-associated virus-mediated gene transfer to hair cells and support cells of the murine cochlea. Mol Ther 2005;11:843–848.
76. Hacker UT, Wingenfeld L, Kofler DM, Schuhmann NK, Lutz S, Herold T, King SB, Gerner FM, Perabo L, Rabinowitz J, McCarty DM, Samulski RJ, Hallek M, Buning H. Adeno-associated virus serotypes 1 to 5 mediated tumor cell directed gene transfer and improvement of transduction efficiency. J Gene Med 2005;7:1429–1438.
77. Harding TC, Dickinson PJ, Roberts BN, Yendluri S, Gonzalez-Edick M, Lecouteur RA, Jooss KU. Enhanced gene transfer efficiency in the murine striatum and an orthotopic glioblastoma tumor model, using AAV-7- and AAV-8-pseudotyped vectors. Hum Gene Ther 2006;17: 807–820.
78. Thorsen F, Afione S, Huszthy PC, Tysnes BB, Svendsen A, Bjerkvig R, Kotin RM, Lonning PE, Hoover F. Adeno-associated virus (AAV) serotypes 2, 4 and 5 display similar transduction profiles and penetrate solid tumor tissue in models of human glioma. J Gene Med 2006;8:1131–1140.
79. Burger C, Gorbatyuk OS, Velardo MJ, Peden CS, Williams P, Zolotukhin S, Reier PJ, Mandel RJ, Muzyczka N. Recombinant AAV viral vectors pseudotyped with viral capsids from serotypes 1, 2, and 5 display differential efficiency and cell tropism after delivery to different regions of the central nervous system. Mol Ther 2004;10:302–317.
80. Davidson BL, Stein CS, Heth JA, Martins I, Kotin RM, Derksen TA, Zabner J, Ghodsi A, Chiorini JA. Recombinant adeno-associated virus type 2, 4, and 5 vectors: transduction of variant cell types and regions in the mammalian central nervous system. Proc Natl Acad Sci USA 2000;97:3428–3432.
81. Shen F, Su H, Liu W, Kan YW, Young WL, Yang GY. Recombinant adeno-associated viral vector encoding human VEGF165 induces neomicrovessel formation in the adult mouse brain. Front Biosci 2006;11:3190–3198.
82. Wang C, Wang CM, Clark KR, Sferra TJ. Recombinant AAV serotype 1 transduction efficiency and tropism in the murine brain. Gene Ther 2003;10:1528–1534.
83. Broekman ML, Comer LA, Hyman BT, Sena-Esteves M. Adeno-associated virus vectors serotyped with AAV8 capsid are more efficient than AAV-1 or -2 serotypes for widespread gene delivery to the neonatal mouse brain. Neuroscience 2006;138:501–510.
84. Wu K, Meyer EM, Bennett JA, Meyers CA, Hughes JA, King MA. AAV2/5-mediated NGF gene delivery protects septal cholinergic neurons following axotomy. Brain Res 2005;1061:107–113.
85. Tenenbaum L, Chtarto A, Lehtonen E, Velu T, Brotchi J, Levivier M. Recombinant AAV-mediated gene delivery to the central nervous system. J Gene Med 2004;6 (Suppl 1): S212–S222.

86. Du L, Kido M, Lee DV, Rabinowitz JE, Samulski RJ, Jamieson SW, Weitzman MD, Thistlethwaite PA. Differential myocardial gene delivery by recombinant serotype-specific adeno-associated viral vectors. Mol Ther 2004;10:604–608.
87. Pacak CA, Mah CS, Thattaliyath BD, Conlon TJ, Lewis MA, Cloutier DE, Zolotukhin I, Tarantal AF, Byrne BJ. Recombinant adeno-associated virus serotype 9 leads to preferential cardiac transduction in vivo. Circ Res 2006;99:e3–e9.
88. Su H, Huang Y, Takagawa J, Barcena A, Arakawa-Hoyt J, Ye J, Grossman W, Kan YW. AAV Serotype-1 mediates early onset of gene expression in mouse hearts and results in better therapeutic effect. Gene Ther 2006;13:1495–1502.
89. Kawamoto S, Shi Q, Nitta Y, Miyazaki J, Allen MD. Widespread and early myocardial gene expression by adeno-associated virus vector type 6 with a beta-actin hybrid promoter. Mol Ther 2005;11:980–985.
90. Thomas CE, Storm TA, Huang Z, Kay MA. Rapid uncoating of vector genomes is the key to efficient liver transduction with pseudotyped adeno-associated virus vectors. J Virol 2004; 78:3110–3122.
91. Wang AY, Peng PD, Ehrhardt A, Storm TA, Kay MA. Comparison of adenoviral and adeno-associated viral vectors for pancreatic gene delivery in vivo. Hum Gene Ther 2004; 15:405–413.
92. Halbert CL, Rutledge EA, Allen JM, Russell DW, Miller AD. Repeat transduction in the mouse lung by using adeno-associated virus vectors with different serotypes. J Virol 2000;74:1524–1532.
93. Chen J, Wu Q, Yang P, Hsu HC, Mountz JD. Determination of specific CD4 and CD8 T cell epitopes after AAV2- and AAV8-hF.IX gene therapy. Mol Ther 2006;13:260–269.
94. Vandenberghe LH, Wang L, Somanathan S, Zhi Y, Figueredo J, Calcedo R, Sanmiguel J, Desai RA, Chen CS, Johnston J, Grant RL, Gao G, Wilson JM. Heparin binding directs activation of T cells against adeno-associated virus serotype 2 capsid. Nat Med 2006; 12:967–971.
95. Zabner J, Seiler M, Walters R, Kotin RM, Fulgeras W, Davidson BL, Chiorini JA. Adeno-associated virus type 5 (AAV5) but not AAV2 binds to the apical surfaces of airway epithelia and facilitates gene transfer. J Virol 2000;74:3852–3858.
96. Halbert CL, Allen JM, Miller AD. Adeno-associated virus type 6 (AAV6) vectors mediate efficient transduction of airway epithelial cells in mouse lungs compared to that of AAV2 vectors. J Virol 2001;75:6615–6624.
97. Teschendorf C, Warrington KH, Jr, Shi W, Siemann DW, Muzyczka N. Recombinant adeno-associated and adenoviral vectors for the transduction of pancreatic and colon carcinoma. Anticancer Res 2006;26:311–317.
98. Chao H, Liu Y, Rabinowitz J, Li C, Samulski RJ, Walsh CE. Several log increase in therapeutic transgene delivery by distinct adeno-associated viral serotype vectors. Mol Ther 2000;2:619–623.
99. Grimm D, Pandey K, Nakai H, Storm TA, Kay MA. Liver transduction with recombinant adeno-associated virus is primarily restricted by capsid serotype not vector genotype. J Virol 2006;80:426–439.
100. Hauck B, Xiao W. Characterization of tissue tropism determinants of adeno-associated virus type 1. J Virol 2003;77:2768–2774.
101. Aikawa R, Huggins GS, Snyder RO. Cardiomyocyte-specific gene expression following recombinant adeno-associated viral vector transduction. J Biol Chem 2002;277: 18979–18985.
102. Girod A, Ried M, Wobus C, Lahm H, Leike K, Kleinschmidt J, Deleage G, Hallek M. Genetic capsid modifications allow efficient re-targeting of adeno-associated virus type 2. Nat Med 1999;5:1052–1056.
103. Wu P, Xiao W, Conlon T, Hughes J, Agbandje-McKenna M, Ferkol T, Flotte T, Muzyczka N. Mutational analysis of the adeno-associated virus type 2 (AAV2) capsid gene and construction of AAV2 vectors with altered tropism. J Virol 2000;74:8635–8647.

104. Ried MU, Girod A, Leike K, Buning H, Hallek M. Adeno-associated virus capsids displaying immunoglobulin-binding domains permit antibody-mediated vector retargeting to specific cell surface receptors. J Virol 2002;76:4559–4566.
105. White SJ, Nicklin SA, Buning H, Brosnan MJ, Leike K, Papadakis ED, Hallek M, Baker AH. Targeted gene delivery to vascular tissue in vivo by tropism-modified adeno-associated virus vectors. Circulation 2004;109:513–519.
106. Muller OJ, Leuchs B, Pleger ST, Grimm D, Franz WM, Katus HA, Kleinschmidt JA. Improved cardiac gene transfer by transcriptional and transductional targeting of adeno-associated viral vectors. Cardiovasc Res 2006;70:70–78.
107. Perabo L, Goldnau D, White K, Endell J, Boucas J, Humme S, Work LM, Janicki H, Hallek M, Baker AH, Buning H. Heparan sulfate proteoglycan binding properties of adeno-associated virus retargeting mutants and consequences for their in vivo tropism. J Virol 2006;80:7265–7269.
108. Shi X, Fang G, Shi W, Bartlett JS. Insertional mutagenesis at positions 520 and 584 of adeno-associated virus type 2 (AAV2) capsid gene and generation of AAV2 vectors with eliminated heparin-binding ability and introduced novel tropism. Hum Gene Ther 2006;17:353–361.
109. Ruffing M, Heid H, Kleinschmidt JA. Mutations in the carboxy terminus of adeno-associated virus 2 capsid proteins affect viral infectivity: lack of an RGD integrin-binding motif. J Gen Virol 1994;75 (Pt 12):3385–3392.
110. Gigout L, Rebollo P, Clement N, Warrington KH, Jr, Muzyczka N, Linden RM, Weber T. Altering AAV tropism with mosaic viral capsids. Mol Ther 2005;11:856–865.
111. Stachler MD, Bartlett JS. Mosaic vectors comprised of modified AAV1 capsid proteins for efficient vector purification and targeting to vascular endothelial cells. Gene Ther 2006;13:926–931.
112. Rabinowitz JE, Bowles DE, Faust SM, Ledford JG, Cunningham SE, Samulski RJ. Cross-dressing the virion: the transcapsidation of adeno-associated virus serotypes functionally defines subgroups. J Virol 2004;78:4421–4432.
113. Kohlbrenner E, Aslanidi G, Nash K, Shklyaev S, Campbell-Thompson M, Byrne BJ, Snyder RO, Muzyczka N, Warrington KH, Jr, Zolotukhin S. Successful production of pseudotyped rAAV vectors using a modified baculovirus expression system. Mol Ther 2005;12: 1217–1225.
114. Virella-Lowell I, Zusman B, Foust K, Loiler S, Conlon T, Song S, Chesnut KA, Ferkol T, Flotte TR. Enhancing rAAV vector expression in the lung. J Gene Med 2005;7:842–850.
115. Ostedgaard LS, Rokhlina T, Karp PH, Lashmit P, Afione S, Schmidt M, Zabner J, Stinski MF, Chiorini JA, Welsh MJ. A shortened adeno-associated virus expression cassette for CFTR gene transfer to cystic fibrosis airway epithelia. Proc Natl Acad Sci USA 2005; 102:2952–2957.
116. Sirninger J, Muller C, Braag S, Tang Q, Yue H, Detrisac C, Ferkol T, Guggino WB, Flotte TR. Functional characterization of a recombinant adeno-associated virus 5-pseudotyped cystic fibrosis transmembrane conductance regulator vector. Hum Gene Ther 2004; 15:832–841.
117. Wagner JA, Messner AH, Moran ML, Daifuku R, Kouyama K, Desch JK, Manley S, Norbash AM, Conrad CK, Friborg S, Reynolds T, Guggino WB, Moss RB, Carter BJ, Wine JJ, Flotte TR, Gardner P. Safety and biological efficacy of an adeno-associated virus vector-cystic fibrosis transmembrane regulator (AAV-CFTR) in the cystic fibrosis maxillary sinus. Laryngoscope 1999;109:266–274.
118. Wagner JA, Nepomuceno IB, Messner AH, Moran ML, Batson EP, Dimiceli S, Brown BW, Desch JK, Norbash AM, Conrad CK, Guggino WB, Flotte TR, Wine JJ, Carter BJ, Reynolds TC, Moss RB, Gardner P. A phase II, double-blind, randomized, placebo-controlled clinical trial of tgAAVCF using maxillary sinus delivery in patients with cystic fibrosis with antrostomies. Hum Gene Ther 2002;13:1349–1359.
119. Flotte TR, Zeitlin PL, Reynolds TC, Heald AE, Pedersen P, Beck S, Conrad CK, Brass-Ernst L, Humphries M, Sullivan K, Wetzel R, Taylor G, Carter BJ, Guggino WB. Phase I trial of

intranasal and endobronchial administration of a recombinant adeno-associated virus serotype 2 (rAAV2)-CFTR vector in adult cystic fibrosis patients: a two-part clinical study. Hum Gene Ther 2003;14:1079–1088.

120. Aitken ML, Moss RB, Waltz DA, Dovey ME, Tonelli MR, McNamara SC, Gibson RL, Ramsey BW, Carter BJ, Reynolds TC. A phase I study of aerosolized administration of tgAAVCF to cystic fibrosis subjects with mild lung disease. Hum Gene Ther 2001; 12:1907–1916.

121. Wagner JA, Moran ML, Messner AH, Daifuku R, Conrad CK, Reynolds T, Guggino WB, Moss RB, Carter BJ, Wine JJ, Flotte TR, Gardner P. A phase I/II study of tgAAV-CF for the treatment of chronic sinusitis in patients with cystic fibrosis. Hum Gene Ther 1998; 9:889–909.

122. Moss RB, Rodman D, Spencer LT, Aitken ML, Zeitlin PL, Waltz D, Milla C, Brody AS, Clancy JP, Ramsey B, Hamblett N, Heald AE. Repeated adeno-associated virus serotype 2 aerosol-mediated cystic fibrosis transmembrane regulator gene transfer to the lungs of patients with cystic fibrosis: a multicenter, double-blind, placebo-controlled trial. Chest 2004;125:509–521.

123. Wagner JA, Nepomuceno IB, Shah N, Messner AH, Moran ML, Norbash AM, Moss RB, Wine JJ, Gardner P. Maxillary sinusitis as a surrogate model for CF gene therapy clinical trials in patients with antrostomies. J Gene Med 1999;1:13–21.

124. Flotte T, Carter B, Conrad C, Guggino W, Reynolds T, Rosenstein B, Taylor G, Walden S, Wetzel R. A phase I study of an adeno-associated virus-CFTR gene vector in adult CF patients with mild lung disease. Hum Gene Ther 1996;7:1145–1159.

125. Grimm D, Zhou S, Nakai H, Thomas CE, Storm TA, Fuess S, Matsushita T, Allen J, Surosky R, Lochrie M, Meuse L, McClelland A, Colosi P, Kay MA. Preclinical in vivo evaluation of pseudotyped adeno-associated virus vectors for liver gene therapy. Blood 2003;102: 2412–2419.

126. Davidoff AM, Gray JT, Ng CY, Zhang Y, Zhou J, Spence Y, Bakar Y, Nathwani AC. Comparison of the ability of adeno-associated viral vectors pseudotyped with serotype 2, 5, and 8 capsid proteins to mediate efficient transduction of the liver in murine and nonhuman primate models. Mol Ther 2005;11:875–888.

127. Wang L, Takabe K, Bidlingmaier SM, Ill CR, Verma IM. Sustained correction of bleeding disorder in hemophilia B mice by gene therapy. Proc Natl Acad Sci USA 1999;96: 3906–3910.

128. Snyder RO, Miao CH, Patijn GA, Spratt SK, Danos O, Nagy D, Gown AM, Winther B, Meuse L, Cohen LK, Thompson AR, Kay MA. Persistent and therapeutic concentrations of human factor IX in mice after hepatic gene transfer of recombinant AAV vectors. Nat Genet 1997;16:270–276.

129. Wang L, Calcedo R, Nichols TC, Bellinger DA, Dillow A, Verma IM, Wilson JM. Sustained correction of disease in naive and AAV2-pretreated hemophilia B dogs: AAV2/8-mediated, liver-directed gene therapy. Blood 2005;105:3079–3086.

130. Wang L, Nichols TC, Read MS, Bellinger DA, Verma IM. Sustained expression of therapeutic level of factor IX in hemophilia B dogs by AAV-mediated gene therapy in liver. Mol Ther 2000;1:154–158.

131. Mount JD, Herzog RW, Tillson DM, Goodman SA, Robinson N, McCleland ML, Bellinger D, Nichols TC, Arruda VR, Lothrop CD, Jr, High KA. Sustained phenotypic correction of hemophilia B dogs with a factor IX null mutation by liver-directed gene therapy. Blood 2002;99:2670–2676.

132. Liu YL, Mingozzi F, Rodriguez-Colon SM, Joseph S, Dobrzynski E, Suzuki T, High KA, Herzog RW. Therapeutic levels of factor IX expression using a muscle-specific promoter and adeno-associated virus serotype 1 vector. Hum Gene Ther 2004;15:783–792.

133. Chao H, Monahan PE, Liu Y, Samulski RJ, Walsh CE. Sustained and complete phenotype correction of hemophilia B mice following intramuscular injection of AAV1 serotype vectors. Mol Ther 2001;4:217–222.

134. Arruda VR, Schuettrumpf J, Herzog RW, Nichols TC, Robinson N, Lotfi Y, Mingozzi F, Xiao W, Couto LB, High KA. Safety and efficacy of factor IX gene transfer to skeletal muscle in murine and canine hemophilia B models by adeno-associated viral vector serotype 1. Blood 2004;103:85–92.
135. Herzog RW, Mount JD, Arruda VR, High KA, Lothrop CD, Jr. Muscle-directed gene transfer and transient immune suppression result in sustained partial correction of canine hemophilia B caused by a null mutation. Mol Ther 2001;4:192–200.
136. Monahan PE, Samulski RJ, Tazelaar J, Xiao X, Nichols TC, Bellinger DA, Read MS, Walsh CE. Direct intramuscular injection with recombinant AAV vectors results in sustained expression in a dog model of hemophilia. Gene Ther 1998;5:40–49.
137. Chao H, Samulski R, Bellinger D, Monahan P, Nichols T, Walsh C. Persistent expression of canine factor IX in hemophilia B canines. Gene Ther 1999;6:1695–1704.
138. Kay MA, Manno CS, Ragni MV, Larson PJ, Couto LB, McClelland A, Glader B, Chew AJ, Tai SJ, Herzog RW, Arruda V, Johnson F, Scallan C, Skarsgard E, Flake AW, High KA. Evidence for gene transfer and expression of factor IX in haemophilia B patients treated with an AAV vector. Nat Genet 2000;24:257–261.
139. Larson PJ, High KA. Gene therapy for hemophilia B: AAV-mediated transfer of the gene for coagulation factor IX to human muscle. Adv Exp Med Biol 2001;489:45–57.
140. Manno CS, Chew AJ, Hutchison S, Larson PJ, Herzog RW, Arruda VR, Tai SJ, Ragni MV, Thompson A, Ozelo M, Couto LB, Leonard DG, Johnson FA, McClelland A, Scallan C, Skarsgard E, Flake AW, Kay MA, High KA, Glader B. AAV-mediated factor IX gene transfer to skeletal muscle in patients with severe hemophilia B. Blood 2003;101:2963–2972.
141. Jiang H, Pierce GF, Ozelo MC, de Paula EV, Vargas JA, Smith P, Sommer J, Luk A, Manno CS, High KA, Arruda VR. Evidence of multiyear factor IX expression by AAV-mediated gene transfer to skeletal muscle in an individual with severe hemophilia B. Mol Ther 2006;14:452–455.
142. Tan M, Qing K, Zhou S, Yoder MC, Srivastava A. Adeno-associated virus 2-mediated transduction and erythroid lineage-restricted long-term expression of the human beta-globin gene in hematopoietic cells from homozygous beta-thalassemic mice. Mol Ther 2001;3:940–946.
143. Johnston J, Tazelaar J, Rivera VM, Clackson T, Gao GP, Wilson JM. Regulated expression of erythropoietin from an AAV vector safely improves the anemia of beta-thalassemia in a mouse model. Mol Ther 2003;7:493–497.
144. Hayashita-Kinoh H, Yamada M, Yokota T, Mizuno Y, Mochizuki H. Down-regulation of alpha-synuclein expression can rescue dopaminergic cells from cell death in the substantia nigra of Parkinson's disease rat model. Biochem Biophys Res Commun 2006;341: 1088–1095.
145. Hadaczek P, Kohutnicka M, Krauze MT, Bringas J, Pivirotto P, Cunningham J, Bankiewicz K. Convection-enhanced delivery of adeno-associated virus type 2 (AAV2) into the striatum and transport of AAV2 within monkey brain. Hum Gene Ther 2006;17:291–302.
146. Forsayeth JR, Eberling JL, Sanftner LM, Zhen Z, Pivirotto P, Bringas J, Cunningham J, Bankiewicz KS. A dose-ranging study of AAV-hAADC therapy in Parkinsonian monkeys. Mol Ther 2006;14:571–577.
147. Sanftner LM, Sommer JM, Suzuki BM, Smith PH, Vijay S, Vargas JA, Forsayeth JR, Cunningham J, Bankiewicz KS, Kao H, Bernal J, Pierce GF, Johnson KW. AAV2-mediated gene delivery to monkey putamen: evaluation of an infusion device and delivery parameters. Exp Neurol 2005;194:476–483.
148. During MJ, Samulski RJ, Elsworth JD, Kaplitt MG, Leone P, Xiao X, Li J, Freese A, Taylor JR, Roth RH, Sladek JR, Jr, O'Malley KL, Redmond DE, Jr. In vivo expression of therapeutic human genes for dopamine production in the caudates of MPTP-treated monkeys using an AAV vector. Gene Ther 1998;5:820–827.
149. Bankiewicz KS, Leff SE, Nagy D, Jungles S, Rokovich J, Spratt K, Cohen L, Libonati M, Snyder RO, Mandel RJ. Practical aspects of the development of ex vivo and in vivo gene therapy for Parkinson's disease. Exp Neurol 1997;144:147–156.

150. Daly TM, Okuyama T, Vogler C, Haskins ME, Muzyczka N, Sands MS. Neonatal intramuscular injection with recombinant adeno-associated virus results in prolonged beta-glucuronidase expression in situ and correction of liver pathology in mucopolysaccharidosis type VII mice. Hum Gene Ther 1999;10:85–94.
151. Fraites TJ, Jr, Schleissing MR, Shanely RA, Walter GA, Cloutier DA, Zolotukhin I, Pauly DF, Raben N, Plotz PH, Powers SK, Kessler PD, Byrne BJ. Correction of the enzymatic and functional deficits in a model of Pompe disease using adeno-associated virus vectors. Mol Ther 2002;5:571–578.
152. Sun B, Zhang H, Franco LM, Brown T, Bird A, Schneider A, Koeberl DD. Correction of glycogen storage disease type II by an adeno-associated virus vector containing a muscle-specific promoter. Mol Ther 2005;11:889–898.
153. Watson G, Bastacky J, Belichenko P, Buddhikot M, Jungles S, Vellard M, Mobley WC, Kakkis E. Intrathecal administration of AAV vectors for the treatment of lysosomal storage in the brains of MPS I mice. Gene Ther 2006;13:917–925.
154. Passini MA, Dodge JC, Bu J, Yang W, Zhao Q, Sondhi D, Hackett NR, Kaminsky SM, Mao Q, Shihabuddin LS, Cheng SH, Sleat DE, Stewart GR, Davidson BL, Lobel P, Crystal RG. Intracranial delivery of CLN2 reduces brain pathology in a mouse model of classical late infantile neuronal ceroid lipofuscinosis. J Neurosci 2006;26:1334–1342.
155. McEachern KA, Nietupski JB, Chuang WL, Armentano D, Johnson J, Hutto E, Grabowski GA, Cheng SH, Marshall J. AAV8-mediated expression of glucocerebrosidase ameliorates the storage pathology in the visceral organs of a mouse model of Gaucher disease. J Gene Med 2006;8:719–729.
156. Cardone M, Polito VA, Pepe S, Mann L, D'Azzo A, Auricchio A, Ballabio A, Cosma MP. Correction of Hunter syndrome in the MPSII mouse model by AAV2/8-mediated gene delivery. Hum Mol Genet 2006;15:1225–1236.
157. Matalon R, Surendran S, Rady PL, Quast MJ, Campbell GA, Matalon KM, Tyring SK, Wei J, Peden CS, Ezell EL, Muzyczka N, Mandel RJ. Adeno-associated virus-mediated aspartoacylase gene transfer to the brain of knockout mouse for Canavan disease. Mol Ther 2003;7:580–587.
158. Leone P, Janson CG, Bilaniuk L, Wang Z, Sorgi F, Huang L, Matalon R, Kaul R, Zeng Z, Freese A, McPhee SW, Mee E, During MJ. Aspartoacylase gene transfer to the mammalian central nervous system with therapeutic implications for Canavan disease. Ann Neurol 2000;48:27–38.
159. McPhee SW, Janson CG, Li C, Samulski RJ, Camp AS, Francis J, Shera D, Lioutermann L, Feely M, Freese A, Leone P. Immune responses to AAV in a phase I study for Canavan disease. J Gene Med 2006;8:577–588.
160. Janson C, McPhee S, Bilaniuk L, Haselgrove J, Testaiuti M, Freese A, Wang DJ, Shera D, Hurh P, Rupin J, Saslow E, Goldfarb O, Goldberg M, Larijani G, Sharrar W, Liouterman L, Camp A, Kolodny E, Samulski J, Leone P. Clinical protocol. Gene therapy of Canavan disease: AAV-2 vector for neurosurgical delivery of aspartoacylase gene (ASPA) to the human brain. Hum Gene Ther 2002;13:1391–1412.
161. Wang Z, Zhu T, Rehman KK, Bertera S, Zhang J, Chen C, Papworth G, Watkins S, Trucco M, Robbins PD, Li J, Xiao X. Widespread and stable pancreatic gene transfer by adeno-associated virus vectors via different routes. Diabetes 2006;55:875–884.
162. Rehman KK, Wang Z, Bottino R, Balamurugan AN, Trucco M, Li J, Xiao X, Robbins PD. Efficient gene delivery to human and rodent islets with double-stranded (ds) AAV-based vectors. Gene Ther 2005;12:1313–1323.
163. Fukuchi K, Tahara K, Kim HD, Maxwell JA, Lewis TL, Accavitti-Loper MA, Kim H, Ponnazhagan S, Lalonde R. Anti-Abeta single-chain antibody delivery via adeno-associated virus for treatment of Alzheimer's disease. Neurobiol Dis 2006;23:502–511.
164. Hara H, Monsonego A, Yuasa K, Adachi K, Xiao X, Takeda S, Takahashi K, Weiner HL, Tabira T. Development of a safe oral Abeta vaccine using recombinant adeno-associated virus vector for Alzheimer's disease. J Alzheimer Dis 2004;6:483–488.

165. Zhang J, Wu X, Qin C, Qi J, Ma S, Zhang H, Kong Q, Chen D, Ba D, He W. A novel recombinant adeno-associated virus vaccine reduces behavioral impairment and beta-amyloid plaques in a mouse model of Alzheimer's disease. Neurobiol Dis 2003;14:365–379.
166. Pachori AS, Melo LG, Zhang L, Solomon SD, Dzau VJ. Chronic recurrent myocardial ischemic injury is significantly attenuated by pre-emptive adeno-associated virus heme oxygenase-1 gene delivery. J Am Coll Cardiol 2006;47:635–643.
167. Woo YJ, Zhang JC, Taylor MD, Cohen JE, Hsu VM, Sweeney HL. One year transgene expression with adeno-associated virus cardiac gene transfer. Int J Cardiol 2005;100: 421–426.
168. Shi W, Hemminki A, Bartlett JS. Capsid modifications overcome low heterogeneous expression of heparan sulfate proteoglycan that limits AAV2-mediated gene transfer and therapeutic efficacy in human ovarian carcinoma. Gynecol Oncol 2006: 103:1054–1062.
169. Koppold B, Sauer G, Buning H, Hallek M, Kreienberg R, Deissler H, Kurzeder C. Efficient gene transfer of CD40 ligand into ovarian carcinoma cells with a recombinant adeno-associated virus vector. Int J Oncol 2005;26:95–101.
170. Subramanian IV, Ghebre R, Ramakrishnan S. Adeno-associated virus-mediated delivery of a mutant endostatin suppresses ovarian carcinoma growth in mice. Gene Ther 2005; 12:30–38.
171. Li ZB, Zeng ZJ, Chen Q, Luo SQ, Hu WX. Recombinant AAV-mediated HSVtk gene transfer with direct intratumoral injections and Tet-On regulation for implanted human breast cancer. BMC Cancer 2006;6:66.
172. Yanamandra N, Kondraganti S, Gondi CS, Gujrati M, Olivero WC, Dinh DH, Rao JS. Recombinant adeno-associated virus (rAAV) expressing TFPI-2 inhibits invasion, angiogenesis and tumor growth in a human glioblastoma cell line. Int J Cancer 2005;115: 998–1005.
173. Watanabe M, Nasu Y, Kashiwakura Y, Kusumi N, Tamayose K, Nagai A, Sasano T, Shimada T, Daida H, Kumon H. Adeno-associated virus 2-mediated intratumoral prostate cancer gene therapy: long-term maspin expression efficiently suppresses tumor growth. Hum Gene Ther 2005;16:699–710.
174. Sun X, Krissansen GW, Fung PW, Xu S, Shi J, Man K, Fan ST, Xu R. Anti-angiogenic therapy subsequent to adeno-associated-virus-mediated immunotherapy eradicates lymphomas that disseminate to the liver. Int J Cancer 2005;113:670–677.
175. Ma H, Liu Y, Liu S, Xu R, Zheng D. Oral adeno-associated virus-sTRAIL gene therapy suppresses human hepatocellular carcinoma growth in mice. Hepatology 2005;42: 1355–1363.
176. Shi J, Zheng D, Liu Y, Sham MH, Tam P, Farzaneh F, Xu R. Overexpression of soluble TRAIL induces apoptosis in human lung adenocarcinoma and inhibits growth of tumor xenografts in nude mice. Cancer Res 2005;65:1687–1692.
177. Zhang W, Singam R, Hellermann G, Kong X, Juan HS, Lockey RF, Wu SJ, Porter K, Mohapatra SS. Attenuation of dengue virus infection by adeno-associated virus-mediated siRNA delivery. Genet Vaccines Ther 2004;2:8.
178. Murphy JE, Zhou S, Giese K, Williams LT, Escobedo JA, Dwarki VJ. Long-term correction of obesity and diabetes in genetically obese mice by a single intramuscular injection of recombinant adeno-associated virus encoding mouse leptin. Proc Natl Acad Sci USA 1997;94:13921–13926.
179. Keen-Rhinehart E, Kalra SP, Kalra PS. AAV-mediated leptin receptor installation improves energy balance and the reproductive status of obese female Koletsky rats. Peptides 2005;26:2567–2578.
180. Shklyaev S, Aslanidi G, Tennant M, Prima V, Kohlbrenner E, Kroutov V, Campbell-Thompson M, Crawford J, Shek EW, Scarpace PJ, Zolotukhin S. Sustained peripheral expression of transgene adiponectin offsets the development of diet-induced obesity in rats. Proc Natl Acad Sci USA 2003;100:14217–14222.

181. Xin KQ, Mizukami H, Urabe M, Toda Y, Shinoda K, Yoshida A, Oomura K, Kojima Y, Ichino M, Klinman D, Ozawa K, Okuda K. Induction of robust immune responses against HIV is supported by the inherent tropism of AAV5 for DC. J Virol 2006;80: 11899–11910.
182. Kuck D, Lau T, Leuchs B, Kern A, Muller M, Gissmann L, Kleinschmidt JA. Intranasal vaccination with recombinant adeno-associated virus type 5 against human papillomavirus type 16 L1. J Virol 2006;80:2621–2630.
183. Blouin V, Brument N, Toublanc E, Raimbaud I, Moullier P, Salvetti A. Improving rAAV production and purification: towards the definition of a scaleable process. J Gene Med 2004;6 (Suppl 1):S223–S228.
184. Booth MJ, Mistry A, Li X, Thrasher A, Coffin RS. Transfection-free and scalable recombinant AAV vector production using HSV/AAV hybrids. Gene Ther 2004;11:829–837.
185. Conway JE, Rhys CM, Zolotukhin I, Zolotukhin S, Muzyczka N, Hayward GS, Byrne BJ. High-titer recombinant adeno-associated virus production utilizing a recombinant herpes simplex virus type I vector expressing AAV-2 Rep and Cap. Gene Ther 1999;6:986–993.
186. Farson D, Harding TC, Tao L, Liu J, Powell S, Vimal V, Yendluri S, Koprivnikar K, Ho K, Twitty C, Husak P, Lin A, Snyder RO, Donahue BA. Development and characterization of a cell line for large-scale, serum-free production of recombinant adeno-associated viral vectors. J Gene Med 2004;6:1369–1381.
187. Urabe M, Ding C, Kotin RM. Insect cells as a factory to produce adeno-associated virus type 2 vectors. Hum Gene Ther 2002;13:1935–1943.
188. Graham FL, Smiley J, Russell WC, Nairn R. Characteristics of a human cell line transformed by DNA from human adenovirus type 5. J Gen Virol 1997;36:59–74.
189. Matsushita T, Elliger S, Elliger C, Podsakoff G, Villarreal L, Kurtzman GJ, Iwaki Y, Colosi P. Adeno-associated virus vectors can be efficiently produced without helper virus. Gene Ther 1998;5:938–945.
190. Zolotukhin S, Potter M, Zolotukhin I, Sakai Y, Loiler S, Fraites TJ, Jr, Chiodo VA, Phillipsberg T, Muzyczka N, Hauswirth WW, Flotte TR, Byrne BJ, Snyder RO. Production and purification of serotype 1, 2, and 5 recombinant adeno-associated viral vectors. Methods 2002;28:158–167.
191. Francis JD, Snyder RO. Production of research and clinical grade recombinant adeno-associated viral vectors. In: Berns KI, Flotte TR (eds) Laboratory techniques in biochemistry and molecular biology. Elsevier, Amsterdam, 2005:19–56.
192. Zen Z, Espinoza Y, Bleu T, Sommer JM, Wright JF. Infectious titer assay for adeno-associated virus vectors with sensitivity sufficient to detect single infectious events. Hum Gene Ther 2004;15:709–715.
193. Clark KR, Liu X, McGrath JP, Johnson PR. Highly purified recombinant adeno-associated virus vectors are biologically active and free of detectable helper and wild-type viruses. Hum Gene Ther 1999;10:1031–1039.
194. Drittanti L, Rivet C, Manceau P, Danos O, Vega M. High throughput production, screening and analysis of adeno-associated viral vectors. Gene Ther 2000;7:924–929.
195. Veldwijk MR, Topaly J, Laufs S, Hengge UR, Wenz F, Zeller WJ, Fruehauf S. Development and optimization of a real-time quantitative PCR-based method for the titration of AAV-2 vector stocks. Mol Ther 2002;6:272–278.
196. Gao G, Qu G, Burnham MS, Huang J, Chirmule N, Joshi B, Yu QC, Marsh JA, Conceicao CM, Wilson JM. Purification of recombinant adeno-associated virus vectors by column chromatography and its performance in vivo. Hum Gene Ther 2000;11:2079–2091.
197. Sommer JM, Smith PH, Parthasarathy S, Isaacs J, Vijay S, Kieran J, Powell SK, McClelland A, Wright JF. Quantification of adeno-associated virus particles and empty capsids by optical density measurement. Mol Ther 2003;7:122–128.
198. Kronenberg S, Kleinschmidt JA, Bottcher B. Electron cryo-microscopy and image reconstruction of adeno-associated virus type 2 empty capsids. EMBO Rep 2001;2:997–1002.
199. Bleker S, Pawlita M, Kleinschmidt JA. Impact of capsid conformation and Rep-capsid interactions on adeno-associated virus type 2 genome packaging. J Virol 2006;80:810–820.

200. Wobus CE, Hugle-Dorr B, Girod A, Petersen G, Hallek M, Kleinschmidt JA. Monoclonal antibodies against the adeno-associated virus type 2 (AAV-2) capsid: epitope mapping and identification of capsid domains involved in AAV-2-cell interaction and neutralization of AAV-2 infection. J Virol 2000;74:9281–9293.
201. Van Vliet K, Blouin V, Agbandje-McKenna M, Snyder RO. Proteolytic mapping of the adeno-associated virus capsid. Mol Ther 2006;14:809–821.

Chapter 3
Delivering Small Interfering RNA for Novel Therapeutics

Patrick Y. Lu and Martin C. Woodle

Abstract The gene silencing capability of RNA interference (RNAi) is being used to study individual gene's biological function and role in biochemical pathways. However, the efficacy of RNAi depends upon efficient delivery of the intermediates of RNAi, small interfering RNA (siRNA) oligonucleotides. The delivery challenge is even greater when the aim is to inhibit the expression of target genes in disease tissues. In vivo delivery of siRNA is complicated and challenging, and recent works on various animal disease models and early successes in human clinical trials are enlightening the tremendous potential of RNAi therapeutics. In this chapter, the latest developments of in vivo delivery of siRNA and the critical issues related to this effort are addressed.

Keywords RNA interference; Small interfering RNA; In vivo delivery; RNAi therapeutics; Nanoparticle; Local delivery; Systemic delivery

1 Introduction

RNA interference (RNAi) has been rapidly adopted for the discovery and validation of gene function through cell culture and animal model studies, using sequence-specific small interfering RNA (siRNA) [1, 2]. These siRNA intermediates (21–23-nt oligos) were found to bind to an RNA induced silencing complex (RISC) and then selectively degrade the complementary single-stranded target RNA in a sequence-specific manner. Not only is siRNA being used to characterize gene function in high throughput screens for potential therapeutic targets, the growing success as a research tool has also stirred up tremendous interest in using siRNA as a therapeutic agent. With many reports on the in vivo application of siRNA inhibitors and the success of several early phase clinical trials, a broad therapeutic application of RNAi therapeutics is coming. One of the major hurdles to realize the RNAi therapeutic potential is to overcome the obstacles to the siRNA delivery locally and systemically to the disease tissues. In this review, we address the challenges and problems for in vivo siRNA delivery for potential therapeutic applications.

From: *Methods in Molecular Biology, Vol. 437: Drug Delivery Systems*
Edited by: Kewal K. Jain © Humana Press, Totowa, NJ

2 Challenges of In Vivo siRNA Delivery

As a potent and specific inhibitor of gene expression, siRNA is being rapidly adopted as the preferred tool for functional genomics research [3, 4]. siRNA oligos are typically used to inhibit an individual gene, though targeting multiple genes or groups of genes are possible by using a combination of multiple siRNA sequences [5, 6]. The success of using siRNA to knockdown gene expression in vitro has led to a growing interest in applications of siRNA inhibitors in vivo for evaluation of the therapeutic potentials of the gene targets of interest, potency of siRNA inhibitors, the route of administration, and the unwanted adverse effects. These applications should eventually provide validated targets for conventional therapeutic modalities such as small molecule and monoclonal antibody inhibitors, as well as validation of siRNA drug lead itself (Fig. 3.1) [3].

One example is using siRNA oligos specifically targeting angiogenesis factors, such as VEGFs, EGF, FGFs, and their receptors, representing the most widely recognized targets that took years to validate. One study using siRNA-mediated downregulation of these proangiogenesis genes, VEGF and VEGF R2, in clinically relevant xenograft tumor models resulted in a significant antitumor efficacy. Thus, the functions of these two targets were further validated rapidly in a matter of weeks [7]. This example demonstrates the power of in vivo target validation with

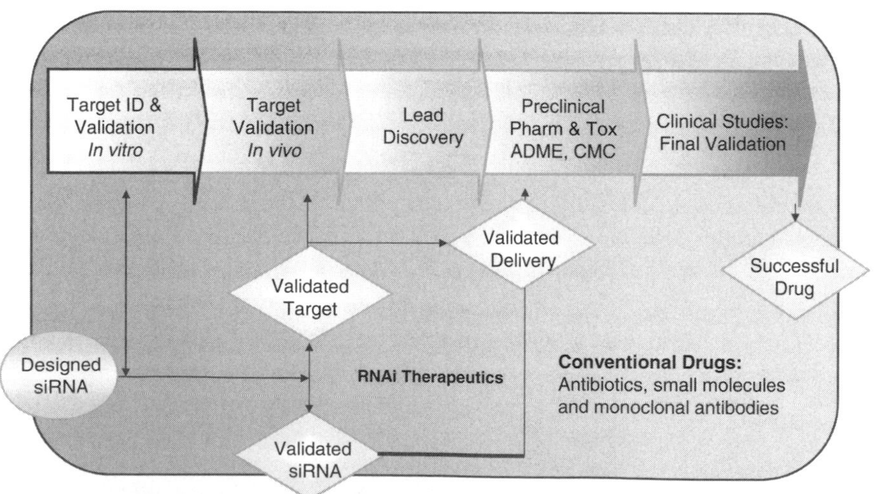

Fig. 3.1 Delivery of siRNA for drug discovery and development. Effective siRNA deliveries in vitro and in vivo are playing very important role for siRNA-based target validation and potential RNAi therapeutics [5]. When targets are validated through in vivo siRNA delivery process, there are already three types of outcomes: validated targets, validated siRNA duplexes, and validated in vivo delivery

siRNA inhibitors. In this case, not only the roles of the proangiogenic factors were validated, the siRNA inhibitors themselves were also validated as potential anticancer drugs.

Delivering siRNA oligos in vivo to animal tissues and keeping them active in the targeted cells are complicated and involve using a physical, chemical, or biological approach and in some cases their combination [7]. Since the main goal of in vivo delivery is to have active siRNA oligos in the target cells, the stability of siRNA oligos in both the extracellular and intracellular environment after a systemic administration is the most challenging issue (Fig. 3.2). The first hurdle is the size of the 21–23-nt double-stranded siRNA oligos: they are relatively small and thus rapidly excreted through urine when administrated into the blood stream, even if those siRNA molecules remain stable through chemical modifications. Second, the double-stranded siRNA oligos are relatively unstable in the serum environment and are potentially degraded by RNase activity within a short period of time. Third, when siRNA is administered systemically, the nonspecific distribution of the oligos throughout the body will significantly decrease the local concentration at the site of disease. In addition, the siRNA oligos need to overcome the blood vessel endothelial wall and multiple tissue barriers in order to reach the target cells. Finally, when siRNA reaches the target cells, cellular uptake of the oligos and intracellular RNAi activity require efficient endocytosis and intact double-stranded oligos.

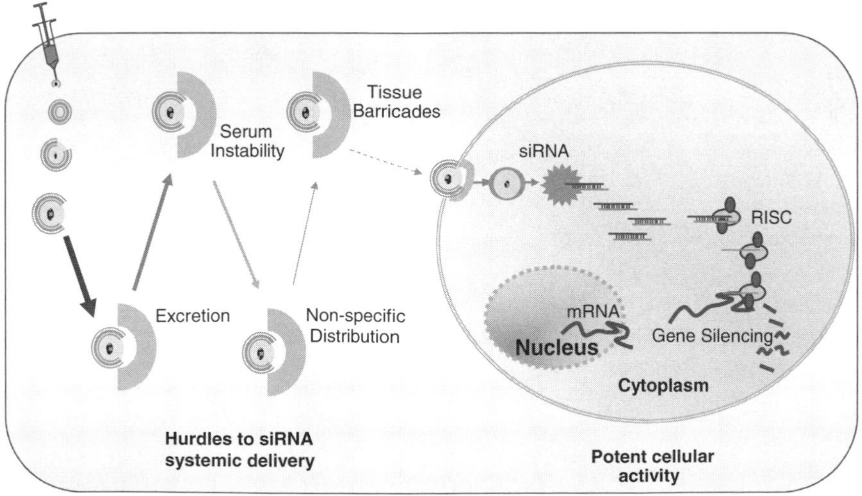

Fig. 3.2 Challenges of systemic in vivo siRNA delivery. The in vivo application, especially systemic delivery of siRNA, is facing challenges from multiple hurdles in the extracellular environment and various barriers for the intracellular uptake. Addressing those issues is critical for efficient in vivo delivery of siRNA in preclinical animal models for drug target validation and potential therapeutics

To increase their stability in both extracellular and intracellular environments, siRNA oligos can be chemically modified by a variety of methods, including change of oligo backbone, replacement of individual nucleotide with nucleotide analogue, and adding conjugates to the oligo. The chemically modified siRNA demonstrated a significant serum resistance and higher stability [8], but it did not solve the problems of excretion through urine and targeted delivery. Therefore, a delivery system capable of protecting siRNA oligos from the urinary excretion and RNase degradation, transporting siRNA oligos through the physical barriers to the target tissue, and enhancing cellular uptake of the siRNA, is the key to the success of in vivo siRNA application.

The accessibility of different tissue types, the presence of various delivery routes, and a variety of pharmacological requirements makes it impossible to have a universal in vivo delivery system suitable to every scenario of siRNA delivery. In terms of in vivo delivery vehicles for siRNA, the "nonviral" carriers are the major type being investigated so far, though some physical and viral delivery approaches are also very effective. The routes of in vivo deliveries are commonly categorized as local or systemic. Some of the delivery vehicles and delivery routes are very effective in animals for target validation but may not be useful for delivery of siRNA therapeutics in humans (Fig. 3.3). Therefore, in vivo siRNA delivery carriers and methods can also be classified as clinically viable and nonclinically viable, according to their suitability for the human use.

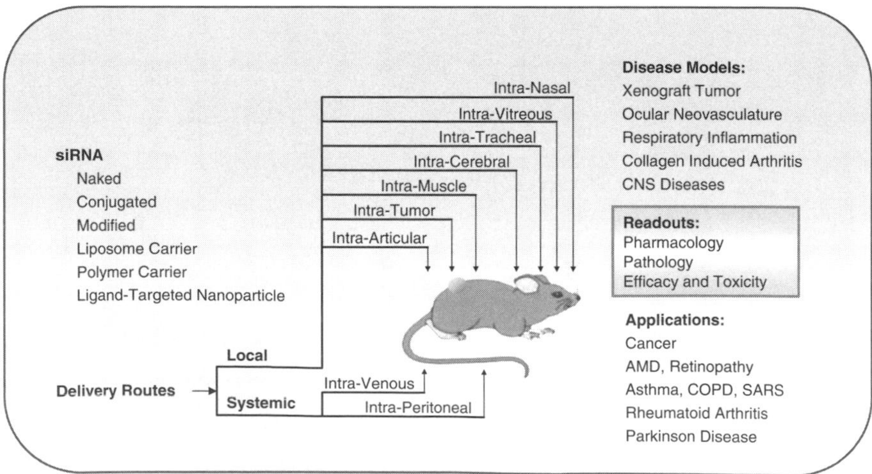

Fig. 3.3 Applications of in vivo siRNA delivery in disease models. Mouse models are widely used for in vivo siRNA delivery studies. siRNA can be delivered by many routes based on the disease types and targeted tissues. The efficacy and toxicity readouts of the siRNA inhibitors from the preclinical models will provide vital information for the in vivo target validation [9]. The clinically viable siRNA delivery provides foundation for designing the administration route and condition of an RNAi therapeutic protocol

3 Delivering siRNA In Vivo Using Nonviral Carriers

Many nonviral carriers are currently used in delivery of either siRNA oligos or DNA-based short hairpin RNA (shRNA) vectors. Because RNAi is a mechanism of action in cytoplasm, achieving effective siRNA delivery is relatively easier than achieving the delivery of DNA-based shRNA systems. Nonviral siRNA delivery to disease tissue usually does not elicit an immune response and it is relatively less toxic to the target cells, which is a great advantage for drug target validation. The possibility of multiple administrations of siRNA makes the therapeutic applications of siRNA very practical.

Cationic lipids and polymers are two major classes of nonviral siRNA delivery carriers and both of them are positively charged and can form complexes with negatively charged siRNA. The siRNA/carrier complex can be condensed into a tiny nanoparticle of size around 100 nm which allows a very efficient cellular uptake of the siRNA agent through the endocytosis process. In a mouse model, the reporter gene silencing and downregulation of TNF-α expression were achieved after intraperitoneal administration of siRNA/lipoplexes [10]. One recent study reported that the use of a cationic derivative of cardiolipin to form lipoplexes with siRNA targeting the c-RAF oncogene led to an inhibition of tumor growth in a sequence-specific manner [11]. In another study, a family of highly branched histidine-lysine (HK) polymer peptides was found to be effective carriers of siRNA [12]. The Raf-1 expression in MDA-MB-435 xenografts was significantly inhibited by intratumoral injection of Raf-1 siRNA complexed with HK polymer (Fig. 3.4) [13]. Another recent study demonstrated a significant inhibition of HER-2 expression and tumor growth through intraperitoneal injection of HER-2 siRNA formulated with polyethylenimine (PEI) [14]. These studies demonstrated that cationic lipids and polymers can enhance siRNA delivery in vivo through systemic routes either intravenously (IV) or intraperitoneally (IP). These carrier-administered siRNA agents were efficiently knocking down the target genes and achieved antitumor efficacies. In contrast, direct intratumoral injection of VEGF siRNA without carrier did not generate any significant antitumor efficacy [15]. In an antiviral study, IV administration of siRNA specific targeting influenza virus RNA genome complexed with PEI was able to inhibit influenza virus production in mice [16].

Some ligand-targeted nucleic acid delivery systems have been developed based on the cationic liposome complex and polymer complex systems. Recent successes of using ligand-targeted complexes to deliver therapeutic siRNA into tumor tissue and liver tissue suggest that targeted systemic delivery for siRNA therapeutics is a possibility [17–19]. We demonstrated an efficient delivery of siRNA to local neovasculature in tumor xenograft or viral infected eye through the systemic administration of a ligand-targeted nanoparticle containing siRNA, targeting VEGF R2, an angiogenesis factor overexpressed in endothelium of new blood vessels. The Arg-Gly-Asp (RGD)-motif peptide ligand is specific to integrin receptor, a marker of the activated endothelial and tumor cells. The ligand-targeted nanoparticle maintains the stability of the siRNA payload, targets to the tumor neovasculature, and

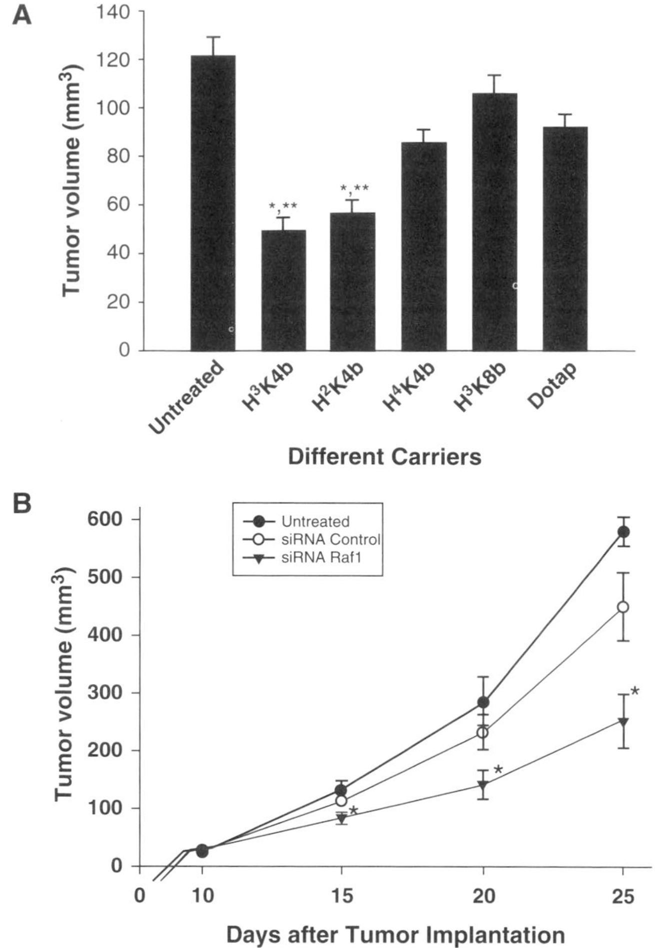

Fig. 3.4 Raf-1 siRNA inhibits tumor growth in vivo. Ten days after the injection of MDA-MB-435 cells into the mammary fat pad, mice with visible tumors were separated into treatment groups. **a** Mice received 4 μg siRNA/tumor with each intratumoral injection every 5 days to determine the optimal carrier of Raf-1 siRNA. **b** To confirm the antitumor efficacy of siRNA Raf-1 with the optimal polymer, mice with tumors were divided into these groups: untreated, β-galactosidase (β-gal) siRNA, and Raf-1 siRNA. The β-gal siRNA and Raf-1 siRNA groups were injected with H3K4b complexed with siRNA every 5 days, three times. $*P < 0.02$, Raf-1 siRNA when compared with untreated; $*P < 0.05$, Raf-1 siRNA when compared with β-gal siRNA [13]

enhances the cellular uptake of the siRNA. As a result, VEGF R2 knockdown and antiangiogenesis effects were observed both in xenografts tumor model and in ocular neovascularization disease model [5, 20]. Clearly, this siRNA nanoparticle is a clinically viable delivery system for various applications of siRNA therapeutics.

It has been reported that siRNA can trigger "off-target effects" [21, 22] and activate cellular interferon pathway [23, 24]. These issues raise concerns for the integrity of target validation studies and the safety and selectivity of the potential siRNA therapeutics. However, majority of these alarming results were found from the in vitro studies and many more studies have shown that siRNA inhibitors are highly specific both in vitro and in vivo [25, 26]. One recent study using systemic delivery of unmodified and unformulated siRNA oligos into mice revealed a lack of interferon response [27]. Two recent reports have revealed that siRNA oligos, containing "5-UGUGU-3" motif, were able to induce a Toll-like receptor-mediated interferon response only when they were delivered in vivo with cationic lipid or polymer carrier, through either intravenous or intraperitoneal administration [28, 29]. In contrast, neither the unformulated siRNA oligos containing 5-UGUGU-3 motif nor siRNA containing no 5-UGUGU-3 motif but with cationic carriers were able to induce the interferon response. For this reason, any 5-UGUGU-3 motif and other potential immunostimulatory motifs should be eliminated from the siRNA oligos if cationic lipid or polymer carriers are going to be used.

4 Delivering siRNA In Vivo Using Local Administration

The choice between local and systemic delivery depends on what tissues and cell types are targeted. For example, skin and muscle can be better accessed using local siRNA delivery, while lung and tumor can be reached efficiently by both local and systemic siRNA deliveries. There is increasing evidence supporting that siRNA can be efficiently delivered to various tissue types, using different approaches (Fig. 3.2).

4.1 Intranasal siRNA Delivery

Airway delivery of siRNA is a very useful method for both target validation and therapeutic development, because of the relevance of respiratory system in various diseases. In a recent study, intranasal delivery of GAPDH-specific siRNA mixed with pulmonary surface active material (InfaSurf) and elastase resulted in lowered GAPDH protein levels in lung, heart, and kidney by ~50–70% at 1 and 7 days after administration, when compared with scrambled siRNA control [30]. Direct delivery of unformulated siRNA into mouse airway led to knockdown of heme oxygenase-1 expression in the lung [31]. Intranasal administration of cationic liposome formulated siRNA specifically targeting the influenza virus RNA genome into mouse lung infected with the influenza virus resulted in a significantly reduced lung virus titer in infected mice and protected animals from lethal challenge [32]. However, in vivo delivery of siRNA with cationic polymer carrier, such as PEI, is often associated with severe toxicity in the host and may induce nonspecific interferon response through the Toll-like receptor pathway, as discussed earlier. Therefore, pulmonary siRNA delivery may require formulations without cationic carriers. Of late, we have successfully delivered siRNA with D5W (5% D-glucose in water) solution into

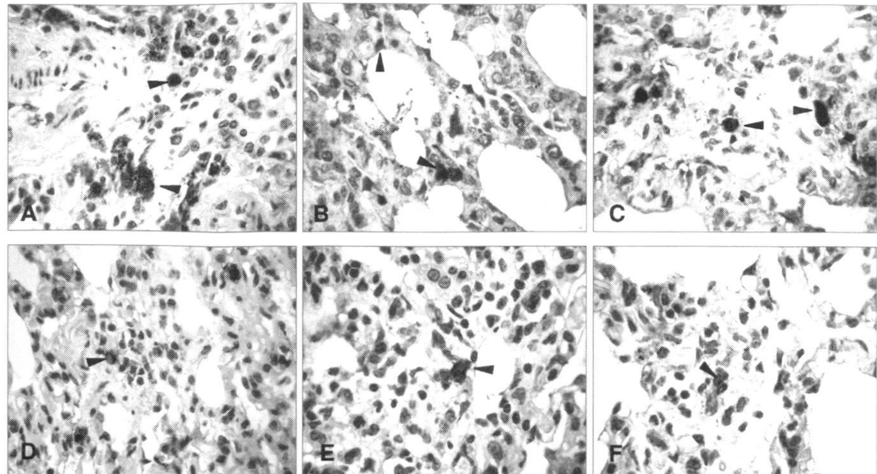

Fig. 3.5 Anti-SARS activity of intranasally delivered siRNA in macaques. The SARS coronavirus (SCV)-specific antigen was detected in alveoli deep into the lungs, including various cell types (original magnification, ×200), confirmed by the specific staining with monoclonal antibodies. **a** The *upper arrow* indicates an SCV-infected type II pneumocyte and the *lower arrow* indicates an infected alveolar macrophage. **b** *Arrows* indicate SCV-infected epithelium-originated type I pneumocytes. **c** *Arrows* indicate SCV-antigen-positive alveolar macrophages. **d–f** *Arrows* indicate SCV-infected cells scattered within the siRNA-treated lungs [33]

mouse and monkey lungs, achieving effective knockdown of SARS coronavirus RNA (Fig. 3.5) [33].

4.2 Intraocular siRNA Delivery

An increasing number of clinical protocols have been approved for treating eye diseases with nucleic acid drugs such as antisense oligonucleotides or RNA aptamers. Delivery of nonformulated siRNA specific to VEGF to the subretinal space in a mouse model of retinal neovascularization resulted in a significant reduction of angiogenesis in the eye. Importantly, this study indicated that chemical protection of the siRNA was not essential, at least in the intravitreous compartment of the eye, in contrast to antisense oligonucleotides or RNA aptamers, which need protection by chemical modification for applications in eye [34]. Using a murine model of herpetic stromal ketatitis that develops from herpes simplex virus corneal infection, we found that subconjunctival administration of siRNA targeting several genes in the VEGF pathway significantly inhibited the corneal angiogenesis and disease symptoms [5]. Subconjunctival delivery of siRNA specific to TGF-β significantly reduced the inflammatory response and matrix deposition in a wound-induced mouse model of ocular inflammation [35]. The evidence also provided clinically viable means for the local delivery of siRNA for gene function validation in various eye disease models. Local delivery of siRNA to the front of the eye subconjunctivally or to the back of the eye intravitreously is highly efficient in silencing target gene expression, and therefore are effective administration routes for target validation for eye disease. However, the frequency and time intervals between repeated deliveries may be the limiting factors of these delivery routes, especially for clinical application of siRNA therapeutics.

4.3 Intracerebral siRNA Delivery

The brain tissue is the foundation of the central nervous system (CNS), obviously a very important biological system and one representing considerable interest for both functional genomics research and therapeutic development [36]. A recent study showed that infusion of chemically protected siRNA oligonucleotides in an aqueous solution directly into the brain was able to selectively inhibit gene expression [37]. Treatment of rats with aqueous siRNA against α(2A)-ARs on days 2–4 after birth resulted in an acute decrease in the levels of α(2A)-AR mRNA in the brainstem into which siRNA was injected [38]. Nonviral infusion of siRNA in brain provided a unique approach to accelerate target validation for neuropsychiatric disorders that involve a complex interplay of gene(s) from various brain regions. For example, infusion of siRNA specific to an endogenous dopamine transporter (DAT) gene in regions (ventral midbrain) far distal to the infusion site resulted in a significant downregulation of DAT mRNA and protein in the brain, and elicited a temporal hyperlocomotor response similar to that obtained upon infusion of GBR-12909, a pharmacologically selective DAT inhibitor [39]. However, the difficulty of performing surgical implantation of an infusion pump delivering high dose of siRNA limits its usefulness as a tool for functional genomics. Another recent study on the use of cationic formulations for siRNA delivery to the brain revealed that delivery was more efficient using a lipid carrier than using a polymer carrier [40]. Electroporation is a physical approach which has been used to introduce DNA into the cells. During the process of electroporation, an electric field pulse induces pores (electropores) in cell membrane that allow DNA molecules to enter the cell. Recently, electroporation procedures have been adopted for local delivery of siRNA. In one study, siRNA introduced into hippocampus region by local electroporation led to a marked reduction in the expression of both the mRNA and protein of the target genes, such as GluR2 and Cox-1, without affecting the expression of other proteins [41].

4.4 Intramuscle siRNA Delivery

The skeletal-muscle tissue is accessible for local siRNA administration. Direct injection of siRNA formulated with cationic lipids or polymers can be considered for local delivery, although inflammation caused by the injection is a common problem. A recent study with nonformulated siRNA delivered by direct injection into mouse muscle, followed by electroporation, demonstrated a significant gene silencing that lasted for 11 days [42]. The electroporation method was also applied in a different study targeting several reporter genes in murine skeletal muscle [43]. A local hydrodynamic approach, in which siRNA in a sufficient volume was rapidly injected into a distal vein of a limb that is transiently isolated by a tourniquet or blood pressure cuff, was tested for siRNA delivery in muscles of animal models and demonstrated a knockdown of both reporter and endogenous gene [44].

4.5 Intratumoral siRNA Delivery

Intratumoral delivery of siRNA is a very attractive approach for functional validation of the tumorigenic genes. We observed inhibition of tumor growth in two human breast cancer xenograft models using intratumoral delivery of VEGF-specific siRNA [45]. It was reported that atelocollagen, a collagen solubilized by protease, can protect siRNA from being digested by RNase when it forms a complex with siRNA. In addition, the siRNA can be slowly released from atelocollagen to efficiently transduce into cells, allowing a long-term target gene silencing [46, 47]. In a mouse xenograft tumor study, after administration of atelocollagen/luc-siRNA complex intratumorally, a reduced luciferase expression was observed. Furthermore, intratumoral injection of atelocollagen/VEGF-siRNA showed an efficient inhibition of tumor growth in an orthotopic xenograft model of a human nonseminomatous germ cell tumor [46]. A similar result was observed in a PC-3 human prostate xenograft tumor model, using the same siRNA delivery approach [47]. Therefore, the atelocollagen-based siRNA delivery method could be a reliable approach to achieve maximal inhibition of gene function in vivo. On the basis of the experience in the successful validation of a group of novel genes for their roles in tumorgenesis using intratumoral delivery of formulated siRNA [2, 4], we believe that intratumoral delivery of siRNA into xenograft tumor models is a very useful platform for in vivo target validation.

5 Delivery of siRNA In Vivo Using Systemic Administration

5.1 Liver-Targeted Systemic Delivery

Some of the first published results showed activity of siRNA in mammals by delivering into mouse liver using the hydrodynamic delivery, a rapid injection of a large volume of aqueous solution into the mouse tail vein creating a high pressure in the vascular circulation that leads to an extensive delivery of siRNA into hepatocytes [48–52]. This procedure allows high efficiency of siRNA uptake and potent siRNA activity in hepatocytes, and thus is a useful tool for functional genomic studies in liver. On the other hand, this procedure is not a clinically viable procedure because of potential damage of liver and other organs, and is limited only to research on liver function and metabolism or liver infectious diseases such as hepatitis [53, 54]. Hepatocyte-specific targeting carriers for siRNA delivery into liver are very attractive approaches for development of siRNA therapeutics for hepatic diseases and are currently under investigation. As one step toward the liver targeting delivery, liver delivery of chemically modified oligonucleotide with cholesterol conjugates was tested, as described in recent publications [8, 55]. However, the data suggested that at least three challenges must be addressed: adequate protection of the siRNA oligonucleotide from serum degradation en route to the liver, protection of the siRNA oligonucleotide from rapid glomerulofiltration by the kidney into the urine, and selective uptake by the target hepatocytes. In addition, the high dose used for intravenous cholesterol-conjugated siRNA delivery indicated a widespread distribution rather than targeting the liver.

5.2 Tumor-Targeted Systemic Delivery

Malignant tumors grow fast and spread throughout the body via blood or the lymphatic system. Metastatic tumors established at distant locations are usually not encapsulated and thus more amenable for systemic delivery. Local siRNA administration methods discussed earlier can meet the requirements for most functional genomics studies by acting on primary tumors or xenograft models which form the basis of most cancer biology research. On the other hand, systemic delivery of siRNA is needed for development of siRNA-based cancer therapeutics.

Systemic siRNA delivery imposes several requirements and greater hurdles than does local siRNA delivery. It requires stable oligonucleotides in the blood and in the local environment to enter the target cells. In addition, the siRNA needs to pass through multiple tissue barriers to reach the target cell. A recent study in pancreas xenografts used the systemic administration of CEACAM6-specific siRNA without protection and formulation. The study demonstrated a significant suppression of primary tumor growth by 68%, compared with that by control siRNA, associated with a decreased proliferation index of the tumor cells, impaired angiogenesis, and increased apoptosis. Treatment with CEACAM6-specific siRNA completely inhibited metastasis and significantly improved survival, without apparent toxicity [56]. Recent results revealed the tumor-targeting siRNA delivery using an RGD peptide ligand directed nanoparticle and its application in antiangiogenic treatment for cancer (Fig. 3.6) by systemic siRNA delivery [20, 57], as reviewed elsewhere recently [58].

5.3 Other Neovasculature-Targeted Systemic Delivery

In addition to targeting tumor neovasculature, we also studied the RGD ligand targeted nanoparticle for targeting ocular neovasculature tissues [5]. The antiangiogenesis efficacy observed in ocular neovascularization models further demonstrated this

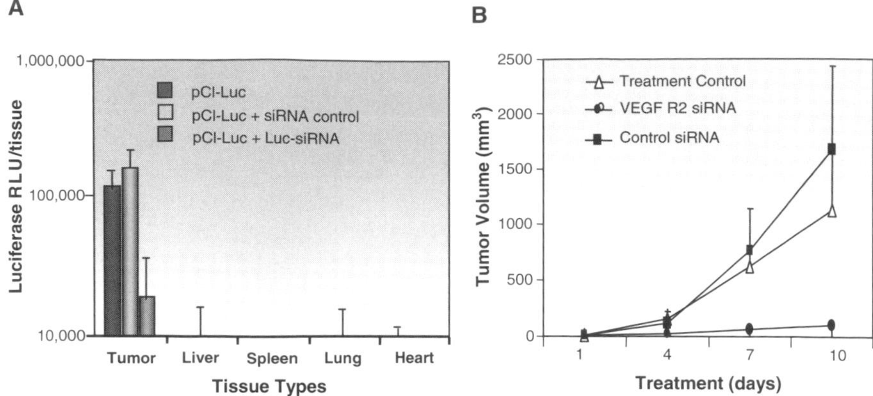

Fig. 3.6 Nanoparticle siRNA delivery for tumor treatment. **a** N2Atumor-bearing mice received a single intravenous injection of 40mg pLuc in RPP-nanoplexes only, with control siRNA or with Luc-specific siRNA. 24h following administration, tissues were assayed for luciferase activity ($n = 5$). **b** Mice were inoculated withN2Atumor cells and left untreated (*open squares*) or treated every 3 days by tail vein injection with RPP-nanoplexes with control siRNA or VEGF R2-specific siRNA at a dose of 40mg per mouse. Treatment was started when the tumors became palpable (>20mm^3). Only VEGF R2-sequence-specific siRNA inhibited tumor growth, whereas treatment with control siRNA did not affect tumor growth rate when compared with untreated controls ($n = 5$)

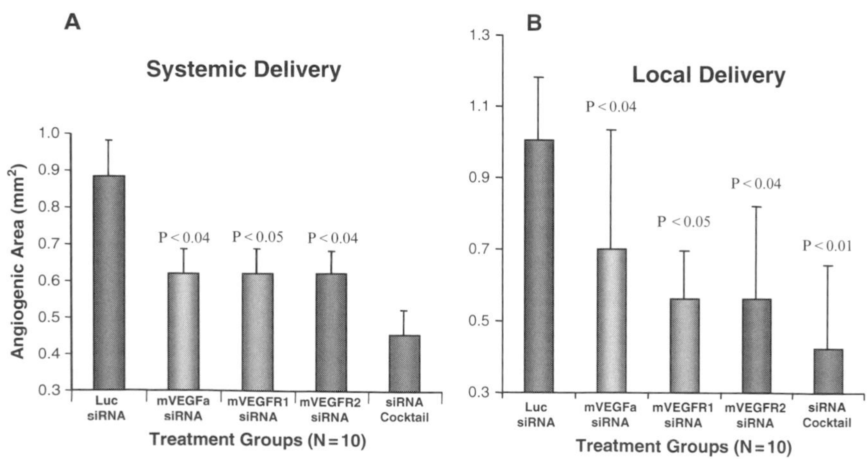

Fig. 3.7 Systemic and local nanoparticle delivery for siRNA therapeutics to treat ocular neovascularization. **a** Systemic delivery of siRNAs against VEGF pathway genes inhibits the CpG ODN-induced angiogenesis. Individual siRNAs or a mixture of total siRNAs against VEGF pathway genes were delivered with RPP nanoparticles 6 and 24h after the CpG ODN induction. **b** Local delivery of siRNAs against VEGF pathway genes inhibits the CpG ODN-induced angiogenesis. Individual siRNAs or a mixture of total siRNAs against VEGF pathway genes were delivered with HKP nanoparticles 6 and 24h after the induction

approach as a clinically viable method for siRNA therapeutics (Fig. 3.7). Using the HSV DNA induced ocular neovascularization model, we demonstrated that siRNA oligos specific for several genes can be combined in the same nanoparticle as a "cocktail" approach to achieve a stronger antiangiogenesis activity inhibiting the disease pathology [5, 59]. This ligand-directed nanoparticle delivery represents a novel and effective approach for a clinically viable systemic administration of siRNA oligos as the dual-targeted RNAi therapeutics.

6 Conclusion

Currently, delivery of siRNA oligos as a therapeutic agent in vivo, through either local or systemic route, is evolving from the target validation tools to the proof of principle for potential RNAi therapeutics. Therefore, examining the utility of each particular siRNA delivery method in vivo requires confirmation of its robustness during the target validation process with repeated testing in the preclinical models. One significant advantage of siRNA, however, is rapidity with which different siRNA sequences and the matching genes can be studied, which is particularly useful for drug target validation. Moreover, developing and optimizing siRNA delivery in various types of animal disease models will be a challenging but worthy effort to accelerate the novel drug discovery process. Ultimately, this effort will be translated into clinically viable administration method for siRNA-based therapeutics to treat cancer, infectious diseases, and many other critical diseases.

References

1. Dykxhoorn, D.M. et al. (2005). Killing the messenger: short RNAs that silence gene expression. *Nat Rev Mol Cell Biol.* 4, 457–467.
2. Lu, P.Y. et al. (2003). siRNA-mediated antitumorigenesis for drug target validation and therapeutics. *Curr Opin Mol Ther.* 5, 225–234.
3. Lu, P.Y. et al. (2005). In vivo application of RNA interference: from functional genomics to therapeutics. *Adv Genet.* 54:117–142.
4. Xie, F.Y. et al. (2004). Delivering siRNA to animal disease models for validation of novel drug targets *in vivo. PharmaGenomics,* July/August, 28–38.
5. Kim, B. et al. (2004). Inhibition of ocular angiogenesis by siRNA targeting vascular endothelial growth factor pathway genes: therapeutic strategy for herpetic stromal keratitis. *Am J Pathol.* 165, 2177–2185.
6. Song, E. et al. (2005). Antibody mediated in vivo delivery of small interfering RNAs via cell-surface receptors. *Nat Biotechnol.* 23(6), 709–717.
7. Lu, P.Y., and Woodle, M.C. (2005). Delivering siRNA *in vivo* for functional genomics and novel therapeutics. In: Appasani, K. (ed.) *RNA interference technology.* Cambridge University Press, London.
8. Soutschek, J. et al. (2004). Therapeutic silencing of an endogenous gene by systemic administration of modified siRNAs. *Nature.* 432, 173–178.

9. Xie, et al. (2006). Harnessing in vivo siRNA delivery for drug discovery and therapeutic development. *Drug Discov Today.* 11(1/2), 67–73.
10. Sorensen, D.R. et al. (2003). Gene silencing by systemic delivery of synthetic siRNAs in adult mice. *J Mol Biol.* 327, 761–766.
11. Chien, P.Y. et al. (2004). Novel cationic cardiolipin analogue-based liposome for efficient DNA and small interfering RNA delivery in vitro and in vivo. *Cancer Gene Ther.* 12, 321–328.
12. Leng, Q. et al. (2005). Highly branched HK peptides are effective carriers of siRNA. *J Gene Med.* 7(7), 977–986.
13. Leng, Q. et al. (2005). Small interfering RNA targeting Raf-1 inhibits tumor growth *in vitro* and *in vivo. Cancer Gene Ther.* Advance online publication, April 1, 2005; doi:10.1038/sj.cgt.7700831
14. Urban-Klein, B. et al. (2005). RNAi-mediated gene-targeting through systemic application of polyethylenimine (PEI)-complexed siRNA in vivo. *Gene Ther.* 12, 461–466.
15. Filleur, S. et al. (2003). SiRNA-mediated inhibition of vascular endothelial growth factor severely limits tumor resistance to antiangiogenic thrombospondin-1 and slows tumor vascularization and growth. *Cancer Res.* 63, 3919–3922.
16. Ge, Q. et al. (2004). Inhibition of influenza virus production in virus-infected mice by RNA interference. *Proc Natl Acad Sci USA.* 101, 8676–8681.
17. Woodle, M.C. et al. (2001). Sterically stabilized polyplex: ligand-mediated activity. *J Control Release.* 74, 309–311.
18. Song, E. et al. (2005). Antibody mediated in vivo delivery of small interfering RNAs via cell-surface receptors. *Nat Biotechnol.* 23(6), 709–717.
19. Morrissey, D.V. et al. (2005). Potent and persistent in vivo anti-HBV activity of chemically modified siRNAs. *Nat Biotechnol.* 23(8), 1002–1007.
20. Schiffelers, R.M. et al. (2004). Cancer siRNA therapy by tumor selective delivery with ligand-targeted sterically stabilized nanoparticle. *Nucleic Acids Res.* 32, e149.
21. Agrawal, S., and Kandimalla, E.R. (2004). Antisense and siRNA as agonists of toll-like receptors. *Nat Biotechnol.* 22, 1533–1537.
22. Jackson, A. et al. (2003). Expression profiling reveals off-target gene regulation by RNAi. *Nat Biotechnol.* 21, 635–637.
23. Kariko, K. et al. (2004). Small interfering RNAs mediate sequence-independent gene suppression and induce immune activation by signaling through toll-like receptor 3. *J Immunol.* 172, 6545–6549.
24. Sledz, C.A. et al. (2003). Activation of the interferon system by short-interfering RNAs. *Nat Cell Biol.* 5, 834–839.
25. Chi, J.T. et al. (2003). Genomewide view of gene silencing by small interfering RNAs. *Proc Natl Acad Sci USA.* 100, 6364–6369.
26. Semizarov, D. et al. (2003). Specificity of short interfering RNA determined through gene expression signatures. *Proc Natl Acad Sci USA.* 100, 6347–6352.
27. Heidel, J.D. et al. (2004). Lack of interferon response in animals to naked siRNAs. *Nat Biotech.* 22, 1579–1582.
28. Hornung, V. et al. (2005). Sequence-specific potent induction of IFN-a by short interfering RNA in plasmacytoid dendritic cells through TLR7. *Nat Med.* 11, 263–270.
29. Judge, A.D. et al. (2005). Sequence-dependent stimulation of the mammalian innate immune response by synthetic siRNA. *Nat Biotech.* 23, 457–462.
30. Massaro, D. et al. (2004). Noninvasive delivery of small inhibitory RNA and other reagents to pulmonary alveoli in mice. *Am J Physiol Lung Cell Mol Physiol.* 287, L1066–L1070.
31. Zhang, X. et al. (2003). Small interfering RNA targeting heme oxygenase-1 enhances ischemia-reperfusion-induced lung apoptosis. *J Biol Chem.* 279, 10677–10684.
32. Tompkins, S.M. et al. (2004). Protection against lethal influenza virus challenge by RNA interference in vivo. *Proc Natl Acad Sci USA.* 101, 8682–8686.
33. Li, B.J. et al. (2005). Prophylactic and therapeutic efficacies of siRNA targeting SARS coronavirus in rhesus macaque. *Nat Med.* 11(9), 944–951.

34. Reich, S.J. et al. (2003). Small interfering RNA (siRNA) targeting VEGF effectively inhibits ocular neovascularization in a mouse model. *Mol Vis.* 9, 210–216.
35. Nakamura, H. et al. (2004). RNA interference targeting transforming growth factor-beta type II receptor suppresses ocular inflammation and fibrosis. *Mol Vis.* 10, 703–711.
36. Buckingham, S.D. et al. (2004). RNA interference: from model organisms towards therapy for neural and neuromuscular disorders. *Hum Mol Genet Spec.* 2, R275–R288.
37. Dorn, G. et al. (2004). siRNA relieves chronic neuropathic pain. *Nucleic Acids Res.* 32, e49.
38. Shishkina, G.T. et al. (2004). Attenuation of alpha (2A)-adrenergic receptor expression in neonatal rat brain by RNA interference or antisense oligonucleotide reduced anxiety in adulthood. *Neuroscience.* 129, 521–528.
39. Thakker, D.R. et al. (2004). Neurochemical and behavioral consequences of widespread gene knockdown in the adult mouse brain by using nonviral RNA interference. *Proc Natl Acad Sci USA.* 101, 17270–17275.
40. Hassani, Z. et al. (2004). Lipid-mediated siRNA delivery down-regulates exogenous gene expression in the mouse brain at picomolar levels. *J Gene Med.* 7, 198–207.
41. Akaneya, Y. et al. (2005). RNAi-induced gene silencing by local electroporation in targeting brain region. *J Neurophysiol.* 93, 594–602.
42. Golzio, M. et al. (2004). Inhibition of gene expression in mice muscle by in vivo electrically mediated siRNA delivery. *Gene Ther.* 12, 246–251.
43. Kishida, T. et al. (2004). Sequence-specific gene silencing in murine muscle induced by electroporation-mediated transfer of short interfering RNA. *J Gene Med.* 6, 105–110.
44. Hagstrom, J.E. et al. (2004). A facile nonviral method for delivering genes and siRNAs to skeletal muscle of mammalian limbs. *Mol Ther.* 10(2), 386–398.
45. Lu, P.Y. et al. (2002). Tumor inhibition by RNAi-mediated VEGF and VEGFR2 down regulation in xenograft models. *Cancer Gene Ther.* 10 (Suppl.), S4.
46. Minakuchi, Y. et al. (2004). Atelocollagen-mediated synthetic small interfering RNA delivery for effective gene silencing in vitro and in vivo. *Nucleic Acids Res.* 32, e109.
47. Takei, Y. et al. (2004). A small interfering RNA targeting vascular endothelial growth factor as cancer therapeutics. *Cancer Res.* 64, 3365–3370.
48. Lewis, D.L. et al. (2002). Efficient delivery of siRNA for inhibition of gene expression in postnatal mice. *Nat Genet.* 32, 107–108.
49. McCaffrey, A.P. et al. (2002). RNA interference in adult mice. *Nature.* 418, 38–39.
50. Song, E. et al. (2003). RNA interference targeting Fas protects mice from fulminant hepatitism. *Nat Med.* 9, 347–351.
51. Zender, L. et al. (2003). Caspase 8 small interfering RNA prevents acute liver failure in mice. *Proc Nat Acad Sci USA.* 100, 7797–7802.
52. Layzer, J.M. et al. (2004). In vivo activity of nuclease-resistant siRNAs. *RNA 10*, 766–771.
53. Giladi, H. et al. (2004). Small interfering RNA inhibits hepatitis B virus replication in mice. *Mol Ther.* 8, 769–776.
54. Sen, A. et al. (2003). Inhibition of hepatitis C virus protein expression by RNA interference. *Virus Res.* 96, 27–35.
55. Lorenz, C. et al. (2004). Steroid and lipid conjugates of siRNAs to enhance cellular uptake and gene silencing in liver cells. *Bioorg Med Chem Lett.* 14, 4975–4977.
56. Duxbury, M.S. (2004). Systemic siRNA-mediated gene silencing: a new approach to targeted therapy of cancer. *Ann Surg.* 240, 667–674.
57. Dubey, et al. (2004). Liposomes modified with cyclic RGD peptide for tumor targeting. *J Drug Target.* 12, 257–264.
58. Lu, P.Y. et al. (2005). Modulation of angiogenesis with siRNA inhibitors for novel therapeutics. *Trend Mol Med.* 11, 104–113.
59. Woodle, M.C., and Lu, P.Y. (2005). Nanoparticles deliver RNAi therapy. Materials Today, 8 (suppl 1), 34–41.

Chapter 4
Catheters for Chronic Administration of Drugs into Brain Tissue

Michael Guarnieri, Benjamin S. Carson, Sr., and George I. Jallo

Abstract Methods to infuse drugs into the parenchyma of the central nervous system (CNS) have been reported as inconsistent or unpredictable. The source of variability appears to be a compromised seal between the tissue and the outer surface of the cannula. Failure of the tissue to seal to the cannula creates a path of least resistance. Rather than penetrate the target area, the drug backflows along the path of the cannula. This artifact can be difficult to detect because drugs enter the systemic circulation and provide some fraction of the intended therapy.

Decreasing the rate of the infusion can reduce backflow. However, this may not be an attractive option for certain therapeutic targets because decreased infusion rates decrease the volume of drug distribution in normal tissue. Cannula design plays a role. Rigid catheters that are fixed to the skull will oppose movements of the brain and break the seal between the catheter and the tissue during chronic infusions. Flexible infusion cannulas, which can be readily made by modifying commercially available brain infusion catheters with plastic tubing, appear to provide consistent infusion results because they can move with the brain and maintain their tissue seal.

Keywords Local delivery; Intraparenchymal therapy; Infusion cannula; Infusion artifacts

1 Introduction

Methods are needed to circumvent the poor solubility of biologically active agents at the blood-brain barrier. The disadvantages of bolus cerebral injections and implanted biodegradable drug crystals have been known for several decades [1]. In 1974, Ott and coworkers described an implantable cannula for chronic drug injections into the hippocampus of laboratory animals [2]. The device consisted of a brass cannula body that was cemented to the animal's skull. A rigid tube was press-fit through the cannula body and extended 3 mm into the hippocampus. Numerous modifications of this basic design have been reported [3–7].

From: *Methods in Molecular Biology, Vol. 437: Drug Delivery Systems*
Edited by: Kewal K. Jain © Humana Press, Totowa, NJ

In 1976, Theeuwes and Yum described an implantable slow-flow (0.5–1 μL/h) osmotic minipump for the local delivery of drugs [8]. Because of their simplicity and reliability [9, 10], osmotic pumps seem suitable for the chronic delivery of chemotherapy. In 1982, Kroin and Penn coupled the minipump to a stainless steel cannula to examine the intracerebral infusion of cisplatin [11]. They used a similar system to demonstrate the efficacy of intratumoral cisplatin and fluorouracil against a 9L rat cerebellar tumor model [12]. Subsequently, efficacy against other CNS tumor models has been observed with infusions of bleomycin [13], phenyl acetate [14], topotecan [15], cyclopentenyl cytosine [16], and carboplatin [17].

We have investigated local therapy for tumors in surgically eloquent areas [18, 19]. However, others and we have encountered inconsistencies in the use of slow-flow infusions (*see* **Note 1**). These systems have been described as unpredictable [20]. Several authors reported that cannulas may be subject to clogging by tissue debris, although specific details are lacking [21, 22]. Reviews of slow-flow infusions typically have focused on their limited capacity for drug distribution, and the steep concentration gradients between the point of delivery and the surrounding brain tissue. Our studies with carboplatin have been consistent with reports showing a radius of distribution for platinum-based drugs of about 5 mm from the point of infusion [11, 17]. The distribution of small and large molecular weight drugs is remarkably different in normal and abnormal rat brain tissue models [23]. The radius of distribution for small molecules can be significantly longer in abnormal tissue [18, 24], probably as a result of convection secondary to edema [25, 26].

Rigid catheters that are fixed to the skull will oppose movements of the brain. We consider that inconsistencies associated with intraparenchymal infusions may be caused by leak-back of the infused drug along the tract of the catheter whenever there is a failure in the seal between the catheter and the tissue. To test this hypothesis, we reviewed studies measuring the distribution of drugs in normal and abnormal tissues when the drugs were infused with a rigid catheter (*see* **Note 2**). The observed distribution patterns were compared with the patterns obtained when drugs were infused using a catheter with a flexible tip. The results show that the inconsistent distribution patterns found with a rigid catheter tip are not seen when flexible-tipped catheters are used (*see* **Note 3**). Infusion rates also have been associated with reflux [27]. All of our studies have been conducted at infusion rates of 10 μL/h or less.

2 Materials

2.1 *Animals*

Rats, dogs, and monkeys were housed according to Johns Hopkins Animal Care and Use Committee policies and federal guidelines. Animals with intracranial (IC) pump implants were housed in individual cages.

1. Fischer rats (females, 180–200 g) were from Charles River.
2. Male Beagle dogs were from Harlan or Charles River.
3. Adult male cynomolgus monkeys (*Macaca fascicularis*) weighing 3–5 kg were obtained from Biologic Research Farms (Houston, TX).

2.2 Supplies

1. Alzet osmotic pumps were from Durect Corporation, Cupertino, CA.
2. Stainless steel brain cannulas were purchased from Plastics One (Roanoke, VA).
3. Modified cannulas were prepared by using a drill with a burr head to cut the 0.28-gauge (0.47 mm outer diameter (OD)) stainless steel proximal cannula tips to a length of 1 mm below the pedestal head. A 3 mm length of 0.38 mm inner diameter (ID), 1.09 mm OD polyethylene tubing (Fisher Scientific, Newark, DE) was fitted over the 1-mm stub of stainless steel tubing and secured with acrylic cement. Flexible tipped assemblies were stored in 70% alcohol overnight (*see* **Note 5**).

2.3 Tumor Cells

1. Rat F98 glioma cells were from R. Goodman, Ohio State University (Columbus, OH). The 9L gliosarcoma line was obtained from the Brain Tumor Research Center, University of California, San Francisco. Cells were maintained in 10% fetal calf serum in DMEM supplemented with penicillin/streptomycin and tested by the Gen-Probe Rapid Detection System (Fisher Scientific) to rule out mycoplasma contamination. Cells were harvested with 0.25% trypsin, counted, and resuspended in DMEM solution before intracranial implantation.
2. Canine tumor cells [28] were from J. Hilton, Johns Hopkins Hospital (Baltimore, MD).

3 Methods

3.1 Surgery

1. Monkeys: After induction of anesthesia, monkeys were placed in a Kopf stereotactic head holder, and a linear incision was made from the inion to the spinous process of C2. A burr hole was made in the midline on the occipital bone 2.5 cm below the inion. The dura was opened with a scalpel blade (no. 11) and the edges of the dura were coagulated by bipolar cautery. The infusion

catheter was then inserted to a pontine target of 1.75 mm anteroposterior, −12.5 mm dorsoventral, and 0 mm mediolateral in a standard stereotaxic atlas [29]. These coordinates were used to determine the resting depth of the catheter tip. We then mapped the linear path of the catheter for those coordinates and inserted the catheter freehand through the cerebellum at a 45° angle between the catheter and the occipital bone to a depth of ~2.25 cm from the surface of the cerebellum (*see* **Note 4**).

In animals 1 and 2, the 3-cm-long catheter was secured to the skull by placing cyanoacrylate in the burr hole. To accommodate independent movement between the skull and the brain, a 2.25-cm-long catheter was passed through the skull in animals 3–5. The tubing was secured to the bone with cyanoacrylate. The body of the pump was placed in a subcutaneous (SC) pocket in the low cervical/high thoracic region and connected to the catheter by the silicon tubing. The tubing was looped into the subcutaneous pocket between the pump and the burr hole (*see* **Note 5**).

2. Dogs: The animal is sedated with acetopromazine (0.2 mg/kg, i.m.) and transferred to animal-operating rooms where an intravenous line will be placed for infusion of D5NS solution. The dogs will receive a prophylactic dose of penicillin G benzathine, 30,000 IU/kg. General anesthesia will consist of sodium thiopental (10–20 mg/kg, i.v.). Sterile conditions are maintained for all procedures. Dogs are orally intubated for subsequent mechanical ventilation. Inhalable halothane and isofluorane are used to maintain general anesthesia. In addition to electrocardiographic monitoring, a catheter is introduced into the femoral artery for continuous monitoring of blood pressure and heart rate. For cell inoculation surgery, dogs are fasted from solid food for 12 h prior to surgery; water is available at all times. After induction of anesthesia, the dog's head is shaved and washed with alcohol and providone-iodine solution.

   An incision is made in the left frontal region commencing at the superior orbital ridge extending 2 cm and then coursing inferiorly. Scalp bleeding is controlled with electrocautery. Underlying connective tissue is removed from the frontal bone to reveal the sagittal suture. A high-speed drill is used to create a craniectomy defect in the left frontal bone that extends ~0.5 cm. A small round curette and small mastoid rongeurs are used to round out the defect. The dura is opened using a blade (no. 15) in a linear fashion. Edges are cauterized using a bipolar cautery. Approximately 30 µL of tumor cell suspension (5 million cells) is slowly (5 min) injected, using a Hamilton 50-µL syringe, ~7 mm into the cortex. The wound is covered with a piece of Gelfoam cut to the size of the defect. The temporalis muscle is closed using a running 3–0 Vicryl suture. The scalp is closed with interrupted 3–0 Prolene sutures. Prior to reversal of general anesthesia, the animal is given one injection of meperidine (2.0 mg/kg, i.m.) to abate postoperative pain and discomfort. Approximately 5 days after the cells have been implanted and an MRI imaging confirms the tumor growth, animals are prepared as described for cell inoculation surgery. The scalp wound is identified, reshaved if necessary, and washed with alcohol and providone-iodine solution. An area between the shoulder blades of approximately 3 cm wide × 10 cm long is shaved and washed with alcohol and providone-iodine. The sutures of the scalp wound are reopened to reveal the initial burr hole. A 5-mm-long stainless steel cannula connected to ~20 cm of silicon tubing is brought to the field. The tip of the tube is inserted to a depth of ~5 mm into the cortex. The silicon tubing connected to the tube is now in the burr hole and is sealed into place with surgical cement. A 5-cm-long incision is made to open a subcutaneous pocket in the prepared area between the shoulder blades. Bleeding is controlled by electrocautery. The distal end of the tubing is passed through a tunnel under the skin created by spreading hemostat blades under the skin caudally toward the area between the shoulder blades. A drug-containing pump is placed in the SC pocket and connected to the distal end of the silicon tubing. The temporalis muscle is closed using a running 3–0 Vicryl suture. The scalp and shoulder are closed with interrupted 3–0 Prolene sutures. Prior to reversal of general anesthesia, the animal is given one injection of meperidine (2.0 mg/kg, i.m.) to abate postoperative pain and discomfort (*see* **Note 4**). Seven to nine days after the pump implant, the animal is lightly anesthetized, and the area over the shoulder blades is painted with alcohol and providone-iodine solution. The wound is opened to expose the body of the pump and the connecting tubing. The tubing is cut and sealed with a knot. The pump is removed. The wound is resutured.

3. Rats: Animals are anesthetized with 0.65 mL of a solution containing ketamine hydrochloride (25 mg/mL), xylazine (2.5 mg/mL), and 14.25% ethyl alcohol in saline. Surgical surfaces are shaved, and washed with 70% ethyl alcohol and Betadine. With the aid of a Zeiss operating microscope, a 2-mm burr hole is made ~2 mm lateral and 1 mm anterior to the bregma. Pump cannulas are placed to a depth of 3 mm in the burr hole (*see* **Note 4**). The hole with the rigid or flexible cannula is sealed with surgical glue. The body of the pump is implanted subcutaneously on the back of the anesthetized rodent slightly posterior to the scapulae in a pocket created by inserting and opening a hemostat into a midscapular incision and thereby spreading the subcutaneous tissue. The pocket is large enough to allow some movement of the pump, i.e., 1 cm longer than the pump. Wounds are closed with 4.0 vicryl.

   In all cases to be reported, pumps were examined at the end of the infusion study to verify that the full content of drug solution had been delivered (*see* **Note 4**).

3.2 Biodistribution Measurements

1. Tissue platinum was assayed by atomic absorption spectroscopy to estimate the distribution of infused carboplatin [11, 18].
2. Doxorubicin levels were measured in clear supernatant solution obtained by centrifugation at 14,000 × *g* of 10% homogenates of 1-mm coronal tissue sections in saline. Fluorescence was measured at an excitation wavelength of 490 nm. Emission was measured at 594 nm [30].

4 Notes

1. Carboplatin distribution patterns in animals infused at 1 μL/h for 7 days with rigid catheters are shown in Table 4.1. The expected distribution pattern is a normal curve with the maximum tissue concentration centered about the infusion point. Regardless of whether the drug was infused into an established tumor, many of the brains infused with a rigid catheter had no detectable tissue levels of platinum. In three experiments with tumor-challenged rats, we found a normal distribution of platinum in the brains of 15 of 28 animals examined. No platinum tissue levels were found in the brains of 13 animals. A similar result was observed in studies with normal animals. The data shows that in three experiments, a normal distribution of platinum was found in only 10 of 17 animal brains examined. No platinum was detected in the brains of two dogs challenged with canine tumor cells and infused at the established tumor site with carboplatin through a rigid catheter that was fixed to the skull.
2. For additional information about the results in Table 4.1, we collected data from the analysis of platinum in a series of studies using rigid and flexible infusion catheters. The summary shown in Table 4.2 demonstrates that the remarkable distribution artifact observed with rigid catheters is not seen in rats infused with flexible catheters.

Table 4.1 Proportion of brains examined having expected CNS tissue distribution of carboplatin when infused with rigid catheter

Species	Tumor[a]	Expected distribution
Rat	–	5/8
Rat	+	4/8
Rat	+	2/8
Rat	–	2/5
Rat	–	3/4
Rat	+	9/12
Dog	+	0/2

[a]Rat tumors include 9L and F98 gliomas; dog tumor is a canine glioma

Table 4.2 Platinum (Pt) distribution patterns in F98-tumor-challenged rat brains after carboplatin infusion

	Infusion via rigid catheter					Infusion via flexible catheter		
Section[a] (mm)	1[b]	2	3	4	5	1	2	3
R3	0.9	0.0	0.0	0.7	0.3	0.4	2.3	0.3
R2	3.1	0.0	0.0	6.2	2.1	1.2	4.7	0.4
R1	10.4	0.0	0.0	10.4	7.1	2.2	5.9	1.4
Center	3.1	0.0	0.0	2.2	7.9	3.4	7.7	1.8
C1	9.0	0.0	0.0	0.6	2.9	2.8	6.9	3.5
C2	2.8	0.0	0.0	0.0	1.3	1.2	4.5	3.1
C3	0.9	0.0	0.0	0.0	0.0	0.6	2.5	1.8
C4	0.5	0.0	0.0	0.0	0.0	0.1	1.8	1.1

[a]Sections are listed as R*x* – rostral, and C*x* – caudal, where *x* is millimeters from the center of the infusion site

[b]Numbers 1–5 indicate ng Pt/mg tissue

Table 4.3 Proportion of brains examined having a normal CNS tissue distribution of drug when infused with a flexible catheter

Species	Tumor	Assay	Normal distribution
Rat	+	Platinum	6/6
Rat	+	Doxorubicin	6/6
Rat	–	Doxorubicin	6/6
Rat	+	TRF-dox	4/4
Rat	–	TRF-dox	4/4
Dog	+	Platinum	6/6
Monkey	–	Platinum	5/6

3. Table 4.3 shows the distribution pattern of carboplatin, doxorubicin, and a doxorubicin-transferrin conjugate infused via a flexible catheter. With the exception of one infusion in a monkey, the drug was found normally distributed in all cases tested. A review of this monkey revealed that the catheter and its tubing were inadvertently loaded with drug solution at surgery. We have assumed that the immediate flow of the drug solution prevented the tissue from sealing about the catheter and that the drug leaked back along the catheter into the subdural space. The pump was empty at the end of the infusion and the tubing was intact.
4. Our current practice of implanting empty tubing and catheters – to allow 2–24 h (depending on the pump flow rate) for the catheter to seal before drug solutions reach the tissue – would seem to promote the potential for blockage. However, pressures within osmotic pumps can reach more than 100 atm. [10]. It is difficult to imagine that tissue debris or a clot could block this pressure. Moreover, such an effect would be easily detectable by the presence of broken tubing, or liquid within the pump at the end of the delivery period. This has not been observed in our experience despite a routine practice of examining the pump and tubing at the end of the delivery period.
5. Osmotic pumps connected to intratumoral catheters have been associated with tissue damage [31, 32]. In more than 5 years of rodent studies, we have never observed tissue damage that was not associated with the wound created by implanting the catheter [17, 33]. The histopathology and radiographic examination of cynomolgus monkeys infused for 3 months with saline were unremarkable [18]. Inflammatory responses caused by toxic drug concentrations at the point of delivery have been documented by radiographic and histopathology studies in monkeys [18]. Nevertheless, the outer diameter of the polyethylene catheter used in this work is 2-fold larger than the commercially available stainless steel model. The larger tubing most likely causes additional damage. Additional studies are needed to determine whether this damage affects survival research with tumor models.

Acknowledgments The authors have received financial support from DURECT Inc. (Cupertino, CA) to examine the toxicity of carboplatin infused into the brainstem of monkeys via an osmotic pump. Financial aid for the present study has been supplied entirely by the Children's Cancer Foundation (Baltimore, MD).

References

1. Routtenberg A (1972) Intracranial chemical injection and behavior: a critical review. *Behav Biol* 7:601–641.
2. Ott T, Schmitt M, Krug M, Matthies H (1974) Intrahippocampal injection of chemicals: analysis of spread. *Pharmacol Biochem Behav* 2:715–718.

3. Crane LA, Glick SD (1979) Simple cannula for repeated intracerebral drug administration in rats. *Pharmacol Biochem Behav* 10:799–800.
4. Rezek M, Havlicek V (1975) Cannula for intracerebral administration of experimental substances. *Pharmacol Biochem Behav* 3:1125–1128.
5. Kokkinidis L, Raffler L, Anisman H (1977) Simple and compact cannula system for mice. *Pharmacol Biochem Behav* 6:595–597.
6. Williams LR, Vahlsing HL, Lindamood T, Varon S, Gage FH, Manthorpe M (1987) A small-gauge cannula device for continuous infusion of exogenous agents into the brain. *Exp Neurol* 95:743–754.
7. Lal S, Lacroix M, Tofilon P, Fuller GN, Sawaya R, Lang FF (2000) An implantable guide-screw system for brain tumor studies in small animals. *J Neurosurg* 92:326–333.
8. Theeuwes F, Yum SI (1976) Principles of the design and operation of generic osmotic pumps for the delivery of semisolid or liquid drug formulations. *Ann Biomed Eng* 4:343–353.
9. White JD, Schwartz MW (1994) Using osmotic minipumps for intracranial delivery of amino acids and peptides. In: Flanagan TR, Emerich DF, Winn SR (eds) *Methods in neurosciences, vol 21*. Academic, San Diego, CA, 187–200.
10. Wright JC, Stevenson CL (1999) Pumps/osmotic. In: Mathiowitz E (ed) *Encyclopedia of controlled drug delivery*. Wiley, New York, 896–915.
11. Kroin JS, Penn RD (1982) Intracerebral chemotherapy: chronic microinfusion of cisplatin. *Neurosurgery* 10:349–354.
12. Penn RD, Kroin JS, Harris JE, Chiu KM, Braun DP (1983) Chronic intratumoral chemotherapy of a rat tumor with cisplatin and fluorouracil. *Appl Neurophysiol* 46:240–244.
13. Kimler BF, Martin DF, Evans RG, Morantz RA, Vats TS (1990) Combination of radiation therapy and intracranial bleomycin in the 9L rat brain tumor model. *Int J Radiat Oncol Biol Phys* 18:1115–1121.
14. Ram Z, Samid D, Walbridge S, Oshiro E, Viola J, Tao-cheng J-H, Shack S, Thibault A, Meyers C, Oldfield E (1994) Growth inhibition, tumor maturation, and extended survival in experimental brain tumors in rats treated with phenyl acetate. *Cancer Res* 54:2923–2927.
15. Pollina J, Plunkett RJ, Ciesielski MJ, Lis A, Barone TA, Greenberg SJ, Fenstermaker RA (1998) Intratumoral infusion of topotecan prolongs survival in the nude rat intracranial U87 human glioma model. *J Neurooncol* 39:217–225.
16. Viola JJ, Agbaria A, Walbridge S, Oshiro EM, Johns DG, Kelly JA, Oldfield EH, Ram Z (1995) *In situ* cyclopentenyl cytosine infusion for the treatment of experimental brain tumors. *Cancer Res* 55:1306–1309.
17. Carson BS, Wu QZ, Tyler B, Sukay L, Raychaudhuri R, DiMeco F, Clatterbuck R, Olivi A, Guarnieri M (2002) New approach to tumor therapy for inoperable areas of the brain: chronic intraparenchymal drug delivery. *J Neurooncol* 60:151–158.
18. Storm PB, Clatterbuck RE, Liu YJ, Johnson RM, Gillis EM, Guarnieri M, Carson BS (2003) A surgical technique for safely placing a drug delivery catheter into the pons of primates: preliminary results of carboplatin infusion. *Neurosurgery* 52:1169–1177.
19. Carson Sr BS, Guarnieri M (2002) Local therapy for brain tumors. *Adv Clin Neurosci* 12:89–99.
20. Groothuis DR (2000) The blood-brain and blood-tumor barriers: a review of strategies for increasing drug delivery. *Neurooncol* 2:45–59.
21. Weingart J, Rhines L, Brem H (2000) Intratumoral chemotherapy. In: Bernstein M, Berger MS (eds) *Neuro-oncology. The essentials*. Thieme, New York, 240–248.
22. Wang PP, Frazier J, Brem H (2002) Local drug delivery to the brain. *Adv Drug Deliv Rev* 54:987–1013.
23. Khan A, Jallo GI, Liu YJ, Carson Sr BS, Guarnieri M (2005) Infusion rates and drug distribution in brain tumor models. *J Neurosurg (Pediatr 1)* 102:53–58.
24. Fung LK, Ewend MG, Aills A, Sipos EP, Thompson R, Watts M, Colvin OM, Brem H, Saltzman WM (1998) Pharmacokinetics of interstitial delivery of carmustine, 4-hydroperoxycyclophosphamide, and paclitaxel from a biodegradable polymer implant in the monkey brain. *Cancer Res* 58:672–684.

25. Morikawa N, Mori T, Abe T, Kawashima H, Takeyama M, Hori S (1999) Pharmacokinetics of etoposide and carboplatin in cerebrospinal fluid and plasma during hyperosmotic disruption of the blood brain barrier and intraarterial combination chemotherapy. *Biol Pharm Bull* 22:428–431.
26. Reulen HJ, Grahm R, Spatz M, Klatzo I (1977) Role of pressure gradients and bulk flow in dynamics of vasogenic brain edema. *J Neurosurg* 46:24–34.
27. Krauze MT, Saito R, Noble C, Tamas M, Bringas J, Park JW, Berger MS, Bankiewicz K (2005) Reflux-free cannnula for convection-enhanced high-speed delivery of therapeutic agents. *J Neurosurg* 103:923–929.
28. Salcman M, Scott EW, Schepp RS, Knipp HC, Broadwell RD (1982) Transplantable canine glioma model for use in experimental neuro-oncology. *Neurosurgery* 11:372–381.
29. Szabo J, Cowan WM (1984) A stereotaxic atlas of the brain of the cynomolgus monkey *(Macaca fascicularis). J Comp Neurol* 222:265–300.
30. Barabas K, Sizensky J, Faulk WP (1992) Transferrin conjugates of adriamycin are cytotoxic without intercalating nuclear DNA. *J Biol Chem* 267:9437–9442.
31. Bear MF, Kleinschmidt A, Gu Q, Singer W (1990) Disruption of experience-dependent synaptic modification in striate cortex by infusion of an NMDA receptor antagonist. *J Neurosci* 10:909–925.
32. Cunningham J, Oiwa Y, Nagy D, Podsakoff G, Colosi P, Bankiewicz KS (2000) Distribution of AAV-TK following intracranial convection-enhanced delivery into rats. *Cell Transplant* 9:585–594.
33. Wu Q, Tyler B, Sukay L, Rhines L, DiMeco F, Clatterbuck RE, Guarnieri M, Carson Sr BS. (2002) Experimental rodent models of brainstem tumors. *Vet Pathol* 39:293–299.

Chapter 5
Transdermal Drug Delivery Systems: Skin Perturbation Devices

Marc B. Brown, Matthew J. Traynor, Gary P. Martin, and Franklin K. Akomeah

Abstract Human skin serves a protective function by imposing physicochemical limitations to the type of permeant that can traverse the barrier. For a drug to be delivered passively via the skin it needs to have a suitable lipophilicity and a molecular weight <500 Da. The number of commercially available products based on transdermal or dermal delivery has been limited by these requirements. In recent years various passive and active strategies have emerged to optimize delivery. The passive approach entails the optimization of formulation or drug carrying vehicle to increase skin permeability. However, passive methods do not greatly improve the permeation of drugs with molecular weights >500 Da. In contrast, active methods, normally involving physical or mechanical methods of enhancing delivery, have been shown to be generally superior. The delivery of drugs of differing lipophilicity and molecular weight, including proteins, peptides and oligonucletides, has been shown to be improved by active methods such as iontophoresis, electroporation, mechanical perturbation and other energy-related techniques such as ultrasound and needleless injection. This chapter details one practical example of an active skin abrasion device to demonstrate the success of such active methods. The in vitro permeation of acyclovir through human epidermal membrane using a rotating brush abrasion device was compared with acyclovir delivery using iontophoresis. It was found that application of brush treatment for 10 s at a pressure of 300 N m^{-2} was comparable to 10 min of iontophoresis. The observed enhancement of permeability observed using the rotating brush was a result of disruption of the cells of the stratum corneum, causing a reduction of the barrier function of the skin. However, for these novel delivery methods to succeed and compete with those already on the market, the prime issues that require consideration include device design and safety, efficacy, ease of handling, and cost-effectiveness. This chapter provides a detailed review of the next generation of active delivery technologies.

Keywords Dermal; Drug delivery; Permeability; Skin; Transdermal

From: *Methods in Molecular Biology, Vol. 437: Drug Delivery Systems*
Edited by: Kewal K. Jain © Humana Press, Totowa, NJ

1 Introduction

1.1 The Skin Barrier

Human skin has a multifunctional role, primary among which is its role as a barrier against both the egress of endogenous substances such as water and the ingress of xenobiotic material (chemicals and drugs). This barrier function of the skin is reflected by its multilayered structure (Fig. 5.1). The top or uppermost layer of the skin known as the stratum corneum (SC) represents the end product of the differentiation process initially started in the basal layer of the epidermis with the formation of keratinocytes by mitotic division. The SC, therefore, is composed of dead cells (corneocytes) interdispersed within a lipid rich matrix. It is the "brick and mortar" architecture and lipophilic nature of the SC, which primarily accounts for the barrier properties of the skin [1, 2]. The SC is also known to exhibit selective permeability and allows only relatively lipophilic compounds to diffuse into the lower layers. As a result of the dead nature of the SC, solute transport across this layer is primarily by passive diffusion [3] in accordance with Fick's Law [4] and no active transport processes have been identified.

Typical delivery systems can be utilised to achieve transdermal drug delivery or dermal drug delivery. The former involves the delivery of drugs through the skin

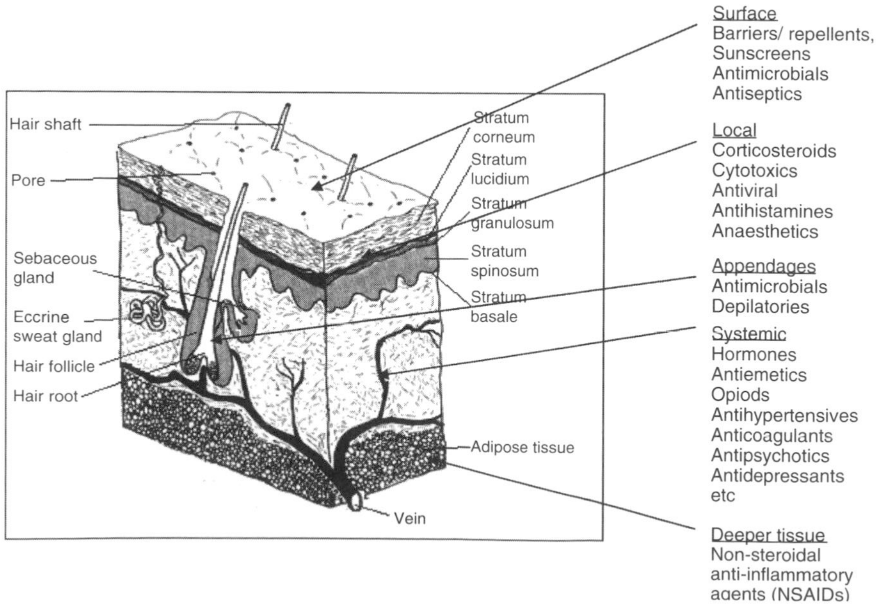

Fig. 5.1 Anatomy and physiology of the skin, showing the potential targets or site of action for cosmetics and drugs

barrier in order that they exert a systemic effect whereas the latter refers to delivery of drugs to particular locations within the skin so that they exert a local effect. This sort of dermal drug delivery approach is commonly used in the treatment of dermatological conditions such as skin cancer, psoriasis, eczema and microbial infections, where the disease is located in the skin. Like many alternative routes of delivery, the skin has both benefits and limitations (Table 5.1) when compared with more conventional methods such as oral drug delivery.

In the last 25 years numerous methods of overcoming the skin barrier have been described, but they can broadly be divided into two main categories defined as either passive or active methods.

Table 5.1 Benefits and limitations associated with cutaneous delivery

Benefits

- The avoidance of first pass metabolism and other variables associated with the GI tract, such as pH, gastric emptying time [5–7]
- Sustained and controlled delivery for a prolonged period of time [8, 9]
- Reduction in side effects associated with systemic toxicity, i.e. minimization of peaks and troughs in blood-drug concentration [7, 10]
- Improved patient acceptance and compliance [11–13]
- Direct access to target or diseased site, e.g. treatment of skin disorders such as psoriasis, eczema and fungal infections [14]
- Ease of dose termination in the event of any adverse reactions, either systemic or local
- Convenient and painless administration [5, 6]
- Ease of use may reduce overall healthcare treatment costs [15, 16]
- Provides an alternative in circumstances where oral dosing is not possible (in unconscious or nauseated patients) [7]

Limitations

- A molecular weight less than 500 Da is essential to ensure ease of diffusion across the SC [17], since solute diffusivity is inversely related to its size
- Sufficient aqueous and lipid solubility, a log P (octanol/water) between 1 and 3 is required for the permeant to successfully traverse the SC and its underlying aqueous layers for systemic delivery to occur [18]
- Intra- and inter-variability associated with the permeability of intact and diseased human skin. This implies that there will be fast, slow and normal skin absorption profiles, resulting in varying biological responses [19, 20]. The barrier nature of intact SC ensures that this route is applicable only for very potent drugs that require only minute concentrations (e.g. 10–30 ng mL^{-1} for nicotine) in the blood for a therapeutic effect [5]
- Pre-systemic metabolism; the presence of enzymes, such as peptidases, esterases, in the skin might metabolise the drug into a form that is therapeutically inactive, thereby reducing the efficacy of the drug [21]
- Skin irritation and sensitization, referred to as the "Achilles heel" of dermal and transdermal delivery. The skin as an immunological barrier may be provoked by exposure to certain stimuli; this may include drugs, excipients or components of delivery devices, resulting in erythema, oedema, etc. [22–25]

2 Passive Methods for Enhancing (Trans)dermal Drug Delivery

The conventional means of applying drugs to skin include the use of vehicles such as ointments, creams, gels and "passive" patch technology. More recently, such dosage forms have been developed and/or modified in order to enhance the driving force of drug diffusion (thermodynamic activity) and/or increase the permeability of the skin. Such approaches include the use of penetration enhancers [26], supersaturated systems [27], prodrugs or metabolic approach [28, 29], liposomes and other vesicles [30–33]. However, the amount of drug that can be delivered using these methods is still limited since the barrier properties of the skin are not fundamentally changed. As such there are still no medicines on the market in the USA that contain a labelled penetration enhancer.

3 Active Methods for Enhancing (Trans)dermal Drug Delivery

These methods involve the use of external energy to act as a driving force and/or act to reduce the barrier nature of the SC in order to enhance permeation of drug molecules into the skin. Recent progress in these technologies has occurred as a result of advances in precision engineering (bioengineering), computing, chemical engineering and material sciences, all of which have helped to achieve the creation of miniature, powerful devices that can generate the required clinical response. The use of active enhancement methods has gained importance because of the advent of biotechnology in the later half of the twentieth century, which has led to the generation of therapeutically active, large molecular weight (>500 Da) polar and hydrophilic molecules, mostly peptides and proteins. However, gastrointestinal enzymes often cause degradation of such molecules and hence there is a need to demonstrate efficient delivery of these molecules by alternative administration routes. Passive methods of skin delivery are incapable of enhancing permeation of such large solutes, which has led to studies involving the use of alternative active strategies such as those discussed here.

3.1 Electroporation

The use of electropermeabilization, as a method of enhancing diffusion across biological barriers, dates back as far as 100 years [34]. Electroporation involves the application of high-voltage pulses to induce skin perturbation. High voltages (≥100 V) and short treatment durations (milliseconds) are most frequently employed. Other electrical parameters that affect delivery include pulse properties such as waveform, rate and number [35]. The increase in skin permeability is suggested to be caused by the generation of transient pores during electroporation [36]. The technology has been successfully used to enhance the skin permeability of molecules with differing lipophilicity and size (i.e. small molecules, proteins, peptides and oligonucleotides), including biopharmaceuticals with a molecular weight greater that 7 kDa, the current limit for iontophoresis [37].

Inovio Biomedical Corporation (San Diego, CA) have developed a prototype electroporation transdermal device, which has been tested with various compounds with a view to achieving gene delivery, improving drug delivery and aiding the application of cosmetics. Other transdermal devices based on electroporation have been proposed by various groups [38–41]; however, more clinical information on the safety and efficacy of the technique is required to assess the future commercial prospects.

3.2 Iontophoresis

This method involves enhancing the permeation of a topically applied therapeutic agent by the application of a low level electric current either directly to the skin or indirectly via the dosage form [42–46]. Increase in drug permeation as a result of this methodology can be attributed to either one or a combination of the following mechanisms: electrorepulsion (for charged solutes), electroosmosis (for uncharged solutes) and electropertubation (for both charged and uncharged).

Parameters that affect design of an iontophoretic skin delivery system include electrode type, current intensity, pH of the system, competitive ion effect and permeant type [35]. The launch of commercialised systems of this technology either has occurred or is currently under investigation by various companies. Extensive literature exists on the many types of drugs investigated using iontophoretic delivery and the reader is referred to the following extensive reviews [35, 44, 47–49]. The Phoresor™ device (Iomed Inc.) was the first iontophoretic system to be approved by the FDA in the late 1970s as a physical medicine therapeutic device. In order to enhance patient compliance, the use of patient-friendly, portable and efficient iontophoretic systems have been under intense development over the years. Such improved systems include the Vyteris and E-TRANS iontophoretic devices. Previous work has also reported that the combined use of iontophoresis and electroporation is much more effective than either technique used alone in the delivery of molecules across the skin. [50–52].

The limitations of ionotophoretic systems include the regulatory limits on the amount of current that can be used in humans (currently set at 0.5 mA cm^{-2}) and the irreversible damage such currents could do to the barrier properties of the skin. In addition, iontophoresis has failed to significantly improve the transdermal delivery of macromolecules of >7,000 Da [53].

3.3 Ultrasound (Sonophoresis and Phonophoresis)

Ultrasound involves the use of ultrasonic energy to enhance the transdermal delivery of solutes either simultaneously or via pre-treatment and is frequently referred to as sonophoresis or phonophoresis. The proposed mechanism behind the increase in skin permeability is attributed to the formation of gaseous cavities within the intercellular lipids on exposure to ultrasound, resulting in disruption of the SC [54]. Ultrasound parameters such as treatment duration, intensity and frequency are all known to affect percutaneous absorption, with the latter being the most important [55]. Although frequencies between 20 kHz – 16 MHz have been reported to enhance skin permeation, frequencies at the lower end of this range (<100 kHz) are believed to have a more significant effect on transdermal drug delivery, with the delivery of macromolecules of molecular weight up to 48 kDa being reported [54, 56, 57].

The SonoPrep® device (Sontra Medical Corporation) uses low-frequency ultrasound (55 kHz) for an average duration of 15 s to enhance skin permeability. This battery-operated hand-held device consists of a control unit, ultrasonic horn with control panel, a disposable coupling medium cartridge and a return electrode. The ability of the SonoPrep device to reduce the time of onset of action associated with the dermal delivery of local anaesthetic from EMLA cream was recently reported [58]. In the study by Kost et al. [58], skin treatment by ultrasound for an average time of 9 s resulted in the attainment of dermal anaesthesia within 5 min, compared with 60 min required for non-treated skin. The use of other small, lightweight novel ultrasound transducers to enhance the in vitro skin transport of insulin has also been reported by a range of workers [56, 59–61].

3.4 Laser Radiation and Photomechanical Waves

Lasers have been used in clinical therapies for decades, and therefore their effects on biological membranes are well documented. Lasers are frequently used for the treatment of dermatological conditions such as acne and to confer "facial rejuvenation" where the laser radiation destroys the target cells over a short frame of time (~300 ns). Such direct and controlled exposure of the skin to laser radiation results in ablation of the SC without significant damage to the underlying epidermis. Removal of the SC via this method has been shown to enhance the delivery of lipophilic and hydrophilic drugs [62–64]. The extent of barrier disruption by laser radiation is known to be controlled by parameters such wavelength, pulse length, pulse energy, pulse number and pulse repetition rate [62].

A hand-held portable laser device has been developed by Norwood Abbey Ltd (Victoria, Australia). In a study involving human volunteers [65], the Norwood Abbey laser device was found to reduce the onset of action of lidocaine to 3–5 min, whilst 60 min was required to attain a similar effect in the control group. The Norwood Abbey system has been approved by the US and Australian regulatory bodies for the administration of a topically applied anaesthetic.

Pressure waves (PW), which can be generated by intense laser radiation, without incurring direct ablative effects on the skin have also been recently found to increase the permeability of the skin [66–68]. It is thought that PW form a continuous or hydrophilic pathway across the skin due to expansion of the lacunae domains in the SC. Important parameters affecting delivery such as peak pressure, rise time and duration have been demonstrated [69, 70]. The use of PW may also serve as a means of avoiding problems associated with direct laser radiation.

Permeants that have been successfully delivered in vivo include insulin [71], 40 kDa dextran and 20-nm latex particles [66]. A design concept for a transdermal drug delivery patch based on the use of PW has been proposed by Doukas and Kollias [68].

3.5 Radio-Frequency

Radio-frequency involves the exposure of skin to high-frequency alternating current (~100 kHz), resulting in the formation of heat-induced microchannels in the membrane in the same way as when laser radiation is employed. The rate of drug delivery is controlled by the number and depth of the microchannels formed by the device, which is dependent on the properties of the microelectrodes used in the device. The Viaderm device (Transpharma Ltd) is a hand-held electronic device consisting of a microprojection array (100 microelectrodes/cm^2) and a drug patch. The microneedle array is attached to the electronic device and placed in contact with the skin to facilitate the formation of the microchannels. Treatment duration takes less than a second, with a feedback mechanism incorporated within the electronic control providing a signal when the microchannels have been created, so as to ensure reproducibility of action. The drug patch is then placed on the treated area. Experiments in rats have shown that the device enhances the delivery of granisetron HCL, with blood plasma levels recorded after 12 h rising 30 times the levels recorded for untreated skin after 24 h [72]. A similar enhancement in diclofenac skin permeation was also observed in the same study [72]. The device is reported not to cause any damage to skin, with the radio-frequency-induced microchannels remaining open for less than 24 h. The skin delivery of drugs such as testosterone and human growth hormone by this device is also currently in progress.

3.6 Magnetophoresis

This method involves the application of a magnetic field which acts as an external driving force to enhance the diffusion of a diamagnetic solute across the skin. Skin exposure to a magnetic field might also induce structural alterations that could contribute to an increase in permeability. In vitro studies by Murthy [73] showed a magnetically induced enhancement in benzoic acid flux, which was observed to increase with the strength of the applied magnetic field. Other in vitro studies using a magnet attached to transdermal patches containing terbutaline sulphate (TS) demonstrated an enhancement in permeant flux which was comparable to that attained when 4% isopropyl myristate was used as a chemical enhancer [74]. In the same work the effect of magnetophoresis on the permeation of TS was investigated in vivo using guinea pigs. The preconvulsive time (PCT) of guinea pigs subjected to magnetophoretic treatment was found to last for 36h, which was similar to that observed after application of a patch containing 4% IPM. This was in contrast to the response elicited by the control (patch without enhancer), when the increase in PCT was observed for only 12h. In human subjects, the levels of TS in the blood were higher but not significantly different from those observed with the patch containing 4% IPM. The fact that this technique can only be used with diamagnetic materials will serve as a limiting factor in its applicability and probably explains the relative lack of interest in the method.

3.7 Temperature ("Thermophoresis")

The skin surface temperature is usually maintained at 32°C in humans by a range of homeostatic controls. The effect of elevated temperature (non-physiological) on percutaneous absorption was initially reported by Blank et al. [75]. Recently, there has been a surge in the interest of using thermoregulation as a means of improving the delivery profile of topical medicaments. Previous in vitro studies [76, 77] have demonstrated a 2–3-fold increase in flux for every 7–8°C rise in skin surface temperature. The increased permeation following heat treatment has been attributed to an increase in drug diffusivity in the vehicle and an increase in drug diffusivity in the skin due to increased lipid fluidity [78]. Vasodilation of the subcutaneous blood vessels as a homeostatic response to a rise in skin temperature also plays an important role in enhancing the transdermal delivery of topically applied compounds [79, 80]. The in vivo delivery of nitroglycerin [79], testosterone, lidocaine, tetracaine [81] and fentanyl [82] from transdermal patches with attached heating devices was shown to increase as a result of the elevated temperature at the site of delivery. However, the effect of temperature on the delivery of penetrants >500Da has not been reported.

The controlled heat-aided drug delivery (CHADD) patch (Zars Inc., Salt Lake City, UT) consists of a patch containing a series of holes at the top surface which regulate the flow of oxygen into the patch. The patch generates heat chemically in a powder-filled pouch by an oxidative process regulated by the rate of flow of oxygen through the holes into the patch [83]. The CHADD technology was used in the delivery of a local anaesthetic system (lidocaine and tetracaine) from a patch (S-Caine®) and found to enhance the depth and duration of the anaesthetic action in human volunteers, when the results obtained in active and placebo groups were compared [84]. Zars Inc., together with Johnson and Johnson, recently submitted an investigational new drug (IND) application to the FDA for Titragesia™ (a combination of CHADD disks and Duragesic Patches, the latter containing fentanyl for treatment of acute pain). Kuleza and Dvoretzky [85] also have described a heat delivery patch or exothermic pad for promoting the delivery of substances into the skin, subcutaneous tissues, joints, muscles and blood stream, which may be of use in the application of drug and cosmetic treatments.

All these studies described employed an upper limit skin surface temperature of 40–42°C, which can be tolerated for a long period (>1h). In heat-patch systems where patient exposure to heat is ≤24h, such an upper limit may be necessary for regulatory compliance. In addition, the issue of drug stability may also need to be addressed when elevated temperatures are used.

Thermopertubation refers to the use of extreme temperatures to reduce the skin barrier. Such perturbation has been reported in response to using high temperatures for a short duration (30 ms), with little or no discomfort, using a novel patch system [86]. These investigators developed a polydimethylsiloxane (PDMS) patch for non-intrusive transdermal glucose sensing via thermal micro-ablation. Ablation was achieved by microheaters incorporated within the patch. The heat pulse is regulated by means of a resistive heater, which ensures that the ablation is limited within the superficial dead layers of the skin. Average temperatures of 130°C are required for ablation to occur within 33 ms, after which SC evaporation results. Other heat-assisted transdermal delivery devices under development include the PassPort® patch (Althea therapeutics) which ablates the SC in a manner similar to the PDMS patch. The exposure of skin to low (freezing) temperatures has been reported to decrease its barrier function [87–89] but has however not been exploited as a means of enhancing skin absorption.

The final group of active enhancement methods entails the use of a physical or mechanical means to breach or bypass the SC barrier.

3.8 Microneedle-Based Devices

One of the first patents ever filed for a drug delivery device for the percutaneous administration of drugs was based on this method [90]. The device as described in the patent consists of a drug reservoir and a plurality of projections extending from the reservoir. These microneedles of length 50–110 μm will penetrate the SC and epidermis to deliver the drug from the reservoir. The reservoir may contain drug, solution of drug, gel or solid particulates, and the various embodiments of the invention include the use of a membrane to separate the drug from the skin and control release of the drug from its reservoir. As a result of the current advancement in microfabrication technology in the past ten years, cost-effective means of developing devices in this area are now becoming increasingly common [91–93].

A recent commercialisation of microneedle technology is the Macroflux® microprojection array developed by ALZA Corporation. The macroflux patch can be used either in combination with a drug reservoir [94] or by dry coating the drug on the microprojection array [95]; the latter being better for intracutaneous immunization. The length of the microneedles has been estimated to be around 50–200 μm and therefore they are not believed to reach the nerve endings in the dermo-epidermal junction. The microprojections/microneedles (either solid or hollow) create channels in the skin, allowing the unhindered movement of any topically applied drug. Clinical evaluations report minimal associated discomfort and skin irritation and erythema ratings associated with such systems are reportedly low [96]. This technology serves as an important and exciting advance in transdermal technology because of the ability of the technique to deliver medicaments with extremes of physicochemical properties (including vaccines, small molecular weight drugs and large hydrophilic biopharmaceuticals) [97–99].

Yuzhakov et al. [93] describe the production of an intracutaneous microneedle array and provide an account of its use (microfabrication technology). Various embodiments of this invention can include a microneedle array as part of a closed loop system "smart patch" to control drug delivery based on feedback information from analysis of body fluids. Dual purpose hollow microneedle systems for transdermal delivery and extraction which can be coupled with electrotransport methods are also described by Trautman et al. [91] and Allen et al. [100]. These mechanical microdevices which interface with electronics in order to achieve a programmed or controlled drug release are referred to as microelectromechanical systems (MEMS) devices.

3.9 Skin Puncture and Perforation

These devices are similar to the microneedle devices produced by microfabrication technology. They include the use of needle-like structures or blades, which disrupt the skin barrier by creating holes and cuts as a result of a defined movement when in contact with the skin. Godshall and Anderson [101] described a method and apparatus for disruption of the epidermis in a reproducible manner. The apparatus consists of a plurality of microprotrusions of a length insufficient for penetration beyond the epidermis. The microprotrusions cut into the outer layers of the skin by movement of the device in a direction parallel to the skin surface. After disruption of the skin, passive (solution, patch, gel, ointment, etc.) or active (iontophoresis, electroporation, etc.) delivery methods can be used. Descriptions of other devices based on a similar mode of action have been described by Godshall [102], Kamen [103], Jang [104] and Lin et al. [105].

3.10 Needleless Injection

Needleless injection is reported to involve a pain-free method of administering drugs to the skin. This method therefore avoids the issues of safety, pain and fear associated with the use of hypodermic needles. Transdermal delivery is achieved by firing the liquid or solid particles at supersonic speeds through the outer layers of the skin by using a suitable energy source. Over the years there have been numerous examples of both liquid (Ped-O-Jet®, Iject®, Biojector2000®, Medi-jector® and Intraject®) and powder (PMED™ device, formerly known as powderject® injector) systems [99]. The latter has been reported to deliver successfully testosterone, lidocaine hydrochloride and macromolecules such as calcitonin and insulin [106, 107, 108].

Problems facing needleless injection systems include the high developmental cost of both the device and dosage form and the inability, unlike some of the other techniques described previously, to programme or control drug delivery in order to compensate for inter-subject differences in skin permeability. In addition, the long-term effect of bombarding the skin with drug particles at high speed is not known, and thus, such systems may not be suitable for the regular administration of drugs. It may however be very useful in the administration of medicaments which do not require frequent dosing, e.g. vaccines.

3.11 Suction Ablation

Formation of a suction blister involves the application of a vacuum [109] or negative pressure to remove the epidermis whilst leaving the basal membrane intact. The cellpatch® (Epiport Pain Relief, Sweden) is a commercially available product based on this mechanism [110]. It comprises a suction cup, epidermatome (to form a blister) and device (which contains morphine solution) to be attached to the skin. This method which avoids dermal invasivity, thereby avoiding pain and bleeding, is also referred to as skin erosion. Such devices have also been shown to induce hyperaemia in the underlying dermis in in vivo studies [111], which was detected by laser Doppler flowmetry and confirmed by microscopy, and is thought to further contribute to the enhancement of dextran and morphine seen with this method.

The disadvantages associated with the suction method include the prolonged length of time required to achieve a blister (2.5 h), although this can be reduced to 15–70 min by warming the skin to 38°C [111, 112]. In addition, although there is no risk of systemic infection when compared with the use of intravenous catheters, the potential for epidermal infections associated with the suction method cannot be ignored even though the effects might be less serious [113].

3.12 Application of Pressure

The application of modest pressure (i.e. 25 kPa) has been shown to provide a potentially non-invasive and simple method of enhancing skin permeability of molecules such as caffeine [114]. These workers attributed the increase in transcutaneous flux to either an improved transapendageal route or an increased partition of the compound into the SC when pressure was applied. This method may also work because of the increased solubility of caffeine in the stratum corneum caused by the increase in pressure.

3.13 Skin Stretching

These devices hold the skin under tension in either a unidirectional or a multidirectional manner [115, 116]. The authors claim that a tension of about 0.01–10 mPa results in the reversible formation of micropathways. The efficiency of the stretching process was demonstrated by monitoring the delivery of a decapeptide (1 kDa) across the skin of hairless guinea pigs by using a microprotrusion array. The results of the study showed that the bi-directional stretching of skin after microprotrusion piercing allowed the pathways to stay open (i.e. delayed closure), thereby facilitating drug permeation to a greater extent ($27.9 \pm 3.3\,\mu g\ cm^{-2}\ h^{-1}$) than in the control group ($9.8 \pm 0.8\,\mu g\ cm^{-2}\ h^{-1}$), where the skin was not placed under tension after microneedle treatment. However, increased skin permeation in the absence of microneedle pre-treatment was not found to occur.

Other methods involving the use of skin stretching with subsequent use of delivery devices based on electrotransport, pressure, osmotic and passive mechanisms have also been suggested, but the value of skin stretching alone without the benefit of a secondary active delivery device remains to be seen.

3.14 Skin Abrasion

These techniques, many of which are based on techniques employed by dermatologists in the treatment of acne and skin blemishes (e.g. microdermabrasion), involve the direct removal or disruption of the upper layers of the skin to enhance the permeation of topically applied compounds. The delivery potential of skin abrasion techniques is not restricted by the physicochemical properties of the drug, and previous work has illustrated that such methods enhance and control the delivery of a hydrophilic permeant, vitamin C [64] vaccines and biopharmaceuticals [117–119]. One current method is performed using a stream of aluminium oxide crystals and motor-driven fraises [64, 120] Sage and Bock [121, 122] also describe a method of pre-treating the skin prior to transdermal drug delivery which consists of a plurality of microabraders of length 50–200 μm. The device is rubbed against the area of interest, to abrade the site, in order to enhance delivery or extraction. The microabraders/ microprotrusions terminate as blunt tips and therefore do not penetrate the SC. The device functions by removing a portion of the SC without substantially piercing the remaining layer. Some of these methods are claimed to offer advantages such as minimal patient discomfort, increased patient compliance, ease of use and less risk of infection when compared with their more “invasive” predecessors such as ablation and the use of hypodermic needles/cannulas to deliver medicaments across the skin.

4 A Practical Example of a Skin Abrasion Device

4.1 Introduction

Abrasion devices are generally expensive and usually require trained personnel to operate them, therefore limiting applicability of the technique. One novel strategy might be to employ a rotating brush to perturb the skin barrier. The potential of such a method was investigated and compared with more established methods of enhancing in vitro skin permeation. Acyclovir is an interesting candidate for use in the development of active enhancement devices as it is poorly absorbed through the skin because of its hydrophilicity [123] This is thought to contribute to the low efficacy of commercial acyclovir formulations due to a delay in it reaching its intended target site in the basal epidermis [124, 125]. Iontophoresis currently serves as one of the most effective skin permeation strategies in enhancing the therapeutic profile of ACV [125–128] and as thus was used as a comparative method in the practical example of this review.

4.2 Methodology

In vitro experiments were conducted using excised human epidermal membrane. A rectangular section (~3 × 2 cm^2) of epidermal sheet was selected and a circular region (~1 cm^2) demarcated. Brush treatment of the skin was performed as previously described [129]. In brief, the sample of epidermal sheet was inserted into the device clamp ensuring the demarcated region was exposed. The clamp was tightened and gently raised by means of the latch (lift) until the demarcated region of the epidermis was in slight contact with the bristles (surface area, ~1 cm^2) of the brush. Pre-defined operational parameters (speed, applied pressure, treatment duration) of interest were then set on the control box for the abrasion process to occur.

Calibrated Franz cells of known surface area (~0.65 cm^2) and receptor volume (~2 mL) were used. The receptor chamber was filled with PBS (pH 7.4) and stirred throughout the duration of the experiment by using a PTFE-coated magnetic flea (5 × 2 × 2 m). The treated membrane (using brush or positive controls) or untreated control was then clamped between the donor and receptor chambers of the Franz cell (with the stratum corneum (SC) facing upwards). A radio-labelled formulation was prepared by spiking Zovirax® cream with ^{3}H-ACV (ethanolic solution). A target finite dose of ~9 ± 1 mg cm^{-2} was applied to the epidermal membrane surface using a previously calibrated positive displacement pipette (Gilson Pipetman®, P20 Anachem UK Ltd) and carefully spread to cover the effective surface area by means of a tared syringe plunger. For iontophoresis the (anodal) treatment protocol employed was as described previously [127, 130]. A 0.4-mA current limit and 10-min treatment duration was maintained to simulate "in use" conditions. A shorter iontophoretic treatment (0.4 mA for 10 min) was also employed, so as to reduce the likelihood of potential damage to the skin as a result of prolonged current exposure. All experiments were conducted in a water bath at 37°C for a minimum period of 4 h with sink conditions being maintained throughout. At certain time intervals 200 μL of the receiver fluid was carefully withdrawn from the receiver fluid. Approximately 4 mL of scintillation cocktail was added to each 200-μL sample and analysis was conducted using scintillation counting.

4.3 Results and Discussion

The skin permeation of ACV (Zovirax cream) applied as a finite dose was promoted to a greater extent as the duration of brush treatment was extended (Fig. 5.2). A significant increase in ACV transport was observed following brush treatment ($p \leq 0.05$). The use of iontophoresis proved generally less effective than employing the rotating brush in enhancing permeation. For example the effect of 10-min anodal iontophoresis on the skin permeation of ACV proved to be comparable to that obtained after application of brush treatment at 300 N m^{-2} for 10 s (Table 5.2). The iontophoretic method in this study employed optimum conditions (electrode type, anode; pH of buffer, 7.4; current intensity,

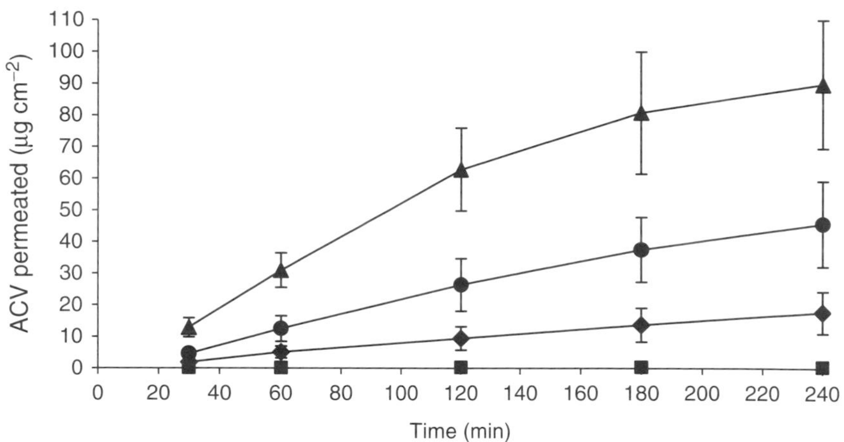

Fig. 5.2 Effect of treatment time on the skin permeation profile of ACV (finite dose) from a topical preparation using rotating device with brush B and constant device parameters (speed, 80 rpm; pressure, 300 N m^{-2}) [(■) untreated, (♦) 10 s, (●) 30 s and (▲) 60 s]. Data represent mean ± SE ($n \geq 4$), and error bars not shown are within size of symbol

Table 5.2 A comparison of the effects of iontophoresis and treatment with brush B at various times on the in vitro skin permeation of ^{3}H-labelled ACV (finite dose)

Treatment type	Amount in receptor (μg cm^{-2}) after 60 min	Enhancement factor
Untreated	0.14 ± 0.08	–
Brush treatment[a] (s)		
10	5.06 ± 1.88*	36.17
30	12.5 ± 4.02*	89.29
60	30.91 ± 5.45*	220.76
Ionotophoresis (anodal)	4.95 ± 2.35*	35.42

Data represent mean ± SE ($n \geq 4$) except otherwise stated

*Significantly different from that of untreated skin ($p \leq 0.05$)

[a]Device parameters (speed, 80 rpm; pressure, 300 N m^{-2}) were maintained constant

≤0.5 mA) which have been previously shown to enhance ACV permeation in vitro [126, 127, 130].

Findings from this present study support the effectiveness of a rotating brush (applied to the skin) in enhancing the cutaneous permeability of acyclovir. The observed enhancement in permeability was a result of the disruption of the cells of the SC, which compromises the principal barrier that skin provides to the absorption of applied compounds. Abrasion devices which allow the controlled removal of only the upper layers of the skin could be an important tool when attempting to generate a standardised skin treatment prior to the topical application of drugs. This prerequisite is a limitation of previous research into this mode of skin penetration enhancement. The use of a rotating brush device as described in this study may serve as an efficient and simple means of overcoming such a limitation. Further in vitro studies are warranted using other solutes to optimize further device parameters, as is an in vivo delivery feasibility study using such a prototype device.

5 The Future

The market for transdermal devices has been estimated at US $2 billion [120] and this figure represents 10% of the overall US $28 billion drug delivery market. Such figures are surprising when it is considered that although the first transdermal patch was granted a licence by the FDA in 1979, only an additional nine drugs have been approved since this time. This short list of "deliverables" highlights the physicochemical restrictions imposed on skin delivery.

Transdermal drug delivery has recently experienced a healthy annual growth rate of 25%, which outpaces oral drug delivery (2%) and the inhalation market (20%) [131]. This figure will certainly rise in the future as novel devices emerge and the list of marketed transdermal drugs increases. The emergence of such devices will increase the use of the skin as a route of administration for the treatment of a variety of conditions.

However, subjective and objective analyses of these devices are required to make sure both scientific, regulatory and consumer needs are met. The devices in development are costlier and more complicated when compared with conventional transdermal patch therapies. As such they may contain electrical and mechanical components which could increase the potential safety risks to patients because of poor operator technique or device malfunction. In addition, effects of the device on the skin must be reversible, since any permanent damage to the SC will result in the loss of its barrier properties and hence its function as a protective organ. Regulatory bodies will also require data to substantiate the safety of the device on the skin for either short- or long-term use. Thus, for any of these novel drug delivery technologies to succeed and compete with those already on the market, their safety, efficacy, portability, user-friendliness, cost-effectiveness and potential market have to be addressed.

References

1. Elias, P.M. (1983) "Epidermal lipids, barrier function and desquamation", *J. Invest. Dermatol.* 80, 44–49.
2. Pirot, F., Kalia, Y.N., Stinchcomb, A.L., Keating, G., Bunge, A. and Guy, R.H. (1997) "Characterization of the permeability barrier of human skin in vivo. *Proc. Nat. Acad. Sci. U.S.A.* 94, 1562–1567.
3. Scheuplein, R.J. and Blank, I.H. (1971) "Permeability of the skin", *Physiol. Rev.* 51, 702–747.
4. Flynn, G., Yalkowsky, S.H. and Roseman, T.J. (1974) "Mass transport phenomena and models", *J. Pharm. Sci.* 63, 479–510.
5. Cleary, G.W. (1993). "Transdermal delivery systems; a medical rationale", In: Shah, V.P. and Maibach, H.I., eds., Topical drug bioavailability, bioequivalence and penetration. Plenum, New York, pp. 17–68.
6. Henzel,, M.R. and Loomba, P.K. (2003) "Transdermal delivery of sex steroids for hormone replacement therapy and contraception. A review of principles and practice". *J. Reprod. Med.* 48, 525–540.
7. Kormic, C.A., Santiago-Palma, J., Moryll, N., Payne, R. and Obbens, E.A. (2003) "Benefit-risk assessment of transdermal fentanyl for the treatment of chronic pain". *Drug Saf.* 26, 951–973.
8. Varvel, J.R., Shafer, S.L., Hwang, S.S., Coen, P.A. and Stanski, D.R. (1989) "Absorption characteristics of transdermally administered fentanyl". *Anaesthesiology.* 70, 928.
9. Yang, S.I., Park, H.Y., Lee. S.H., Lee, S.J., Han, O.Y., Lim, S.C., Jang, C.G., Lee, W.S., Shin, H.Y., Kim, J.J. and Lee, S.Y, (2004) "Transdermal eperisone elicits more potent and longer-lasting muscle relaxation than oral operisone". *Pharmacology.* 71, 150–156.
10. Cramer, M.P. and Saks, S.R. (1994) "Translating safety, efficacy and compliance into economic value for controlled release dosage forms". *Pharmacoeconomics.* 5, 482–504.
11. Payne, R., Mathias, S.D., Pasta, D.J., Wanke, L.A., Williams, R. and Mahmoud, R. (1998) "Quality of life and cancer pain: satisfaction and side effects with transdermal fentanyl versus oral morphine". *J. Clin. Oncol.* 16, 1588–1593.
12. Jarupanich, T., Lamlertkittikul, S. and Chandeying, V. (2003) "Efficacy, safety and acceptability of a seven-day, transdermal estradiol patch for estrogen replacement therapy". *J. Med. Assoc. Thai.* 86, 836–845.
13. Archer, D.F., Cullins, V., Creasy, D.W. and Fisher, A.C. (2004) "The impact of improved compliance with a weekly contraceptive transdermal system (Ortho Evra) on contraceptive efficacy". *Contraception.* 69, 189–195.
14. Long, C. (2002) "Common skin disorders and their topical treatment". In: Walters, K.A., ed., Dermatological and transdermal formulations. Marcel Dekker, New York, pp. 41–60.
15. Whittington, R. and Faulds, D. (1994) "Hormone replacement therapy: I. A pharmacoeconomic appraisal of its therapeutic use in menopausal symptoms and urogenital estrogen deficiency". *Pharmacoeconomics.* 5, 419–445. Review. Erratum in: *Pharmacoeconomics* (1995), 8, 244.
16. Frei, A., Andersen, S., Hole, P. and Jensen, N.H. (2003) "A one year health economic model comparing transdermal fentanyl with sustained-release morphine in the treatment of chronic non-cancer pain". *J. Pain Palliat. Care. Pharmacother.* 17, 5–26.
17. Bos, J.D. and Meinardi, M.M. (2000) "The 500 Dalton rule for skin penetration of chemical compounds and drugs". *Exp. Dermatol.* 9, 165–169.
18. Yano, T., Nagakawa, A., Tsuji, M. and Noda, K. (1986) "Skin permeability of various non-steroidal anti-inflammatory drugs in man". *Life Sci.* 39, 1043–1050.
19. Southwell, S., Barry, B.W. and Woodford, R. (1984) "Variations in permeability of human skin within and between specimens". *Int. J. Pharm.* 18, 299–309.
20. Larsen, R.H., Nielsen, F., Søresen, J.A., Nielsen, J.B. (2003) "Dermal penetration of fentanyl: inter- and intraindividual variations". *Pharmcol. Toxicol.* **93**, 244–248.

21. Steinsträsser, I. and Merkle, H.P. (1995) "Dermal metabolism of topically applied drugs: pathways and models reconsidered". *Pharm. Acta. Helv.* 70, 3–24.
22. Hogan, D.J. and Maibach, H.I. (1990) "Adverse dermatologic reactions to transdermal drug delivery systems". *J. Am. Acad. Dermatol.* 22, 811–814.
23. Carmichael, A.J. (1994) "Skin sensitivity and transdermal drug delivery. A review of the problem". *Drug Saf.* 10, 151–159.
24. Toole, J., Silagy, S., Maric, A., Fath, B., Quebe-Fehling, E., Ibarra de Palacios, P., Laurin, L. and Giguere, M. (2002) "Evaluation of irritation and sensitisation of two 50 microg/day oestrogen patches". *Maturitas.* 43, 257–263.
25. Murphy, M. and Carmichael, A.J. (2000) "Transdermal drug delivery systems and skin sensitivity reactions, Incidence and management". *Am. J. Clin. Dermatol.* 1, 361–368.
26. Williams, A.C. and Barry, B.W. (2004). "Penetration enhancers", Adv. *Drug Deliv. Rev.* 56, 603–618.
27. Pellet, M., Raghavan, S.L., Hadgraft, J. and Davis, A.F. (2003) "The application of supersaturated systems to percutaneous drug delivery". In: Guy, R.H. and Hadgraft, J., eds., Transdermal drug delivery. Marcel Dekker, Marcel Dekker, New York, pp. 305–326.
28. Tsai, J.C., Guy, R.H., Thornfeldt, C.R., Gao, W.N., Feingold, K.R. and Elias, P.M. (1996) "Metabolic approaches to enhance transdermal drug delivery. 1. Effect of lipid synthesis inhibitors", *J. Pharm. Sci.* 85, 643–648.
29. Elias, P.M., Feingold, K.R., Tsai, J., Thornfeldt, C. and Menon, G. (2003) "Metabolic approach to transdermal drug delivery". In: Guy, R.H. and Hadgraft, J., eds., Transdermal drug delivery. Marcel Dekker, pp. 285–304.
30. Schreier, H. and Bouwstra, J. (1994) "Liposomes and niosomes as topical drug carriers: dermal and transdermal drug delivery", *J.Control. Release.* 30, 1–15.
31. Cevc, G. (1996) "Transfersomes, liposomes and other lipid suspensions on the skin: permeation enhancement, vesicle penetration, and transdermal drug delivery". *Crit. Rev. Ther. Drug Carrier Syst.* 13(3–4), 257–388.
32. Cevc, G. (2003). "Transferosomes: innovative transdermal drug carriers". In: Rathbone, M.J., Hadgraft, J. and Roberts, M.S., eds., Modified release drug delivery technology. Marcel Dekker, New York, pp. 533–560.
33. Godin, B. and Touitou, E. (2003) "Ethosomes: new prospects in transdermal delivery", *Crit. Rev. Ther. Drug. Carrier Syst.* 20, 63–102.
34. Helmstädter, A. (2001) "The history of electrically assisted transdermal drug delivery (iontophoresis)", *Pharmazie.* 56, 583–587.
35. Banga, A.K., Bose, S. and Ghosh, T.K. (1999) "Iontophoresis and electroporation: comparisons and contrasts", *Int. J. Pharm.* 179, 1–19.
36. Weaver, J.C., Vaughan, T.E. and Chizmadzhev, Y.A. (1999) "Theory of electrical creation of aqueous pathways across skin transport barriers", *Adv. Drug Deliv. Rev.* 35, 21–39.
37. Denet, A.R., Vanbever, R. and Préat, V. (2004) "Skin electroporation for topical and transdermal delivery", *Adv. Drug Deliv. Rev.* 56, 659–674.
38. Pliquett, U., Vaughan, T. and Weaver, J. (1999) "Apparatus and method for electroporation of tissue". US Pat. 5,983,131.
39. Zhang, L., Hofmann, G.A. and Rabussay, D. (2001) "Electrically assisted transdermal method and apparatus for treatment of erectile dysfunction", US Pat. 6,266,560.
40. Sugibayash, K., Kubo, H. and Mori, K. (2002) "Device and electrode for electroporation". Patent WO 01/28624.
41. Wong, T.W., Chen, C.H., Huang, C.C., Lin, C.D. and Hui, S.W. (2006) "Painless electroporation with a new needle-free microelectrode array to enhance transdermal drug delivery". *J. Control. Release.* 110(3), 557–565.
42. Wang, Y., Allen, L.V., Li, C. and Tu, Y. (1993) "Iontophoresis of hydrocortisone across hairless mouse skin: investigation of skin alteration", *J. Pharm. Sci.* 82, 1140–1144.
43. Turner, N.G., Kalia, Y.N. and Guy, R.H. (1997) "The effect of current on skin barrier function in vivo: recovery kinetics post iontophoresis", *Pharm. Res.* **14**, 1252–1255.

44. Banga, A.K. (1998). "Electrically assisted transdermal and topical drug delivery". Taylor and Francis, London.
45. Guy, R.H., Kalia, Y.N., Delgado-Charro, M.B., Merino, V., López, A. and Marro, D. (2000) "Iontophoresis: electrorepulsion and electroosmosis". *J. Control. Release.* 64, 129–132.
46. Subramony, J.A., Sharma, A. and Phipps, J.B. (2006) "Microprocessor controlled transdermal drug delivery". *Int. J. Pharm.* 317(1), 1–6.
47. Tyle, P. (1986) "Iontophoretic devices for drug delivery". *Pharm. Res.* 3, 318–326.
48. Kalia, Y.N., Naik, A., Garrison, J. and Guy, R.H. (2004) "Iontophoretic drug delivery". *Adv. Drug Del. Rev.* 56, 619–658.
49. Priya, B., Rashmi, T. and Bozena, M. (2006) "Transdermal iontophoresis". *Expert Opin. Drug Deliv.* 3(1), 127–138.
50. Bommannan, D.B., Tamada, J., Leung, L. and Potts, R.O. (1994) "Effects of electroporation on transdermal iontophoretic delivery of luteinizing-hormone-releasing hormone (LHRH) in vitro". *Pharm. Res.* 11, 1809–1814.
51. Chang, S.L., Hofmann, G.A., Zhang, L., Deftos, L.J. and Banga, A.K. (2000) "The effect of electroporation on iontophoretic transdermal delivery of calcium regulating hormones". *J. Control. Release.* 66, 127–133.
52. Badkar, A.V. and Banga, A.K. (2002) "Electrically enhanced transdermal delivery of a macromolecule". *J. Pharm. Pharmacol.* 54, 907–912.
53. Kanikkannan, N. (2002) "Iontophoresis based transdermal delivery systems". *Biodrugs.* 16, 339–347.
54. Mitragotri, S., Blankschtein, D. and Langer, R. (1996) "Transdermal delivery using low frequency sonophoresis", *Pharm. Res.* 13, 411–420.
55. Mitragotri, S. (2004) "Low frequency sonophoresis", *Adv. Drug. Deliv. Rev.* 56, 589–601.
56. Mitragotri, S., Blankschtein, D. and Langer, R. (1995) "Ultrasound mediated transdermal protein delivery", *Science.* 269, 850–853.
57. Liu, H., Li, S., Pan, W., Wang, Y., Han, F. and Yao, H. (2006) "Investigation into the potential of low-frequency ultrasound facilitated topical delivery of Cyclosporin A". *Int. J. Pharm.* 326(1–2), 32–8.
58. Kost, J., Katz, N., Shapiro, D., Herrmann, T., Kellog, S., Warner, N. and Custer, L. (2003) "Ultrasound skin permeation pre-treatment to accelerate the onset of topical anaesthesia". *Proc. Int. Symp. Bioact. Mater.*
59. Tachibana, K. (1992) "Transdermal delivery of insulin to alloxan-diabetic rabbits by ultrasound exposure". *Pharm. Res.* 9, 952–954.
60. Boucaud, A., Garrigue, M.A., Machet, L., Vaillant, L. and Patat, F. (2002) "Effect of sonication parameters on transdermal delivery of insulin to hairless rats", *J. Control. Release.* 81, 113–119.
61. Smith, N.B., Lee, S., Maione, E., Roy, R., McElligott, S. and Shung, K.K. (2003) "Ultrasound mediated transdermal transport of insulin in vitro through human skin using novel transducer designs", *Ultrasound. Med. Biol.* 29, 311–317.
62. Jacques, S.L., McAuliffe, D.J., Blank, I.H. and Parrish, J.A. (1988) "Controlled removal of human stratum corneum by pulsed laser to enhance percutaneous transport". US Pat 4, 775, 361.
63. Lee, W.R., Shen, S.C., Lai, H.H., Hu, C.H. and Fang, J.Y. (2001) "Transdermal drug delivery enhanced and controlled by erbium: YAG laser: a comparative study of lipophilic and hydrophilic drugs". *J. Control. Release.* 75, 155–166.
64. Lee, W.R., Shen, S.C., Wang, K.H., Hu, C.H. and Fang, J.Y. (2003) "Lasers and microdermabrasion enhance and control topical delivery of vitamin C". *J. Invest. Dermatol.* 121, 1118–1125.
65. Baron, E.D., Harris, L., Redpath, W.S., Shapiro, H., Herzel, F., Morley, G., Bar, O.D. and Stevens, S.R. (2003) "Laser assisted penetration of topical anaesthesia". *Arch. Dermatol.* 139, 1288–1290.
66. Lee, S., McAuliffe, D.J., Flotte, T.J., Kollias, N. and Doukas, A.G. (1998) "Photomechanical transcutaneous delivery of macromolecules". *J. Invest. Dermatol.* **111**, 925–929.

67. Lee, S., Kollias, N., McAuliffe, D.J., Flotte, T.J. and Doukas, A.G. (1999) "Topical drug delivery in humans with a single photomechanical wave". *Pharm. Res.* 16, 514–518.
68. Doukas, A.G., Kollias, N. (2004) "Transdermal delivery with a pressure wave". *Adv. Drug. Deliv. Rev.* 56, 559–579.
69. Mulholland, S.E., Lee, S., McAuliffe, D.J. and Doukas, A.G. (1999) "Cell loading with laser generated stress waves: the role of stress gradient". *Pharm. Res.* 16, 514–518.
70. Lee, S., McAuliffe, D.J., Flotte, T.J., Kollias, N. and Doukas, A.G. (2001) "Permeabilization and recovery of the stratum corneum in vivo: the synergy of photomechanical waves and sodium lauryl sulphate". *Lasers Surg. Med.* 29, 145–150.
71. Lee, S., McAuliffe, D.J., Mulholland, S.E. and Doukas, A.G. (2001) "Photomechanical transdermal delivery; the effect of laser confinement". *Lasers Surg. Med.* 28, 344–347.
72. Sintov, A., Krymbeck, I., Daniel, D., Hannan, T., Sohn, Z. and Levin, G. (2003) "Radiofrequency microchanneling as a new way for electrically assisted transdermal delivery of hydrophilic drugs". *J. Control. Release.* 89, 311–320.
73. Murthy, S.N. (1999) "Magnetophoresis: an approach to enhance transdermal drug diffusion". *Pharmazie.* 54, 377–379.
74. Murthy, S.N. and Hiremath, R.R. (2001) "Physical and chemical permeation enhancers in transdermal delivery of terbutaline sulphate". *AAPS PharmSciTech.* 2, 1–5.
75. Blank, I.H., Scheuplein, R.J. and Macfarlane, D.J. (1967) "Mechanism of percutaneous absorption: III. The effect of temperature on the transport of non-electrolytes across the skin". *J. Invest. Dermatol.* 49, 582–589.
76. Clarys, P., Alewaeters, K., Jadoul, A., Barel, A., Mandas, O.R. and Preat, V. (1998) "In vitro percutaneous penetration through hairless rat skin: influence of temperature, vehicle and penetration enhancers". *Eur. J. Pharm. Biopharm.* 46, 279–283.
77. Akomeah, F., Nazir, T., Martin, G.P. and Brown, M.B. (2004) "Effect of heat on the percutaneous absorption and skin retention of 3 model penetrants". *Eur. J. Pharm. Sci.* 21, 337–345.
78. Ogiso, T., Hirota, T., Masahiro, I., Hino, T. and Tadatoshi, T. (1998) "Effect of temperature on percutaneous absorption of terodiline and relationship between penetration and fluidity of stratum corneum lipids". *Int. J. Pharm.* 176, 63–72.
79. Klemsdal, T.O., Gjesdal, K. and Bredesen, J.E. (1992) "Heating and cooling of the nitroglycerin patch application area modify the plasma level of nitroglycerin". *Eur. J. Clin. Pharmacol.* 43, 625–628.
80. Hull, W. (2002) "Heat enhanced transdermal drug delivery: a survey paper". *J. Appl. Res. Clin. Exp. Ther.* 2, 1–9.
81. Shomaker, T.S., Zhang, J. and Ashburn, M.A. (2001) "A pilot study assessing the impact of heat on transdermal delivery of testosterone". *J. Clin. Pharmacol.* 41, 677–682.
82. Ashburn, M.A., Ogden, L.L., Zhan, J., Love, G. and Bastsa, S.V. (2003) "Pharmacokinetics of transdermal fentanyl delivered with and without controlled heat". *J. Pain.* 4, 291–297.
83. Stanley, T., Hull, W. and Rigby, L. (2001) "Transdermal drug patch with attached pocket for controlled heating device". US Pat. 6,261,595.
84. Shomaker, T.S., Zhang, J., Love. G., Basta, S. and Ashburn, M.A. (2000) "Evaluating skin anaesthesia after administration of a local anaesthetic system consisting of an S-Caine™ patch and a controlled heat-aided drug delivery (CHADD™) patch in volunteers". *Clin. J. Pain.* 16, 200–204.
85. Kuleza, J. and Dvoretzky, I. (2001) "Multipurpose drug and heat therapy system". Patent WO 01/58408.
86. Paranjape, M., Garra, J., Brida, S., Schneioder, T., White, R. and Currie, J. (2003) "A PDMS dermal patch for non-intrusive transdermal glucose sensing". *Sens Actuators* A. 104, 195–204.
87. Kasting, G.B. and Bowman, L.A. (1990) "Electrical analysis of fresh excised human skin: A comparison with frozen skin". *Pharm. Res.* **7**, 1141–1146.
88. Yazdanian, M. (1994) "Effect of freezing on cattle skin permeability". *Int. J. Pharm.* **103**, 93–96.
89. Babu, R.J., Kanikkannan, N., Kikwai, L., Ortega, C., Andega, S., Ball, K., Yim, S. and Singh, M. (2003) "The influence of various methods of cold storage on the permeation of melatonin and nimesulide". *J. Control. Release.* 86, 49–57.
90. Gerstel, M.S. and Place, V.A. (1976) "Drug delivery device". US Pat. 3,964,482.

91. Trautman, J., Cormier, M.J., Kim, H.L. and Zuck, M.G. (2000) "Device for enhancing transdermal agent flux". US Pat. 6,083,196.
92. Trautman, J., Wong, P.S., Daddona, P.E., Kim, H.L. and Zuck, M.G. (2001) "Device for enhancing transdermal agent flux". US Pat. 6,322,808 B1.
93. Yuzhakov, V.V., Sherman, F.F., Owens, G.D. and Gartstein, V. (2001) "Apparatus and method for using an intracutaneous microneedle array". US Pat. 6,256,533.
94. Lin, W.Q., Cormier, M., Samiee, A., Griffin, A., Johnson, B., Teng, C.L., Hardee, G.E. and Daddona, P. (2001) "Transdermal delivery of antisense oligonucleotides with microprojection patch (Macroflux®) Technology". *Pharm. Res.* 18, 1789–1793.
95. Matriano, J.A., Cormier, M., Johnson, J., Young, W.A., Buttery, M., Nyam, K. and Daddona, P. (2002) "Macroflux Technology: a new and efficient approach for intracutaneous immunization". *Pharm. Res.* 19; 63–70.
96. Kaushik, S., Hord, A.H., Denson, D.D., McAllister, D.V., Smitra, S., Allen, M.G. and Prausnitz, M.R. (2001) "Lack of pain associated with microfabricated microneedles". *Anesth. Analg.* 92, 502–504.
97. Prausnitz, M.R. (2004) "Microneedles for transdermal drug delivery". *Adv. Drug Deliv. Rev.* 56, 581–587.
98. Martanto, W., Davis, S.P., Holiday, N.R., Wang, J., Gill, H.S., Prausnitz, M.R. (2004) "Transdermal delivery of insulin using microneedles in vivo". *Pharm. Res.* 21, 947–952.
99. Giudice, E.L. and Campbell, J.D. (2006) "Needle-free vaccine delivery". *Adv. Drug Deliv. Rev.* **58**(1), 68–89.
100. Allen, M.G., Prausnitz, M.R., McAllister, D.V. and Cross, F.P.M. (2002) "Microneedle devices and methods of manufacture and use thereof". US Pat. 6,334,856.
101. Godshall, N. and Anderson, R. (1999) "Method and apparatus for disruption of the epidermis". US Pat. 5,879,326.
102. Godshall, N. (1996) "Micromechanical patch for enhancing delivery of compounds through the skin". Patent WO 9637256.
103. Kamen, D. (1998) "System for delivery of drugs by transport". Patent WO 98/11937.
104. Jang, K. (1998) "Skin perforating apparatus for transdermal medication". US Pat. 5,843,114.
105. Lin, W.Q., Theeuwes, F. and Cormier, M. (2001) "Device for enhancing transdermal flux of sampled agents". Patent WO 01/43643.
106. Muddle, A.G., Longridge, D.J., Sweeney, P.A., Burkoth, T.L. and Bellhouse, B.J. (1997) "Transdermal delivery of testosterone to conscious rabbits using powderject (R): a supersonic powder delivery system". *Proc. Int. Symp. Control. Release. Bioact. Mat.* 24, 713.
107. Longbridge, D.J., Sweeney, P.A., Burkoth, T.L. and Bellhouse, B.J. (1998) "Effects of particle size and cylinder pressure on dermal powderject® delivery of testosterone to conscious rabbits". *Proc. Int. Symp. Control. Rel Bioact. Mat.* 25, 964.
108. Burkoth, T.L., Bellhouse, B.J., Hewson, G., Longridge, D.J., Muddle, A.J. and Sarphie, D.J. (1999) "Transdermal and transmucosal powdered delivery". *Crit. Rev. Ther. Drug Carrier Syst.* 16, 331–384.
109. Bernabei, G.F., "Method and apparatus for skin absorption enhancement and transdermal drug delivery". US Pat. 7,083,580.
110. Svedman, P. (1995) "Transdermal perfusion of fluids". US Pat. 5,441,490.
111. Svedman, P., Lundin, S., Höglund, P., Hammarlund, C., Malmros, C. and Panzar, N. (1996) "Passive drug diffusion via standardized skin mini-erosion; methodological aspects and clinical findings with new device". *Pharm. Res.* 13, 1354–1359.
112. Svedman, P. and Svedman, C. (1998) "Skin mini-erosion sampling technique: feasibility study with regard to serial glucose measurement". *Pharm. Res.* **15**, 883–888.
113. Down, J. and Harvey, N.G. (2003) "Minimally invasive systems for transdermal drug delivery". In: Guy, R.H. and Hadgraft, J., eds., Transdermal drug delivery. Marcel Dekker, New York, pp. 327–360.
114. Treffel, P., Panisset, F., Humbert, P., Remoussenard, O., Bechtel, Y. and Agache, P. (1993) "Effect of pressure on in vitro percutaneous absorption of caffeine". *Acta. Derm. Venereol (Stockh).* 73, 200–202.

115. Cormier, M., Trautman, J., Kim, H.L., Samiee, A.P., Ermans, A.P., Edwards, B.P., Lim, W.L. and Poutiatine, A. (2001) "Skin treatment apparatus for sustained transdermal drug delivery". Patent WO 01/41864 A1.
116. Neukermans, A.P., Poutiatine, A.I., Sendelbeck, S., Trautman, J., Wai, L.L., Edwards, B.P., Eng, K.P., Gyory, J.R., Hyunok, K.L., Lin, W.Q. and Cormier M (2001) "Device and method for enhancing microprotrusion skin piercing". Patent WO 0141863.
117. Mikszta, J.A., Britingham, J.M., Alarcon, J., Pettis, R.J. and Dekker, J.P. (2001) "Applicator having abraded surface coated with substance to be applied". Patent WO 01/89622 A1.
118. Mikszta, J.A., Britingham, J.M., Alarcon, J., Pettis, R.J. and Dekker, J.P. (2003) "Topical delivery of vaccines". US Pat. 6,595,947 B1.
119. Lee, W.R., Tsai, R.Y., Fang, C.L., Liu, C.J., Hu, C.H. and Fang, J.Y. (2006) "Microdermabrasion as a novel tool to enhance drug delivery via the skin: an animal study". *Dermatol. Surg.* 32(8), 1013–1022.
120. Barry, B.W. (2001) "Novel mechanisms and devices to enable successful transdermal drug delivery". *Eur. J. Pharm. Sci.* 14, 101–114.
121. Sage, B.H. and Bock, C.R. (2003) "Method and device for abrading skin". US Pat. 2003/199811.
122. Sage, B.H. and Bock, C.R. (2003) "Device for abrading skin". Patent EP 1,086,719 A1.
123. Seth, A.K., Misrad, A., Umrigar, D. and Vora, N. (2003) "Role of acyclovir gel in herpes simplex: clinical implications". *Med. Sci. Monit.* 9, PI93–P198.
124. Parry, G.E., Dunn, P., Shah, V.P. and Pershing, L.K. (1992) "Acyclovir bioavailability in human skin". *J. Invest. Dermatol.* 98, 856–863.
125. Stagni, G., Ali, M.E. and Weng, D. (2004) "Pharmacokinetics of acyclovir in rabbit skin after i.v.-bolus, ointment and iontophoretic administrations". *Int. J. Pharm.* 274, 201–211.
126. Volpato, N., Santi, P. and Colombo, P. (1995) "Iontophoresis enhances the transport of acyclovir through nude mouse skin by electrorepulsion and electroosmosis". *Pharm. Res.* 12, 1623–1627.
127. Volpato, N.M., Nicoli, S., Laureri, C., Colombo, P. and Santi, P. (1998) "*In vitro* acyclovir distribution in human skin layers after transdermal iontophoresis". *J. Control. Rel.* 50, 291–296.
128. Goldberg, D. (2005) "Iontophoretic based drug delivery". *Innov. Pharm. Technol.* 16, 68–72.
129. Brown, M.B. and Martin, G.P. (2005) "Dermal drug delivery system". World Patent No. WO 2005058226.
130. Santi, P., Nicoli, S., Colombo, G., Bettini, R., Artusi, M., Rimondi, S. et al. (2003) "Post iontophoresis transport of ibuprofen lysine across rabbit ear skin". *Int. J. Pharm.* 266, 69–75.
131. Grosh, S. (2000) "Transdermal drug delivery – opening doors for the future". *Euro. Pharm. Contractor* (Nov.). 4, 30–32.

Chapter 6
Controlling the Release of Proteins/Peptides via the Pulmonary Route

Sunday A. Shoyele

Abstract The inhalation route is seen as the most promising non-invasive alternative for the delivery of proteins; however, the short duration of activity of drugs delivered via this route brought about by the activities of alveolar macrophages and mucociliary clearance means there is a need to develop controlled release system to prolong the activities of proteins delivered to the lung. Polymeric materials such as (D,L)-poly(lactic glycolic acid) (PLGA), chitosan and poly(ethylene glycol) (PEGs) have been used for controlled release of proteins. Other systems such as liposomes and microcrystallization have also proved effective.

This chapter gives a more detailed understanding of these techniques and the manufacture of the delivery systems.

Keywords Proteins; Controlled release; Polymers; Liposomes; Microcrystals; Pulmonary

1 Introduction

Formulation of proteins for either local lung or systemic delivery has continued to pose some challenges to drug formulators because of their "fragile" nature. Although the primary structure of proteins is made up of covalent bonds, the higher order structures (secondary, tertiary, and sometimes quaternary) are made up of relatively weak physical interactions (electrostatic, hydrogen bonding, van der Waal's forces and hydrophobic interactions) and not of the much stronger covalent bonding [1]. Owing to these weak interactions, proteins can easily undergo conformational changes, which can lead to a reduction of their biological activity. Formulation of proteins for therapeutic use will thus depend on the physical and chemical stability of such molecules, since the loss of the native conformation may result in reduction or complete loss of biological activity.

The vulnerability of proteins to gastrointestinal enzymes and first pass metabolism in the liver when administered orally makes oral administration of proteins quite challenging. Other routes such as transdermal, buccal, nasal and ocular have been

From: *Methods in Molecular Biology, Vol. 437: Drug Delivery Systems*
Edited by: Kewal K. Jain © Humana Press, Totowa, NJ

investigated without much success. The main issues involved with these routes are variable bioavailability [2] and safety of the enhancers used in the formulations [3].

The inhalation route, however, offers potential possibilities for the delivery of proteins for systemic activity. The route offers enormous absorptive surface area in the range of 35–140 m^2 [4], very thin diffusion path to the blood stream, elevated blood flow, relatively low metabolic activity as well as avoidance of first pass hepatic metabolism [5]. These advantages, coupled with the fact that oral inhalation is well accepted by the general population in most societies, make pulmonary delivery of proteins quite appealing.

Despite the obvious advantages of the pulmonary route, the relatively short duration of clinical effects of drugs delivered via this route may mean multiple daily dosing. This is borne out of the fact that drugs easily cross the thin epithelium of the alveoli into systemic circulation and drugs left in the peripheral airways are easily cleared by both alveolar macrophages and mucociliary apparatus. There is therefore a need for the development of controlled/sustained release methods for pulmonary delivery of proteins/peptides so as to encourage patient compliance.

2 Controlled Release Strategies

Various controlled release techniques for pulmonary delivery of proteins have been studied (Table 6.1). The major successful ones include the following:

1. Use polymeric materials
2. Microcrystallization
3. Liposomes

2.1 Use of Polymeric Materials

Some of the most promising systems for the controlled release of proteins and peptides involve encapsulation or entrapment in biocompatible polymeric materials. The most widely used polymers to date are poly(ethylene glycol) (PEG), (D,L)-poly(lactic glycolic acid) (PLGA), poly(lactic acid) (PLA) and chitosan. Polymers could be attached to the protein to increase the overall molecular weight of the system and so reduce the rate of absorption across the epithelium of the alveoli, or the protein could be encapsulated in the polymeric system and slowly released into systemic circulation.

2.1.1 Attaching PEG to Proteins

Attaching PEG to a protein such as insulin first involves understanding where the reactive functional groups are and where the active centre of the protein is. This is important so as not to block the biological activity of the protein by sterically hindering

Table 6.1 Table showing the effect of excipient/delivery vehicle on the release pattern of protein/peptide via the inhalation route

Controlled release strategy	Method of design	Mean particle diameter (μm)	Molecule of interest	Excipients/delivery vehicle	Evidence of controlled release	Reference
Polymeric entrapment/ encapsulation	Solvent evaporation	8.20	Insulin	PLGA	96-h systemic glucose level suppression, compared with 4 h for uncoated insulin	6
	Solvent evaporation	2.53	Insulin	Cyclodextrin/PLGA	Prolonged release in diabetic rats, compared with other insulin formulations	7
	Gas anti-solvent (GAS)	13.8	Deslorelin	PLGA	7-day in vivo drug release	8
	Solvent evaporation		Insulin	PLGA	48-h systemic glucose level reduction in guinea pigs, compared with 6 h for *uncoated*	9
Polymeric conjugation	Sonication	3	Insulin	Calcium phosphate/ PEG	12-h presence in serum, compared with 6 h for unconjugated	10
	Spray drying	3.3*	Insulin	PEG	12-h presence in serum, compared with 2 h for unconjugated	11
Microcrystal	Seed zone	3.0	Insulin	Rhombohedral? Rhombus microcrystals	Increased T_{max} from 2 h (solution) to 5 h (microcrystal) with prolonged hypoglycaemic effect	12
Liposomes	Thin film hydration	<200 nm**	Detirelix	DSPC/DSPG/ cholesterol	Increased half-life of Detirelix from 8.2 (free) to 21.6 h (liposomes)	13

PLGA (D,L)-poly(lactic glycolic acid), *PEG* poly(ethylene glycol), *DSPC* distearoyl-L-α–phosphatidylcholine, *DSPG* distearoyl-L-α–phosphatidylglycerol)
*Aerodynamic diameter, ** Liposome vesicle size

access to the active site by the polymer. Once this has been identified, the reactive functional groups close to the active site are protected by attaching t-Boc to them in a suitable organic solvent such as dimethyl sulphoxide-triethylamine (DMSO-TEA) mixture. The reaction mixture is then extensively dialyzed and lyophilized (*see* **Note 1**).

The t-Boc protein can then be attached to methoxy PEG-succinimidyl propionate in DMSO-TEA mixture. The mPEG-Boc-protein solution can then be extensively dialyzed and lyophilized [14]. Quantitative deprotection of the lyophilized product can then be achieved by reaction with trifluoroacetic acid (TFA) at 0°C. The conjugate can now be purified by using reverse phase-high performance liquid chromatography (RP-HPLC).

Following the production of the PEG-protein conjugate, there is a need to prepare the powder for inhalation. This can be done by spray drying the solution of the conjugate containing appropriate stabilizing excipients (such as surfactants and polyols). High-quality particles for inhalation can be achieved by using the appropriate parameters (inlet temperature, outlet temperature, pump rate, flow rate, etc.).

2.1.2 Preparation of Polymeric Microspheres

PLGA is the most widely used polymer for encapsulating proteins for pulmonary delivery. The most commonly used method for preparing protein-encapsulated PLGA microspheres is the solvent evaporation technique based on the formation of a double emulsion (w/o/w). Incorporation of protein into the microspheres could be done by two methods [15, 16].

The first method involves preparing the microspheres first by solvent evaporation and then loading the protein into the microspheres. The emulsion is prepared by adding about 1–2 ml of deionised water to about 5 ml of methylene chloride containing certain amount of PLGA which forms the oil phase. This mixture is sonicated for about 60 s to form a water-in-oil (w/o) emulsion. The resulting w/o emulsion is then quickly added to about 200 ml of deionised water containing about 0.5% (w/o) of poly(vinyl alcohol) and stirred at about 1,500 rpm for about 2 min to allow evaporation of the methylene chloride and hardening of the microspheres. The hardened microspheres can now be washed three times with excess deionised water and freeze dried. The porous microspheres formed can then be loaded with the protein of choice by suspending a specific amount of the microspheres in a solution of the protein in buffer and gently shaken for about 2 h. The protein-loaded microspheres can now be separated by centrifugation and freeze dried.

The second method involves incorporating the protein in the microspheres during the preparation of the double emulsion by dissolving the protein in the initial water phase before mixing with the oil phase. Other processes as earlier-described are then performed.

A more recent approach is the formation of a solid-in-oil-in-water emulsion (s/o/w) before solvent evaporation [18]. This involves the formation of solid protein particles by either spray drying or spray freeze-drying, followed by the dispersion of certain amount of the particles formed in about 5 ml of PLGA methylene chloride solution by ultrasonication for about 1 min. The PLGA/protein mixture is then dispersed into about 0.5% PVA aqueous solution with agitation using a stirrer at 250 rpm for about 4 h to allow the methylene chloride to evaporate. The solid particles formed can then be collected by filtration, rinsed three times with deionised water and then freeze dried.

2.2 Microcrystallization

Microcrystals of proteins with mean diameter <3 μm can be prepared for sustained release by the "seed zone" method. The sustained release effect could be attributed to the decreased solubility of the microcrystals [12]. This method has been used in the making of insulin microcrystals of rhombohedra shape without aggregates. Following intratracheal instillation of the insulin microcrystal suspension (32 U/kg) to rats, the blood glucose levels were reduced and hypoglycaemia was prolonged for 13 h when compared with the unmodified insulin solution [12].

Apart from the seed zone method, a conventional seeding technique can also be used for preparing microcrystals for inhalation. This method involves the creation of seeds by suspending protein particles in a suitable buffer solution and centrifuging this suspension at about 10,000 rpm for about 10 min. The supernatant can then be stored at 4°C, to be used as a seed solution. Protein particles can then be dissolved in another buffer solution but at a much lower pH to facilitate the dissolution of the protein. The amount of protein is then slowly increased to achieve supersaturation. The solution is then filtered and the initially prepared seed solution is added to the filtrate. The mixture is sealed and incubated at 37°C. The only issue with this conventional seeding method is that crystals above 5 μm could be formed, which are not suitable for pulmonary delivery (*see* **Note 2**).

Although microcrystals of proteins seem to be a promising sustained release technique, the large molecular weight and flexibility of most proteins could mean that not all therapeutic proteins would be amenable to this technique.

2.3 Liposomes

Liposomes are artificial, spherical vesicles consisting of amphiphilic lipids (mostly phospholipids), enclosing an aqueous core. Depending on the processing conditions and the chemical composition, liposomes can either be unilamellar or multilamellar.

Liposomes are mostly prepared by the thin film hydration method in which a thin film is produced by dissolving the phospholipids in suitable organic solvent (mostly chloroform or ethanol) and then evaporating the solvent in a rotary evaporator under vacuum. Hydration of this thin lipid film with an aqueous solution of the protein, followed by physical shaking, leads to formation of the liposomes [13]. The size of the vesicles can be reduced by sonication. Unencapsulated protein can be removed by centrifugation and separation of the supernatant. The liposomes formed are then dried by either lyophilisation or spray drying following their suspension in an isotonic aqueous solution.

Liposomes can be classified according to the number of lamellae and size:

1. Small unilamellar vesicle (SUV)
2. Large unilamellar vesicle (LUV)
3. Multilamellar vesicle (MLV)
4. Multivesicular vesicle (MVV)

SUVs have a diameter of 20 to ~100 nm while MLVs, LUVs and MVVs range in size from a few hundred nanometres to several micrometers. An average membrane of a liposome (phospholipid bilayer) measures about 7 nm.

Large liposomes are formed when phospholipids are hydrated at temperature above their phase transition temperature (T_c). Although MLVs are normally formed when lipid films are hydrated below T_c, these can be transformed into small vesicles by using high pressure homogenisation.

The fact that liposomes can be formed from a variety of lipids makes them quite versatile, having a wide range of physicochemical properties depending on the types of lipids used. These physicochemical properties such as liposomes size, surface charge, method of preparation and bilayer fluidity affect their drug release properties. It has been observed that the vesicle size and the number of bilayers are major factors in determining the circulating half-life and extent of drug encapsulation [17]. Liposomes less than 0.1 μm are generally less rapidly opsonised than are larger liposomes (>0.1 μm), which translates to having a longer half-life (*see* **Note 3**).

Niven et al. [18] have demonstrated that small liposomes also have a slower release rate than do large multilamellar vesicles following nebulisation. It has also been suggested that liposomes of 50–200 nm diameter are optimal for clinical applications, as they tend to avoid phagocytosis by macrophages and still trap useful drug loads [19].

The T_c of lipids used in the preparation of the liposomes also has significant effect on the release rate of encapsulated drugs. Lipids have a characteristic T_c which depends on the length and saturation of the fatty acid chains and can vary from 20 to 90°C [20]. Below the T_c, lipids are in a rigid, well-ordered arrangement (gel phase) and above the T_c, in a liquid crystalline state (fluid phase) (*see* **Note 4**). Incorporation of lipids with high T_c ($T_c > 37°C$) makes the bilayer of liposomes less fluid at the physiological temperature and less leaky. T_c also appears to influence uptake of liposomes by macrophages, with lipids with high T_c having lower uptake [17]. Cholesterol is an example of lipids with high T_c and is mostly incorporated into the lipid bilayer to increase stability of the liposomes (*see* **Note 5**).

3 Notes

1. During lyophilisation of any protein formulation, there is a need to include a lyoprotectant such as polyols to prevent freeze-drying-induced denaturation.
2. It is important to note that a difference exists between the physical diameter and aerodynamic diameter of a particle. A particle may have a particle size in the range 1–5 μm but aerodynamic diameter outside this inhalable diameter range.
3. There is a need to avoid surfactants in formulations containing liposomes, although low levels (up to 1%) of non-ionic high hydrophilic-lipophilic balance (HLB) surfactants are usually well tolerated.
4. After production, formulations containing liposomes should normally be stored below 25°C (see Sect. 2.3).
5. Liposomes should normally be added to a formulation at temperature below 37°C to avoid phase transition of the phospholipids.

References

1. Crommelin, D., van Winden, E., Mekking, A. (2002) Delivery of pharmaceutical proteins. In: Aulton, M. E. (ed) Pharmaceutics: The science of dosage form design. Churchill Livingstone, Edinburgh, pp. 544–553.
2. Shen, W. C., Wan, J., Ekrami, H. (1992) Means to enhance penetration (3). Enhancement of polypeptide and protein absorption by macromolecular carriers via endocytosis and transcytosis. Adv. Drug Deliv. Rev. **8**, 93–113.
3. Hilsted, J., Madsbad, S., Hvidberg, A., Rasmussen, M. H., Krarup, T., Ipsen, H., Hansen, B., Pedersen, M., Djurup, R., Oxenboll, B. (1995) Intranasal insulin therapy: the clinical realities. Diabetologia **38**, 680–684.
4. Hollinger, M. A. (1985) Respiratory pharmocology and toxicology. Saunders, Philadelphia, pp. 1–20.
5. Shoyele, S. A., Slowey, A. (2006) Prospects of formulating proteins/peptides as aerosols for pulmonary drug delivery. Int. J. Pharm. **314**, 1–8.
6. Edwards, D. A., Hanes, J., Caponetti, G., Hirach, J., Ben-Jebria, A. (1997) Large porous particles for pulmonary drug delivery. Science **276**, 1868–1871.
7. Aquiar, M. M. G., Rodrigues, J. M., Cunha, A. S. (2004) Encapsulation of insulin-cyclodextrin complex in PLGA microspheres: a new approach to prolonged pulmonary insulin delivery. J. Microencapsul. **21**, 553–564.
8. Koushik, K., Kompella, U. B. (2004) Preparation of large porous deslorelin-PLGA microparticles with reduced residual solvent and cellular uptake using supercritical CO_2 process. Pharm. Res. **21**, 524–535.
9. Courrier, H. M., Butz, N., Vandamme, Th. F. (2002) Pulmonary drug delivery systems: recent developments and prospects. Crit. Rev. Ther. Drug Carrier Syst. **19**, 425–498.
10. Garcia-Contreras, L., Morcol, T., Bell, S. J. D., Hickey, A. J. (2003) Evaluation of novel particles as pulmonary delivery systems for insulin in rats. AAPS PharmSci. **5**(2), Article 9.
11. Leach, C. L., Patton, J. S., Perkins, K. M., Kuo, M., Bueche, B., Guo, L., Bentley, M. D., (2002) PEG-insulin delivered by the pulmonary route provides prolonged systemic activity compared with insulin alone. Paper presented at 2002 AAPS meeting and exposition, Toronto, Ont., Canada, Nov. 10–14, 2002.
12. Kwon, J. H., Lee, B. H., Lee, J. J., Kim, C. W. (2004) Insulin microcrystal suspension as a long acting formulation for pulmonary delivery. Eur. J. Pharm. Sci. **22**, 107–116.

13. Bennett, D. B., Tyson, E., Mah, S., de Groot, J. S., Hedge, J. S., Jerao, S., Teitelbaum, Z. (1994) Sustained delivery of detirelix after pulmonary administration of liposomal formulations. J. Control. Release **32**, 27–35.
14. Hinds, K. D., Kim, S. W. (2002) Effects of PEG conjugation on insulin properties. Adv. Drug Deliv. Rev. **54**, 505–530.
15. Kim, H. K., Chung, H. J., Park, T. J. (2006) Biodegradable polymeric microspheres with open/closed pores for sustained release of human growth hormone. J. Control. Release **112**, 167–174.
16. Wang, J., Chua, K. M., Wang, C. H. (2004) Stabilization and encapsulation of human immunoblobulin G into biodegradable microspheres. J. Control. Release **271**, 92–101.
17. Sharma, A., Sharma, U. S. (1997) Liposomes in drug delivery: progress and limitations. Int. J. Pharm. **154**, 123–140.
18. Niven, R. W., Speer, M., Schreier, H. (1991) Nebulization of liposomes. II. The effects of size and modelling of solute release profiles. Pharm. Res. **8**, 217–221.
19. Allen T. M. (1998) Liposomal drug formulations. Rationale for development and what we can expect for the future. Drug **56**, 747–756.
20. Labiris, N. R., Dolovich, M. B. (2002) Pulmonary drug delivery. Part II: The role of inhalation delivery devices and drug formulations in therapeutic effectiveness of aerosolized medications. Br. J. Clin. Pharmacol. **56**, 600–612.

Chapter 7
Engineering Protein Particles for Pulmonary Drug Delivery

Sunday A. Shoyele

Abstract Pulmonary delivery of proteins requires particles for delivery to be in the aerodynamic size range 1–5 µm for deep lung deposition. However, the traditional particle size reduction technique of jet-milling normally used for inhalation is not suitable for processing these protein particles because of their lability brought about by the weak physical interactions making up their higher order structures. Advanced techniques such as spray drying, spray freeze drying and the use of supercritical fluid technology have been developed to produce particles in the suitable size range and morphology for deep long deposition without altering the native conformation of these biomolecules. Judicious use of excipients and operating conditions are some of the factors needed for a successful particle design.

Keywords Pulmonary; Protein; Spray-drying; Spray freeze-drying; Supercritical fluids; Aerodynamic

1 Introduction

The emergence of recombinant DNA technology in the late 1970s suggests that proteins could be developed under cGMP conditions for therapeutic uses. Proteins and peptides such as insulin, salmon calcitonin, leuprolide, interleukins and interferons are now being prescribed for various diseased conditions. Delivery of these proteins via the pulmonary route as an alternative to the invasive injections has been found to be the most promising non-invasive route [1].

However, pulmonary delivery of proteins or peptides either for local or systemic activity comes with the challenge of designing the particles for optimal delivery to the lungs. For proteins or peptides to be absorbed from the lungs, an aerodynamic diameter (d_{ae}) of 1–5 µm must be achieved [2] in order for these particles to be deposited in the peripheral airways (alveoli) where systemic absorption occurs. Particles greater than 5 µm are deposited in the oropharyngeal region by inertial impaction while particles between 1 and 5 µm are deposited in the peripheral region

From: *Methods in Molecular Biology, Vol. 437: Drug Delivery Systems*
Edited by: Kewal K. Jain © Humana Press, Totowa, NJ

(bronchioles and alveoli) by gravitational sedimentation. Particles less than 0.5 µm are deposited on the walls of alveoli by Brownian diffusion [3] although there is a possibility that these particles are exhaled during normal tidal breathing.

Unlike the physical diameter, the d_{ae} is a concept incorporating the size, shape and density of particles [2] and so the pharmaceutical performance of a powder or droplets for pulmonary delivery would be defined by the mass median aerodynamic diameter (MMAD) of the particles. MMAD is the equivalent aerodynamic diameter in which 50% of the powder mass falls below [4]. Therefore, the MMAD is representative of the aerodynamic particle size of an aerosol formulation. An MMAD of <5 µm is desirable for deep lung delivery.

Furthermore, for dosage consistency, there is a need for an aerosol formulation to be monodispersed [4]. The particle size distribution of an aerosol is defined by its geometric standard deviation (GSD). The GSD is the ratio of particle diameters at 84% and 50% cumulative mass of particles or the ratio of the particle diameters at 50% and 16% cumulative mass of particles when the cumulative mass of particles is plotted against the equivalent diameter on a log-probability scale following particle size analysis using either impactors or laser diffraction instruments.

Protein particles would therefore need to be designed to meet the above criteria for effective delivery into the lungs. Engineering protein particles to meet these criteria comes with the challenge of retaining the conformational structure and hence the biological activity of these labile molecules. Because of the marginal stability of the higher order structures (secondary, tertiary and sometimes quaternary) of proteins, they are prone to various degradative reactions that ultimately impact their physical and chemical stabilities [5], and so there is a need to maintain the conformational stability of these molecules while optimising lung delivery through particle design.

In summary, the important features needed for effective aerosol delivery include narrow particle size distribution, low surface energy and optimal aerodynamic particle size [2].

To achieve these features without necessarily reducing the biological activity of therapeutic proteins, various techniques have been adapted. This chapter discusses the various methods used, highlighting their benefits and challenges.

2 Milling

Milling has been the 'traditional' method of size reduction for pulmonary delivery; however, the most popular milling method is jet milling.

2.1 *Jet Milling*

Jet milling uses inter-particle collision and attrition to achieve the micronisation of particles [6]. Coarse particles are fractured into smaller ones in the gas stream due to the inter-particle collisions. Although jet milling can typically produce particles in the size-range 1–20 μm, lack of control over parameters such as size, shape, morphology and surface properties of jet milled particles coupled with the high energy input, which can facilitate chemical degradation of proteins constitute limitations to the use of this technique in pulmonary delivery of proteins.

Further limitation to this technique is the fact that particles fed into the jet mill have to be coarse enough to allow free flow into the mill while fine enough to avoid blocking the hopper and pipe-work in the mill. To get around this limitation proteins or peptides that are not available as crystals are lyophilized to prepare the perfect type of particles fed into the mill. However, lyophilization could lead to degradation of proteins unless a lyoprotectant is used. Lyoprotectants protect proteins against freeze concentration, which occurs because of crystallization of water at sub-zero temperatures [7]. Proteins could also be vulnerable to the ice–liquid interface that occurs during lyophilization and so there might be a need for a suitable surfactant as well.

2.2 *Media Milling*

This technique involves the use of non-CFC propellants as a medium during ball milling. The balls used are always abrasive resistant pearl and there is a need for a cryostat to keep the temperature at about −60°C.

This method could either be used as a one-step process for the preparation of protein or peptide suspension for pressurised metered dose inhaler or the suspension could be poured into a flask and rotated at room temperature until the propellant evaporates. The powder derived from this could be used for the preparation of a dry powder for inhalation, which makes this technique versatile.

3 Spray Drying

Spray drying has been successful in the design of particles for pulmonary delivery (Fig. 7.1). It involves atomising a feed solution containing proteins or peptides through a nozzle using either heated compressed air or nitrogen (*see* **Note 1**). The atomized droplets dry rapidly because of their surface area and intimate contact with the drying gas. The temperature experienced by the droplets is lower than the temperature of the drying gas because of evaporative cooling. Parameters such as flow rate and pump rate, aspiration rate, heat and concentration of the feed solution influence the drying time, morphology and quality of the product from a spray drying process. The challenge involved in spray drying is the maintenance of the physico-chemical stability of the protein processed. Proteins encounter three different types of degradative steps during spray drying (*see* **Note 2**):

- High shear rate during atomization
- Adsorption at air–liquid interface during drying
- Increased molecular mobility due to high temperature of drying

To stabilize proteins against these degradative steps, excipients have proven to be quite effective. For instance, surfactants such as polysorbates, polyvinyl alcohol and dipamitoyl phosphatidyl choline have been used to prevent the adsorption of proteins at the air–liquid interface during spray drying [8, 9].

At conditions above the glass transition temperature (T_g) of an amorphous protein powder, due to the loss of the glassy structure, protein molecules tend to be molecularly mobile and get involved in various degradative reactions such as oxidation. Polysaccharides such as trehalose, mannitol and sucrose with a relatively high T_g can be used to raise the T_g of the system and so maintain the stability of the protein (*see* **Note 3**). However, there is a need for the polysaccharides to be amorphous so that they can be homogenously distributed between protein molecules for them to be effective. Reducing sugars such as lactose are not advisable to be used as they have been found to form Schiff-bases with the primary amine group on lysine residues in proteins.

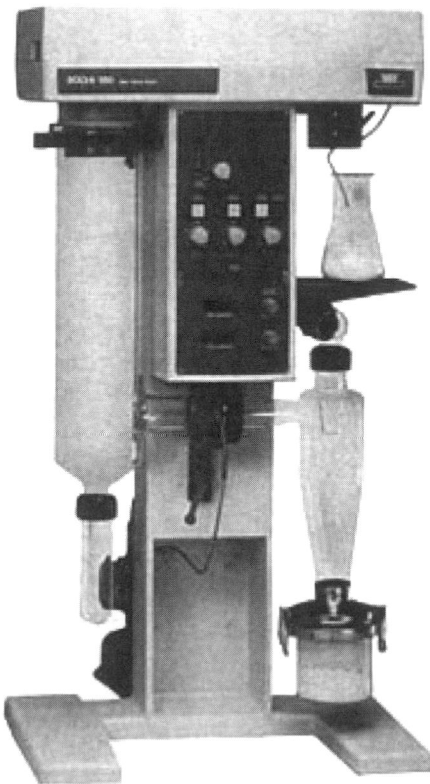

Fig. 7.1 An example of a Buchi 190 Spray Dryer

3.1 CO_2 – Assisted Nebulization with a Bubble Dryer™ (CAN-BD)

Supercritical or subcritical CO_2 has been applied for the atomization of proteins during spray drying. This helps to minimize thermal decomposition associated with normal spray drying.

CO_2 – Assisted Nebulization with a Bubble Dryer involves dissolving the drug in water or alcohol or a mixture of both. The solution produced is then mixed with CO_2 by pumping both through a low volume tee to generate emulsion. The emulsion expands on passing through a nozzle into a drying chamber held at atmospheric pressure to generate micro-droplets, which are dried by heated nitrogen [10]. Typically, particles produced by this method are spherical and below 3 μm in diameter.

4 Spray–Freeze Drying

Spray freeze drying combines atomisation with lyophilisation. It involves spraying a solution containing the protein or peptide of interest into a vessel containing a cryogenic liquid such as nitrogen, oxygen or argon. Nitrogen is mostly used because of its cost effectiveness and its low boiling point (−195.8°C). Droplets generated from spraying are quickly frozen as they make contact with the liquid nitrogen. Lyophilising these frozen droplets results in porous spherical particles suitable for inhalation.

The stresses associated with freezing and drying constitutes limitation to the use of this technique. During atomisation, the protein is exposed to air–liquid interface, which may lead to aggregation. Further limitations to this technique are that it is time consuming and the nature of the process has safety issues with regard to the cryogenic fluids (*see* **Note 4**).

Nevertheless, this method has been used to produce aerodynamic particles of rhDNase, cetrorelix and lysozyme [11–13].

4.1 Spray-Freezing into Liquid (SFL)

Since it has been established that the exposure of protein droplets to liquid–gas interface during spray drying and spray-freeze drying is the major cause of denaturation during particle engineering, a novel technique has been developed, which tries to reduce or exclude exposure to this interface. Although SFL is just a slight modification to spray freeze drying, a huge improvement in terms of product stability could be achieved. Unlike in spray-freeze drying where the protein solution is sprayed on the liquid nitrogen, this process involves spraying the solution below the surface of the cryogenic fluid to produce rapidly frozen particles, which are subsequently lyophilized. The ultra rapid freezing prevents phase separation of the drug and excipients and also prevents crystalline growth in frozen water. The avoidance of exposure to cold vapour experienced in spray freeze drying leads to production of more stable protein particles.

5 Supercritical Fluid Technology

Supercritical fluids (SFs) are gases and liquids at temperature and pressure above their critical points (T_c, critical temperature; P_c, critical pressure). SFs exist as a single phase with several advantageous properties of both liquids and gases. They have density values that enable appreciable solvation power and the viscosity of solute in SF is lower than in liquids. Furthermore, solutes have higher diffusivity, which allows high mass transfer [14]. CO_2 is the most widely used SF for pharmaceutical applications because of its low critical temperature (31.2°C) and pressure (7.4 MPa); it is non-flammable, non-toxic and inexpensive. Depending on the nature of the process, $SFCO_2$ can serve either as a solvent or anti-solvent. To increase the salvation power of an $SFCO_2$ an organic modifier such as acetone or ethanol could be added to the fluid.

SF can be applied to protein particle design (Fig. 7.2) in three broad ways

- Precipitation from supercritical solution composed of $SFCO_2$ and protein
- Precipitation from gas saturated solutions
- Precipitation from saturated solutions using SF as anti-solvent

In the first method, the drug is dissolved in the SF and followed by rapid expansion of the SF solution across a heated orifice to cause a reduction in the density of the solution and reducing the salvation power of the SF, which leads to the precipitation of the drug [16]. This process is termed the 'rapid expansion of supercritical solution' (RESS).

The RESS process depends on sufficient solubility of the material to be processed in the SF while the solubility of the material depends on the density of the SF, the drug's chemical structure and the SF-drug contact time. The morphology and the size distribution of the particles formed could be influenced by the pre-expansion concentration of the solute in SF and the expansion conditions (e.g. temperature and pressure). The higher the pre-expansion concentration, the narrower the particle size distribution. The expansion conditions depend on the temperature, the geometry and size of the nozzle.

The limitations facing the use of RESS for proteins include high temperature needed for the rapid expansion, which could destroy proteins, poor predictive control of particle size and morphology, scale up limited by particle aggregation and nozzle blockage caused by expansion cooling. The solubility of the drug in $SFCO_2$ also constitutes a major limitation.

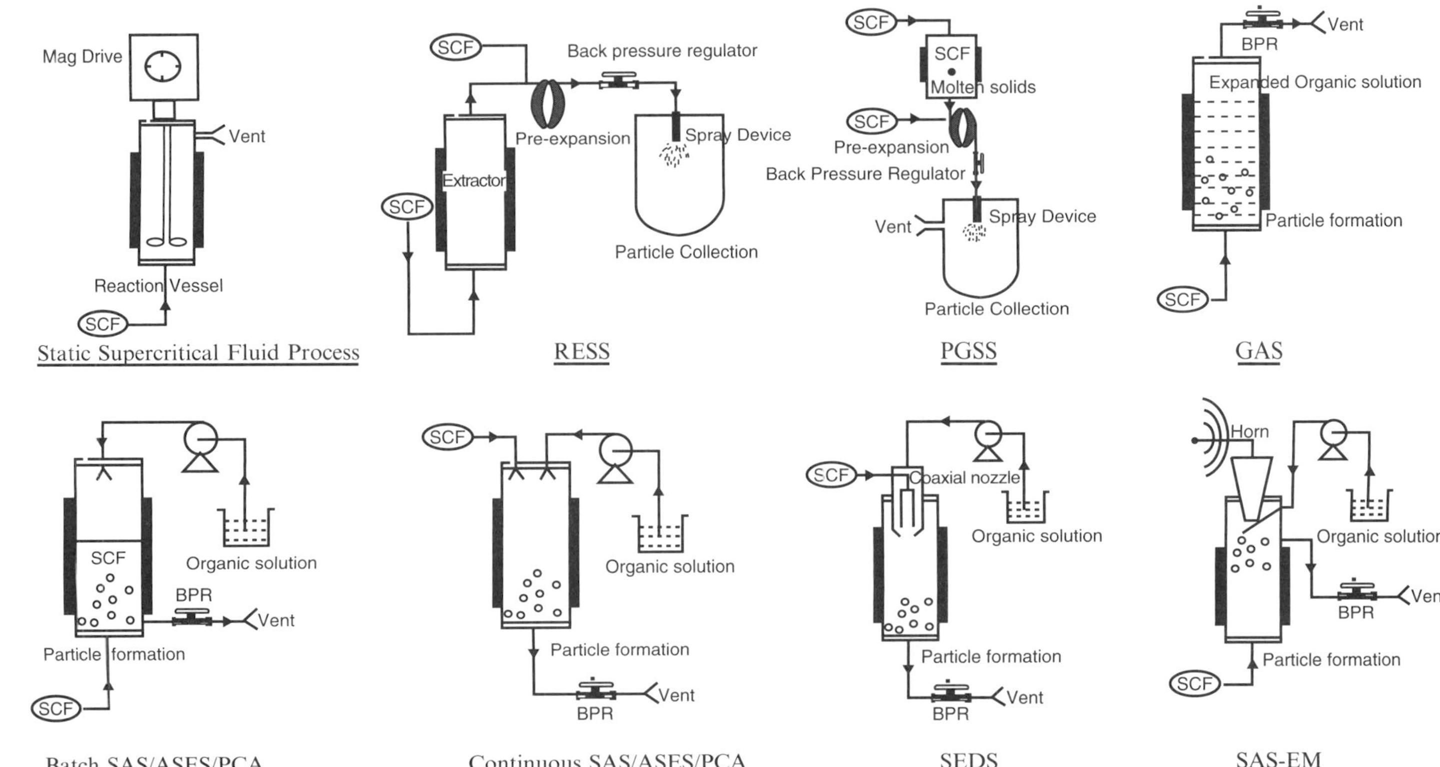

Fig. 7.2 Schematic of various applications of SF technology in particle design, reprinted from ref. 15 with permission from Elsevier

The second method is quite harsh but similar to RESS process as they both involve use of $SFCO_2$ as a solvent rather than an anti-solvent. This process involves dissolving the SF in molten solute and the resulting supercritical solution fed via an orifice into a chamber to allow rapid expansion under ambient conditions [17]. The dissolved gas decreases the viscosity of the molten compound and so the gas saturated liquid phase is expanded to generate particles from materials that are not necessarily soluble in SF. The presence of the CO_2 allows the material to melt at temperature significantly lower than the normal melting or glass transition temperature.

Precipitation from saturated solutions using SF as antisolvent, takes advantage of the limited solvation power of $SFCO_2$ for proteins. This method utilizes a similar concept to the use of anti-solvent in solvent based crystallization processes. The high solubility of $SFCO_2$ in organic solvents leads to volume expansion when the fluids make contact. This leads to reduction in solvent density and subsequent fall in salvation capacity. This leads to super-saturation, solute nucleation and particle formation.

Different variants of the use of SF as anti-solvent have been developed and all have been successful at producing aerodynamic and stable protein particles. These include:

- Gaseous antisolvent (GAS)
- Aerosol solvent extraction system (ASES)
- Solution enhanced dispersion by SF (SEDS)
- Precipitation by compressed antisolvent (PCA)

5.1 Gaseous Antisolvent

This process involves adding SF to a particle formation vessel containing the protein solution of interest. The protein precipitates during the dissolution of SF in the solvent. Generally, the SF is introduced through the bottom of the vessel and bubbled through the protein solution to achieve better mixing of the solvent and anti-solvent. Once the protein has precipitated out, the solvent-SF can be removed. The protein particles can then be washed with a sequence of SF washes. Following the wash, the pressure in the vessel is released and the protein powder removed.

It is important that the solvent in which the protein is dissolved has a high solvent power for the protein, preferably be soluble in the SF of choice and be compatible with the protein. Organic solvents such as dimethyl sulphoxide and dimethyl formaldehyde have been used because they meet the criteria of salvation of the protein and solubility in $SFCO_2$. However, toxicity and compatibility of these solvents with protein molecules remain an issue [18].

Although $SFCO_2$ remains the promising SF for pharmaceutical use, alternatives such as ammonia and ethane have been studied. While ammonia produced completely denatured proteins, ethane produced results comparable to $SFCO_2$ in terms of particle size and improved biological activity of insulin. The limitation facing the use of SF-ethane is its high flammability.

Studies have shown that the type of solvent used has effect on the particle size, morphology and biological activity of the molecule [19, 20]. The principal disadvantage of this process is the lack of control over particle formation. This has been observed to be true in batch operating conditions because the level of saturation is not maintained.

5.2 Aerosol Solvent Extraction System

Unlike in GAS method where the SF is pumped into the protein solution, the opposite happens in the ASES process. The CO_2 is pumped into a high pressure vessel until the system reaches the desired fixed conditions (pressure and temperature). The protein solution is then sprayed via an atomization device into the vessel containing $SFCO_2$. Precipitated particles are then collected on a filter at the bottom of the vessel. This process retains the antisolvent concept of GAS but this happens at the droplet level. This offers a favourable higher antisolvent to solvent ratio, an increased surface area and mass transfer rate, and hence, an acceleration of the drying process. The rate is particularly fast when the operating pressure reaches the mixture critical pressure. The process is then controlled by mixing of miscible fluids rather than mass transfer over the interface of the droplets in the spray [21]. There is a need for special attention for mass transfer when the miscibility of the fluids is poor especially in systems containing water and CO_2. The mass transfer can be improved by increasing the drying medium to solvent ratio, decreasing droplet size or relative velocity between the solvent and drying medium. ASES enables production of protein particles with narrow size distribution, uniform shape and desired physico-chemical characteristics.

5.3 Solution Enhanced Dispersion by SF

To minimise particle agglomeration often observed in ASES and other antisolvent based techniques and to reduce drying times, increased mass transfer rates are required [14]. This has been successfully achieved by the SEDS process. This process involves introducing the dry solution and the SF into a particle formation vessel (where temperature and pressure are controlled) through a co-axial nozzle with a mixing chamber. The higher velocity of the SF allows the production of very small sizes, while the mixing of solvent with the SF inside the mixing chamber leads to increased mass transfer of SF into the solvent and vice versa [16]. A high mass transfer allows a faster nucleation and a smaller particle size with agglomeration. Because of low miscibility of $SFCO_2$ in water, making the use of SEDS quite challenging for proteins or peptides particle design, this process has been optimised by the use of co-axial three component nozzle. Aqueous solution of protein, organic solvent and $SFCO_2$ are simultaneously introduced to increase the miscibility of $SFCO_2$ in the protein solution.

SEDS is a more controllable and reproducible technique compared with other antisolvent based SF process.

5.4 *Precipitation by Compressed Antisolvent*

The PCA process is a slight modification of the ASES process. It is a one-step technique used to produce solvent free particles with a narrow size distribution at mild operating conditions [22]. It involves feeding $SFCO_2$ into a precipitator so as to pressurize the precipitator to a desired value. The CO_2 flow rate into the precipitator is then fixed. The solution of protein is fed into the precipitator through a nozzle. The antisolvent effect of the $SFCO_2$ leads to the precipitation of the protein particles.

Increase in the pressure has been found to help produce a dryer product as the extraction of the liquid solvent to the supercritical phase takes place faster because of a higher solubility of the solvent in the CO_2. Furthermore, as pressure increases, the atomisation of the solution produces smaller droplets and the drying time decreases.

6 Microcrystallization

Most of the earlier discussed processes lead to the production of amorphous protein particles. However, the ordered arrangement of protein molecules in crystalline state makes them more active and stable than their amorphous counterparts. Thermodynamically, the absence of crystallinity in an amorphous protein causes energy content to be higher than in the crystalline state, leading to lower stability and higher reactivity [16].

Amorphous protein particles are cleared rapidly from systemic circulation and are more susceptible to hydrolytic and enzymatic degradation because of their higher reactivity [23]. There is therefore a high desire for crystalline protein as a fine pharmaceutical ingredient. Crystallization of proteins involves a one step process and results in high purity. It can improve protein handling during processing, storage and delivery. It can also offer sustained release of the therapeutic agent for an effective duration by changing the dissolution characteristics.

The relatively high molecular weight of proteins coupled with their flexibility make normal antisolvent crystallization difficult. In a situation where crystallization is achieved, the large particle size and wide size distribution means these particles are not favourable towards pulmonary delivery. Although milling has been applied to size reduce crystallized protein to respirable size, milling may lead to a high energy input leading to particles of reduced crystallinity and may contain disordered regions, thereby introducing regions of reduced stability.

Nevertheless, the concept of Microcrystallization has been used to overcome milling induced disorder in crystalline powders. This method involves the use of pH to achieve a super-saturated status for a protein in solution. A protein normally has a peculiar pH at which it achieves super-saturation in an aqueous medium and if this pH can be achieved in the presence of a stabilizer (normally a polymer such as PEG) and the system stored at temperature around 16°C overnight, microcrystals roughly spherical and about 1–2 µm in diameter can be achieved [23].

Microcrystalline protein particles can also be formed by a process called the 'seed zone' method [24]. Unlike the conventional seeding method in which the protein crystals are suspended in a buffer solution at a pH in which they are least soluble, centrifuged and the supernatant used as a seed solution in subsequent crystallization of the protein from solution, this seed zone method is a one step method than can be easily industrialised. It first involves recognising the seed zone of the protein by dissolving the protein in a buffer at acidic pH and slowly increasing the pH by adding an alkaline solution. A pH is reached when the protein instantly crystallizes to form a turbid solution, and by increasing the pH of the system, the system forms a clearer solution with few crystals. At this pH, seeds are formed and when the system is brought back to its super-saturation pH, protein crystals of peculiar morphology with particle size fit for pulmonary delivery can be achieved. Insulin particles of less than 3 µm have been formed using this method [24].

7 Notes

1. Protein solutions should be minimally shaken as it has been found that shaking leads to exposure of protein to interfaces that aid their degradation.
2. In preparing protein solutions prior to any of the techniques highlighted above, it is important that the protein is dissolved in a buffer solution at a pH of optimal biological activity.
3. In choosing excipients for protein stabilization, reducing sugars such as lactose and glucose should be avoided as they form Schiff's base degradation products with proteins.
4. Care should be taken when handling liquid nitrogen during the spray drying process as liquid nitrogen normally splashes during atomisation process.

References

1. Shoyele, S. A. and Slowey, A. (2006) Prospects of formulating protein/peptides as aerosols for pulmonary drug delivery. Int. J. Pharm. **314**, 1–8.
2. Shoyele, S. A, and Cawthorne, C. (2006) Particle engineering techniques for inhaled biopharmaceuticals. Adv. Drug Deliv. Rev. (in press).
3. Taylor, K. (2002) Pulmonary drug delivery, in Aulton, M. E. (ed.), Pharmaceutics, the Science of Dosage Form Design (). Churchill Livingston, Edinburgh, 473–498.

4. Okamoto, H., Todo, H., Lida, K., and Danjo, K. (2002) Dry powder for pulmonary delivery of peptides and proteins. KONA 20, 71–83.
5. Lai, M. C and Topp, E. M. (1999) Solid state chemical stability of proteins and peptides. J. Pharm. Sci. 489–500.
6. Van Vlack, L. H. (1980) Elements of Material Science and Engineering, Addison Wesley, Reading, 185–208.
7. Hinrichs, W. L. J., De Smelt, N. N., Demeester, J., and Frijlink, H. W. (2005) Inulin is a promising cryo and lyoprotectant for PEGylated lipoplexes. J. Control. Rel. 103, 465–479.
8. Johnson, K. A. (1997) Preparation of peptide and protein powder for inhalation, Adv. Drug Del. Rev. 26, 3–15.
9. Coldron, V., Vanderbist, F., Verbeeck, R., Mohammed, A., Lison, D., Preat, V., and Vanbever, R. (2003) Systemic delivery of parathyroid hormone (1–34) using inhalation dry powder in rats. J. Pharm. Sci. 92, 938–950.
10. Sievers, R. E., Huang, E. T. S., Villa, J. A., Engling, G., Brauer, and P. R. (2003) Micronization of water-soluble or alcohol soluble pharmaceuticals and model compounds with a low temperature Bubble Dryer®. J. Supercritical Fluids 26, 9–16.
11. Constantino, H. R., Firouzabadian, L., Hogeland, K., Wu, C., Banganski, C., Cordova, C. M., Carrasquillo, K. G., Griebenow, Zale, S. E., and Tracy, W. A. (2000) Protein spray freeze drying. Effect of atomization conditions on particle size and stability. Pharm. Res. 17, 1374–1384.
12. Zijlstra, G. S., Hinrichs, W. L. J., De Boer, A. H., and Frijlink, H. W. (2004) The role of particle engineering in relation to formulation and deaggragation principle in the development of a dry powder formulation of cetrorelix. Eur. J. Pharm. Sci. 23, 139–149.
13. Yu, Z., Johnston, K. P., and Williams III, R. O. (2006) Spray freezing into liquid: Influence of atomization on protein aggregation and biological activity. Eur. J. Pharm. Sci. 27, 9–18.
14. York, P. (1997) Strategies for particle design using supercritical fluid technologies. PSST 2, 430–440.
15. Vemavarapu, C., Mollan, M. J., Lodaya, M., Needham, T. E. (2005) Design and process aspects of laboratory scale SCF particle formation system. Int. J. Pharm. 292, 1–16.
16. Pasquali, I., Bettini, R., and Gordono, F. (2006) Solid state chemistry and particle engineering with supercritical fluids in pharmaceutics. Eur. J. Pharm. Sci. 27, 299–310.
17. Tandya, A., Dehghani, F., and Foster, N. R. (2006) Micronization of cyclosporine using dense gas technique. J. Supercritical Fluids 27, 272–278.
18. Yeo, S. D., Lim, G. B., Debenedetti, P. D., and Bernstein, H. (1993) Formation of microparticulate protein powder using a supercritical fluid anti-solvent. Biotechnol. Bioeng. 41, 341–346.
19. Thiering, R., Dehghani, F., Foster, N. R. (2000) Micronization of model proteins using compressed CO2. Proceedings of the 5th International Symposium on Supercritical Fluids, Atlanta.
20. Yeo, S. D., Debenechi, P. G., Radosz, M., and Schmidt, H. W. (1993) Supercritical anti-solvent process for substituted para-linked aromatic polyamides: phase equilibrium and morphology study. Macromolecules 26, 6207–6210.
21. Jovanovic, N., Bouchard, A., Hofland, G. W., Witkamp, G., and Crommelin, Jiskoot, W. (2004) Stabilization of proteins in dry powder formulation using supercritical fluid technology. Pharm. Res. 21, 1955–1969.
22. Bodmeier, R., Wang, H., Dixon, D. J., Mawson, S., and Johnston, K. P. (1995) Polymeric microspheres prepared by spraying into compressed CO2. Pharm. Res. 12, 1211–1217.
23. Lee, M. J., Kwon, J. –H., Shin, J. –S., and Kim, C. W. (2005) Microcrystallization of α-Lactalbumin, J. Crystal Growth 282, 434–437.
24. Kwon, J. –H., Lee, B. –H., Lee, J. –J., and Kim, C. –W. (2004) Insulin microcrystal suspension as a long lasting formulation for pulmonary delivery. Eur. J. Pharm. Sci. **22**, 107–116.

Chapter 8
2B-Trans™ Technology: Targeted Drug Delivery Across the Blood–Brain Barrier

Pieter J. Gaillard and Albertus G. de Boer

Abstract Drug delivery across the blood–brain barrier (BBB) is a major obstacle for the development of effective treatments of many central nervous system disorders. Sophisticated cell culture models of the BBB have helped us to identify, characterize, and validate a novel targeted drug delivery technology, designated 2B-Trans™, for the receptor-mediated uptake and transport of drugs across the BBB. This paper describes in great detail how such a BBB cell culture model should be prepared and handled, and applied for the use of targeted drug delivery across the BBB.

Keywords Blood–brain barrier; Brain capillary endothelial cells; Astrocytes; Diphtheria toxin receptor; CRM197; Drug targeting

1 Introduction

Brain drug delivery is limited by the blood–brain barrier (BBB) [1, 2]. Particularly biopharmaceutical drugs poorly pass the BBB and generally do not reach the brain in sufficient concentrations to be effective. Since almost every neuron is perfused by its own capillary, the most effective way of delivering biopharmaceutical drugs is achieved by targeting them to endogenous transport receptors on these capillaries [3]. In fact, the total length of capillaries in the human brain is impressive (~600 km), with a large surface area (~20 m^2) for effective exchange of drugs. In this paper we describe the methods used for the in vitro validation of the receptor for diphtheria toxin (DTR), the membrane-bound precursor of heparin-binding epidermal growth factor (HB-EGF), as a newly identified endogenous transport receptor for the targeted delivery of drugs across the BBB [4]. Specifically, we used CRM197, a nontoxic and human applicable mutant protein of diphtheria toxin, as the receptor-specific carrier protein. By means of our dynamic cell culture model of the BBB [5] we were able to demonstrate the receptor-specific uptake and transport efficacy of CRM197 conjugated to a 40 kDa enzyme (horseradish peroxidase, HRP, serving

From: *Methods in Molecular Biology, Vol. 437: Drug Delivery Systems*
Edited by: Kewal K. Jain © Humana Press, Totowa, NJ

as a "model" biopharmaceutical drug). Collectively, these results indicate that CRM197 may indeed be developed into a new, safe, and effective brain drug delivery carrier protein for human applications.

2 Materials

2.1 Dynamic Co-Culture Model of the BBB

2.1.1 Coating of Cell Culture Plates, Flasks, and Filters

1. All disposable cell culture plastics are from Corning Costar BV (Schiphol, The Netherlands).
2. Collagen type IV (Sigma-Aldrich BV, Zwijndrecht, The Netherlands, Cat. C-5533, nonsterile) coating solution is prepared in a 100 µg/ml stock solution in 0.1% v/v acetic acid (J.T. Baker, Deventer, The Netherlands, Cat. 6001) solution. The acetic acid solution is prepared in Milli-Q water (from Milli-Q system, Millipore, Etten-Leur, The Netherlands) and sterilized over 0.2-µm filter (Corning Costar BV). The collagen stock solution should not be sterile filtered and can be stored at 4°C no longer than 2 months. Collagen working solution of 10 µg/ml is prepared by dilution in 0.1% v/v acetic acid, kept at 4°C, and can be used for no longer than 2 weeks.
3. Fibronectin (Boehringer Mannheim BV, Almere, The Netherlands, Cat. 1080 938) coating solution is prepared in a 1 mg/ml stock solution in sterile Milli-Q water and stored in 1 ml aliquots at −20°C in 100-ml bottles. Fibronectin working solution of 10 µg/ml is prepared by dilution in phosphate buffered saline (PBS, Cambrex, Verviers, Belgium, Cat. 17-516F), kept at 4°C, and can be used for no longer than 6 weeks.
4. Poly-D-lysine (Sigma-Aldrich BV, Cat. P-7280) coating solution is prepared in a 1 mg/ml stock solution in sterile Milli-Q water and stored in 1 ml aliquots at −20°C in 100-ml bottles. Poly-D-lysine working solution of 10 µg/ml is prepared by dilution in PBS, kept at 4°C, and can be used for no longer than 6 weeks.

2.1.2 Isolation of Astrocytes

1. Newborn (no older than 5 days) Wistar rat pups are obtained from Harlan B.V. (Zeist, The Netherlands).
2. Dulbecco's modified Eagle's medium (DMEM) is supplemented (DMEM + S) with 10% (v/v) heat inactivated (30 min at 56°C) fetal calf serum (FCS, Cambrex, Cat. BE14-603F (*see* **Note 1**)). The DMEM, formulated with high d-glucose (4.5 g/l), $NaHCO_3$ (3.7 g/l), and HEPES (25 mM), contained extra MEM nonessential amino acids (Cambrex, Cat. 13-114E), 2 mM L-glutamine (Cambrex, 17–605E), streptomycin sulfate (0.1 g/l), and penicillin G sodium (100,000 U/l) mixture (Cambrex, 17–602E).

3. Fully (50 mM) HEPES buffered DMEM (i.e., without $NaHCO_3$) is prepared with DMEM from Cambrex (Cat. 15-604D) as described earlier.
4. 0.25% trypsin–0.02% EDTA is from Sigma-Aldrich BV (Cat. T-4049).
5. Nylon mesh filters of 120 and 45 μm are hand-cut from fabric obtained at Merrem & la Porte BV (Zaltbommel, The Netherlands) and placed in filter holders from Millipore (Cat. SX00 047 00).

2.1.3 Isolation of Bovine Brain Capillaries

1. Bovine (calf) brains are obtained at the slaughterhouse from freshly killed animals (*see* **Note 2**).
2. Nylon mesh filters of 200 and 150 μm are hand-cut from fabric obtained at Merrem & la Porte BV and placed in filter holders from Millipore (Cat. SX00 047 00).
3. Collagenase type 3 (Gibco, Breda, The Netherlands, Cat. 17102-013) is prepared in a 2,000 U/ml stock solution in PBS, sterilized over 0.2-μm filter and stored at –20°C.
4. Trypsin TRL (Worthington, Cat. LS003702) is prepared in a 900 U/ml stock solution in PBS, sterilized over 0.2-μm filter, and stored at –20°C.
5. DNAse I (Worthington, Cat. LS002138) is prepared in a 3,400 U/ml stock solution in PBS, sterilized over 0.2-μm filter, and stored at –20°C.
6. Digest mix for one calf brain homogenate is freshly prepared by mixing 2 ml of Collagenase stock, 2 ml Trypsin stock, and 1 ml DNAse stock in 15 ml of DMEM + S.
7. Freeze mix is prepared by adding 10% DMSO (J.T. Baker, Cat. 7033) to 90% FCS.

2.1.4 Differential Seeding of Brain Capillaries and Culture of Bovine Brain Capillary Endothelial Cells

1. Heparin solution (Sigma, Cat. H-3149) is prepared in a 0.5% (w/v) stock solution in PBS, sterilized over 0.2-μm filter, and kept at 4°C.
2. Growth medium is freshly prepared with either 50% DMEM + S, 50% astrocyte conditioned medium (ACM, see below) and 125 μg/ml heparin (GM+), or with 100% DMEM + S and 125 μg/ml heparin (GM–).

2.1.5 Preparation of the In Vitro BBB on Filter Inserts

1. 24-wells Transwell polycarbonate filters are used with a surface area of 0.33 cm^2 and pore-size of 0.4 μm (Corning Costar, Cat. 93413), collagen coated as described below, with astrocytes on the bottom of the filter (as described below).
2. EDTA solution (J.T. Baker, Cat. 1073) is prepared in a 0.02% (w/v) solution in PBS, sterilized over 0.2-μm filter and kept at 4°C.

3. Trypsin-EDTA solution for endothelial cells (500 BAEE units porcine trypsin and 180 μg EDTA per ml) from Sigma (Cat. T-4299) is aliquoted in 5 ml and stored at −20°C.
4. 8-(4-chlorophenylthio (CPT))-cAMP (Sigma, Cat. C-3912) is prepared as a 25 mM stock solution in Milli-Q water (not PBS), sterilized over 0.2-μm filter, aliquoted in 0.5 ml, and stored at −20°C.
5. RO-20-1724 (Calbiochem, Cat. 557502) is prepared as a 35 mM stock solution in DMSO, aliquoted in 25 μl and stored at −20°C.
6. Differentiation medium is freshly prepared by addition of 12.5 μl CPT-cAMP stock solution per ml DMEM + S and 0.5 μl RO-20-1724 stock solution per ml DMEM + S.

2.2 2B-Trans™ Technology

2.2.1 Characterization of Transport Receptor on the BBB In Vitro

1. Used proteins for these studies are: diphtheria toxin (DT, Sigma-Aldrich BV, Cat. D-0564), nontoxic mutant protein of diphtheria toxin CRM197 ("cross-reactive material" 197, or [Glu^{52}]-Diphtheria toxin from Sigma-Aldrich BV, Cat. D-2189) and soluble recombinant human heparin-binding EGF-like growth factor (HB-EGF, carrier free from R&D Systems, Cat. 259-HE/CF).

2.2.2 Receptor Targeted Cell Uptake Studies on the BBB In Vitro

1. Used proteins for these studies are: CRM197 (Sigma-Aldrich BV), HB-EGF (R&D Systems), CRM197, and bovine serum albumin fraction V (BSA, ICN Biomedicals, Cat. 160069).
2. For conjugation of proteins to HRP, a HRP conjugation kit from Alpha Diagnostic International (San Antonio, TX, Cat. 80220) is used according to the manufacturer's instructions.
3. Conjugates are purified on a HiPrep 16/60 column packed with Sephacryl S-200 HR matrix from Amersham Biosciences (Cat. 17-1166-01).
4. HRP activity is determined with TMB liquid substrate (Sigma-Aldrich BV, Cat. T-8665), with H_2SO_4 (J.T. Baker) to stop the reaction, and HRP type VI-A (Sigma-Aldrich BV, Cat. P-6782) for a standard curve (0–2 ng/ml) and absorption is read at 450 nm.
5. Cells are lysated by an aqueous solution of 0.1% Na-deoxycholate (Sigma-Aldrich BV, Cat. D-6750).
6. Cellular protein content is determined using Bio-Rad DC reagents (Bio-Rad Laboratories, Veenendaal, The Netherlands, Cat. 500-0111) and BSA for a standard curve (0–400 μg/ml). Absorption is read at 690 nm.

2.2.3 Receptor Targeted Transcytosis Studies across the BBB In Vitro

1. Materials are used as described earlier.

3 Methods

3.1 Dynamic Co-Culture Model of the BBB

3.1.1 Coating of Cell Culture Plates, Flasks, and Filters

1. Culture flasks for seeding bovine capillaries are coated by adding 6 ml collagen (10 μg/ml) per 250-ml culture flask (T75) for at least 2 h at room temperature, followed by three washes with PBS and addition of 5 ml fibronectin (10 μg/ml) for at least 30 min. Coated flasks should be aspirated and used instantly (*see* **Note 3**).
2. Transwell filters or wells for seeding bovine brain capillary endothelial cells (BCECs) are coated by adding collagen (10 μg/ml) for at least 2 h at room temperature of 100 μl per 24-well filter or 96-well plate. Filters or wells are rinsed 3 times with PBS before seeding the cells.
3. Culture flasks for seeding rat astrocytes are coated by adding 6 ml Poly-d-Lysine (10 μg/ml) per 250-ml culture flask (T75) and left overnight on a shaker at room temperature. Flasks are aspirated and dried for 3 h (not longer). After drying, the flasks are rinsed 3 times with PBS before seeding the cells.

3.1.2 Isolation of Astrocytes

1. All procedures should use sterile instruments and gloves, and medium of 37°C unless otherwise stated.
2. Use six rat pups, 4–5-days old (2 pups/flask). Clean them and keep them in a petri dish. Decapitate each rat with scissors and dip each head in 70% alcohol.
3. Remove the brains with a scalpel and forceps and transport them to a petri dish, containing ice-cold fully HEPES buffered DMEM.
4. Isolate the cortex with forceps and remove the meninges. Break up the cortex with a 5-ml pipette in a 50-ml tube with 14 ml ice-cold fully HEPES buffered DMEM, add 1 ml 0.25% trypsin to the suspension, and incubate for 25 min at 37°C in a water bath.
5. Add DMEM + S to stop the trypsinization, to a total volume of 50 ml and spin the cell suspension for 8 min at 2,000 rpm (720g).
6. Resuspend the pellet in 10 ml DMEM + S, then filter the cell suspension through a 120-μm mesh using a filter holder mounted with a 20-ml syringe and rinse with 10 ml DMEM + S. Subsequently, filter the cell suspension through a 45-μm mesh and rinse with 10 ml DMEM + S and seed the cells in three 250-ml culture flasks (noncoated), 10 ml cell suspension each.

7. Culture cells at 37°C, 10% CO2, and change the medium only after 3 days (not before and do not move the flasks). After 6 days, change the medium and shake the culture flasks overnight on a shaker at 80 rpm at room temperature (use closed caps on the flasks or close the filter caps with parafilm), then wash the cells with PBS, and add fresh DMEM + S, 10 ml per flask.
8. Culture at 37°C, 10% CO2 until confluence (usually after 10 days), then wash the cells with PBS 2 times and passage with 5 ml 0.25% trypsin (shake and tap until most of the cells detach or use a cell scraper, time: 5–10 min). Add 10 ml DMEM + S to stop the trypsinization and pool the cells.
9. Spin the cell suspension for 5 min at 1,000 rpm (180g) and resuspend the pellet in 9 ml DMEM + S and plate the cells in nine Poly-d-Lysine coated flasks containing 10 ml DMEM + S; 1 ml to each.
10. Change the medium every other day, until confluence (usually over the weekend), then collect ACM every other day for 2 weeks, sterile filter, and store in 150-ml flasks at –20°C (weekdays, 10 ml to each flask; weekends, 15 ml).
11. For collecting cells for the co-culture after 2 weeks of ACM collection, the cells are washed with PBS 2 times and passage with 5 ml 0.25% trypsin (shake, tap, and scrape until most of the cells detach, time: ±15 min), pooled into ±150 ml DMEM + S to stop the trypsinization, and the cell suspension centrifuged for 5 min at 1,000 rpm (180g).
12. Cells are resuspended in 18 ml FCS and 2 ml DMSO is added and aliquoted into 20 cryovials (1 ml each). Vials are frozen overnight in –80°C freezer and stored in liquid nitrogen until use.

3.1.3 Isolation of Bovine Brain Capillaries

1. Obtain one calf brain freshly slaughtered from a local slaughterhouse and keep on ice in 300-ml sterile PBS while transporting to the laboratory.
2. Remove and discard both cerebellum and brain stem, then separate the hemispheres and transfer them to a large petridish with ice-cold PBS.
3. Carefully, but thoroughly, remove the meninges from the brain, with your hands, and wash the hemispheres carefully with ice-cold PBS, also between the folders, then transfer the hemispheres to a clean petridish containing ice-cold DMEM + S.
4. Use a sterile razor blade to slice off pieces of the grey matter, leaving the white matter behind. Keep the brain moist by repeatedly bathing it with DMEM + S (cold). Transfer the pieces into a 500-ml flask containing ±50 ml DMEM + S (on ice).
5. Homogenize small aliquots of tissue in a 40-ml Wheaton homogenizer, first with pestle B then pestle A: fill the homogenizer 1/3 with brain pieces and add DMEM + S (cold) to fill it up. Homogenize with pestle B until all large pieces are disrupted (±8 strokes). Continue with pestle A until you have a homogenous suspension (±8 strokes). Transfer the suspension to a 500-ml flask. Repeat this step until all brain tissue is homogenized.

6. Filter the homogenate through a 150-μm mesh, 30–40 ml per filter, using a 20-ml syringe and a filter holder. Rinse with 10 ml DMEM + S (cold). Transfer the mesh with capillaries to a petridish containing 10 ml DMEM + S (cold) and flush the material off the mesh with the pipette. Transfer the suspension to 50-ml tubes; fill the tubes to a maximum of 40 ml and keep them on ice. Use a new piece of 150-μm mesh for the next aliquot. Discard the filtrate.
7. Spin the tubes 5 min at 1,500 rpm (405g) and take white matter off and repeat this step. Resuspend and pool the pellets in 20-ml digest mix. Incubate for 1 h at 37°C (shake once or twice).
8. Filter the suspension through a 200-μm mesh, and then wash with 30 ml DMEM + S (to a total volume of 50 ml). Spin 5 min at 1,500 rpm (405g) and take white matter off. Resuspend in 9 ml FCS, add 1 ml DMSO, and put into cryovials (1 ml each). Freeze the vials in the −80°C freezer overnight. After 1 day, store the vials in liquid nitrogen until use.

3.1.4 Differential Seeding of Brain Capillaries and Culture of BCECs

1. Thaw one or two (depending on the speed of outgrowth) vial(s) with brain capillaries per 2 (or 3) coated culture flasks in a 37°C waterbath and wipe the outside of the vial with 70% ethanol.
2. Transfer the content of the vial(s) to 30 ml DMEM + S (flush the vial) and spin for 5 min at 1,000 rpm (180g). Aspirate the supernatant and resuspend in DMEM + S, 10 ml per culture flask.
3. Leave flasks in the incubator (37°C, 10% CO2) for 2–4 h for capillaries to attach, and then change to 10 ml growth medium (GM+). After 2 or 3 days change the medium (GM+).
4. Passage to filters (see below) at day 4 or 5 after plating (depending on the day of seeding).

3.1.5 Preparation of the In Vitro BBB on Filter Inserts

1. For the preparation of co-cultures with astrocytes on the bottom of the filter, start with collagen coated filters (washed 3 times with PBS) and put them upside down in a large petridish.
2. Pipette 1 ml of DMEM + S to 12 wells of the 24-well plate and put it in the incubator (together with the filters in the petridish).
3. Thaw 1 vial of astrocytes from the liquid nitrogen (other sources of astrocytes were also used (*see* **Note 4**)), transfer the content to 30 ml DMEM + S, and spin for 5 min at 1,000 rpm (180g). Resuspend in 2.2 ml DMEM + S (this is enough for 48 filters) and pipette 40 μl of the suspension (containing ~45.000 cells) on the bottom of each filter. Put the petridish into the incubator for 8 min, then transfer the filters back into the plate, and incubate the cells at 37°C, 10% CO2 for 2 or 3 days.

4. On the day the BCECs can be passaged onto the filters (at about 80% confluence, which should be at day 4 or 5 after plating (see earlier)), change the medium on the filters to GM- (or GM + for monolayers), 1 ml each, and leave in the incubator at 37°C, 10% CO2 for at least 2 h.
5. Rinse the BCECs with PBS (10 ml), then with 0.02% EDTA (5 ml), and add 4 ml endothelial cell trypsin, then observe continuously under the microscope. To stop the trypsinization (after maximum 2 min) add ±7 ml DMEM + S and pipette the contents of the flask into a 50-ml tube; total volume ± 30 ml DMEM + S. This procedure leaves the majority of pericytes still adhered to the substratum. Repeat this step for the other flasks (or passage them simultaneously (maximum, three flasks) with the first flask). Centrifuge for 5 min at 1,000 rpm (180g) and resuspend in 1 ml (or more, depending on the amount of cells) GM- or GM + medium, then count the cells and seed the cell suspension to each filter (30.000 cells per filter) (other sources of endothelial cells were also used (*see* **Note 5**)).
6. After 2 or 3 days, change the medium with DMEM + S or DMEM + S with cAMP: remove the medium basolateral and apical, but leave ~50 µl on the cells (otherwise the quality of the cells reduces significantly), then add 200 µl to the apical side and add 800 µl to the basolateral side. Incubate the cells at 37°C, 10% CO2, and use for experiment the second or third day. See Fig. 8.1 for a schematic drawing of the co-culture model of the BBB.
7. BCEC monolayers are cultured accordingly, but with 50% (v/v) ACM added to the culture medium (GM+).
8. For cell uptake experiments without the use of expensive filter inserts, BCECs are seeded in collagen coated plates at 15.000 cells per well of a 96-well

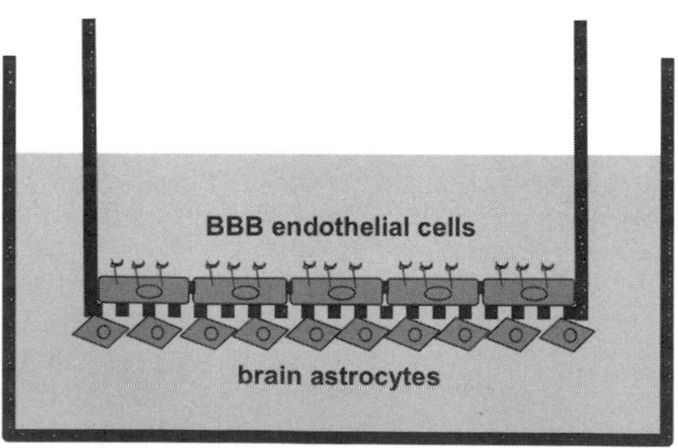

Fig. 8.1 Schematic drawing of the in vitro BBB model. Astrocytes are seeded on the bottom of the collagen-coated filter at a density of 45.000 cells per filter, allowed to adhere for 8 min, and cultured for 2 or 3 days. BBB endothelial cells (BCECs) are seeded at a density of 30.000 cells per filter. BCEC-astrocyte co-cultures are cultured to tight monolayers in growth medium for the first 2 or 3 days and on differentiation medium for the last 2 or 3 days. Transport or transendothelial electrical resistance (TEER), drug transport or receptor characterization experiments are performed after a total of 9 or 10 days after the brain capillaries are seeded

plate in 200 µl GM+ or 30.000 BCECs per well of a 48-well plate in 1 ml GM+. It is not necessary to change the medium after 2 days; use the plates after 4–5 days.

9. BBB functionality is assessed by transendothelial electrical resistance (TEER) across the filters using an electrical resistance system (ERS) with a current-passing and voltage-measuring electrode (Millicell-ERS, Millipore Corporation, Bedford, MA). TEER ($\Omega \bullet cm^2$) is calculated from the displayed electrical resistance on the readout screen by subtraction of the electrical resistance of a collagen coated filter without cells and a correction for filter surface area. TEER across collagen coated filters with only astrocytes on the bottom is close to zero. Effects on TEER are normalized for control treated filters and represented as such. Additional guidelines on how to perform TEER and drug transport experiments are also explained (*see* **Note 6**).

3.2 2B-Trans™ Technology

3.2.1 Characterization of Transport Receptor on the BBB In Vitro

1. Apical exposure to 100 ng/ml up to 50 µg/ml of DT decreases TEER across BCEC-astrocyte co-cultures in a concentration- and time dependent manner (Fig. 8.2a), while concentrations as low as 1 ng/ml are toxic after an overnight incubation period (data not shown). These results indicate that DT is effectively taken up from the apical site (i.e., blood site) by BCECs in which it can exert its toxic effects.
2. Apical exposure to 100 ng/ml DT, which is preincubated with soluble HB-EGF (1 h at room temperature), acting as a noncompetitive antagonist for the DTR by binding to the receptor-binding domain of DT, the toxic effect of DT on BCEC-astrocyte co-cultures decreases in a concentration dependent manner (Fig. 8.2b). In fact, a preincubation of 100 ng/ml DT with 10 µg/ml of soluble HB-EGF completely prevents the DT-induced toxic effect on BCECs, even after an overnight assessment. These results indicate that DT-uptake in BCECs is effectively blocked by previous specific binding of DT to its soluble receptor, making it unable to exert its toxic effects within the BCECs.
3. After BCECs are preincubated with CRM197, the nontoxic mutant protein of DT, acting as a competitive antagonist at the DTR by binding to the receptor-binding domain for DT, the toxic effect of apical exposure to 100 ng/ml DT on BCEC-astrocyte co-cultures is reduced (Fig. 8.2c). These results indicate that DT-uptake in BCECs is effectively antagonized by previous specific binding of CRM197 to the DTR, making it less available for DT to exert its toxic effects within the BCECs. Also note the absence of toxic effects of CRM197 on the BCEC-astrocyte co-cultures (Fig. 8.2c), which is consistent up to the highest tested concentration of 50 µg/ml (data not shown).

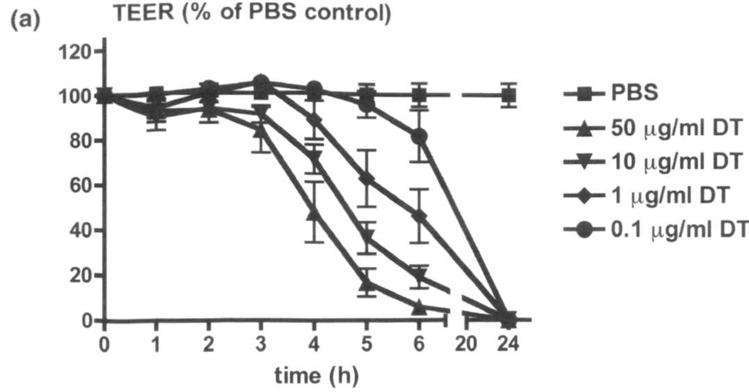

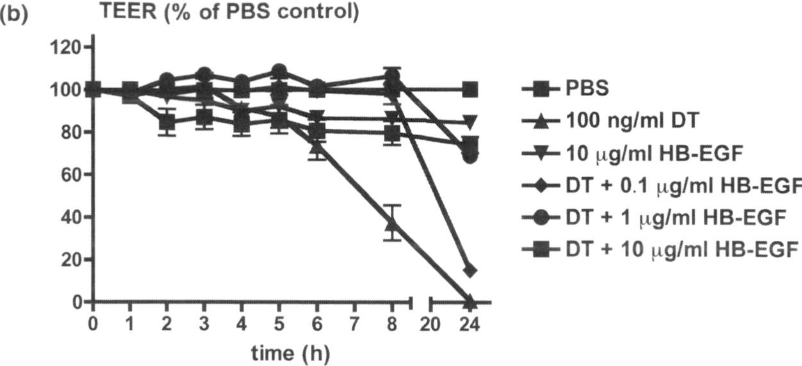

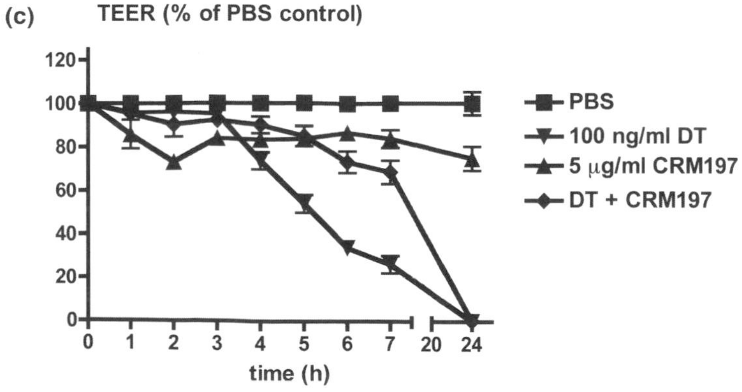

Fig. 8.2 Experiments confirming the functional expression of the specific transport receptor for diphtheria toxin (DT) on the in vitro BBB model. Panel (**a**) is a diagram showing the effect on TEER (expressed as mean % of control) across BCEC-astrocyte co-cultures apically exposed to various concentrations (100 ng/ml up to 50 µg/ml) of DT. Panel (**b**) is a diagram showing the effect of 100 ng/ml DT, which was preincubated with various concentrations of soluble HB-EGF (0.1–10 µg/ml), acting as a noncompetitive antagonist for the DT receptor (DTR) by binding to the receptor-binding domain of DT, before it was exposed to the apical side of the filter. Panel (**c**) is a diagram showing the effect of 100 ng/ml DT on cells that were pretreated for 1 h with 5 µg/ml CRM197, the nontoxic mutant protein of DT, acting as a competitive antagonist at the DTR by binding to the receptor-binding domain for DT

3.2.2 Receptor Targeted Cell Uptake Studies on the BBB In Vitro

1. BCECs cultured as monolayers in 96-wells plates are incubated with HRP-conjugated CRM197 and BSA (corresponding to a concentration of 5 μg/ml of un-conjugated HRP), and to CRM197-HRP conjugate, which is preincubated with 10 μg/ml soluble HB-EGF for 1 h at room temperature, acting as a noncompetitive antagonist for DTR-mediated uptake by binding to the receptor-binding domain of CRM197. After 24 h, HRP activity in cell lysates is detected using a standard colorimetric assay with the appropriate calibration curves (other conjugates were also used (*see* **Note 7**)).
2. In these conditions the CRM197-HRP conjugate is preferably taken up by the BCECs when compared with BSA-HRP conjugate (see Fig. 8.3). In addition, this specific uptake of the CRM197-HRP conjugate is completely inhibited, as compared with the a-specific uptake of BSA-HRP-conjugate, by the preincubation with soluble HB-EGF (see Fig. 8.3). These results indicate that CRM197 conjugated to a cargo of 40 kDa is specifically taken up by BCECs via a DTR-mediated uptake process.

3.2.3 Receptor Targeted Transcytosis Studies across the BBB In Vitro

1. Transcytosis experiments are performed after treatment of the cells with 8-4-CPT-cAMP and RO-20-1724 in complete HEPES buffered DMEM + S for the last 2 or 3 days so as to dramatically increase tightness (i.e., reduce paracellular leakiness).

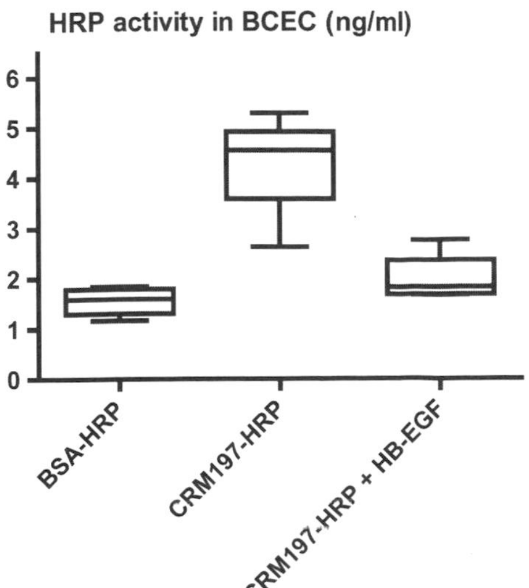

Fig. 8.3 Experiments illustrating the receptor-specific cell uptake of CRM197 conjugates. The diagram shows HRP activity in BCEC lysates after exposure to HRP-conjugated BSA (acting as an inert control conjugate), HRP-conjugated CRM197 and HRP-conjugated CRM197 which was pre-incubated with 10 μg/ml soluble HB-EGF (acting as a noncompetitive antagonist for the uptake receptor (DTR) by binding to the receptor-binding domain of CRM197)

2. The average TEER across the BCEC-astrocyte co-cultures increases from 149.8 ± 5.4 Ω•cm^2 (mean ± standard error, n = 18) to 834 ± 77 Ω•cm^2 (mean ± standard error, n = 24) after treatment with 8-4-CPT-cAMP and RO-20-1724. No difference in DT sensitivity is observed between cells un-treated and cell treated as such (data not shown).
3. To determine specific and active transcytosis across the in vitro BBB, HRP conjugated to CRM197 or BSA is added to the apical side of the filter inserts at 37°C and 4°C in fully HEPES buffered DMEM + S. For the 4°C arm of the experiment, filters are allowed to cool down in the refrigerator for 1 h before the transport experiment is started. This cooling procedure has no effect on TEER across the BCEC-astrocyte co-cultures when it is performed in fully HEPES buffered DMEM + S and when the electrode for the TEER measurement is allowed to cool down to 4°C as well. Directly after the conjugates are added, the filter insert is transferred into a fresh well containing warm (or cold) 250 µl HEPES buffered DMEM + S. Every hour, up to 4 h in total, this procedure is repeated in order to prevent possible re-endocytosis of HRP-conjugated proteins by the abluminal side of the BCECs. Cumulated HRP activity of transcytosed HRP into the basolateral compartment is detected using a standard colorimetric assay with the appropriate calibration curves.
4. After BCEC-astrocyte co-cultures are incubated with HRP-conjugated proteins, the CRM197-HRP conjugate is preferably transcytosed across the BCECs when compared with BSA-HRP conjugates (see Fig. 8.4). At 4°C, the level of transport for the CRM197-HRP conjugate is identical to the BSA-HRP conjugate at 37°C and 4°C (see Fig. 4). These results indicate that CRM197, even when conjugated to a protein cargo of 40 kDa, is specifically and actively transcytosed across the BBB (in vivo validation of this conjugate was also done (*see* **Note 8**)).

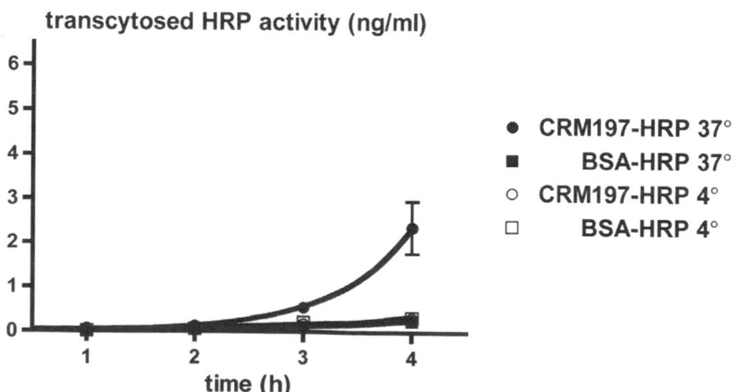

Fig. 8.4 Experiments illustrating the active and receptor-specific transcytosis of CRM197 conjugates across the in vitro BBB model. The graphs depict the active and selective accumulated transcytosis of HRP-conjugated CRM197 (lines with circles), when compared with HRP-conjugated BSA (lines with squares), at 37°C (lines with filled symbols) and 4°C (lines with open symbols)

4 Notes

1. Even from a single supplier, we have had batches of FSC that produced very tight BCEC monolayers and filters with very leaky BCEC monolayers. In these cases, even astrocyte co-cultures were unable to increase tightness. We therefore routinely test about five samples of different batches from different suppliers on their ability to produce high resistance monolayers before we acquire large quantities of that specific batch. By these means we largely avoid a major source of variability in the quality of the BBB cell culture model.
2. We have used calf brains from a local slaughterhouse mainly because of the size of the calf brain (resulting in high yield of brain capillaries) and the tight BCEC monolayers that could ultimately be obtained from primary cultures of calf brains. Alternatively, one can also use brains from other large animals, like porcine brains [6], especially when bovine materials are difficult to obtain because of BSE measures. Rodent brains, on the other hand, usually have a low yield (and thus require many animals) and typically give rise to leaky monolayers, making such models not suitable for the assessment of drug transport and effect measurements. For many obvious reasons, primary cultures of BCECs from human and primate sources are more difficult to obtain on a regular basis, so these are also not readily used for drug transport and effect screening purposes. See also Note 5 for other sources of endothelial cells.
3. Collagen coated flasks can also be prepared a few days in advance and kept on PBS at 4°C. The fibronectin coated flasks should, however, always be prepared directly before seeding of the cells or brain capillaries.
4. As a convenient alternative source of astrocytes we have also tried to make use of commercially available astrocytes (both rat cortical astrocytes and normal human astrocytes from Cambrex, Cat. R-CXAS-520 and CC-2565, respectively) in our experimental setting. Unfortunately, irrespective of the culture media we have used (i.e., our own DMEM + S or the proposed media of the supplier), we have never observed any of the BBB inducing properties on our bovine BCECs (nor on any other source of endothelial cells, see also Note 5) as we observe with our own astrocyte isolation protocol as described in this and earlier papers [5, 7–9].
5. Alternative sources of endothelial cells:
   - 5.1. Although more convenient than primary cultured cells, the currently available human [10] and rodent [11] cell lines are still too leaky to use for drug transport and effect screening purposes.
   - 5.2. Should one be unable or unwilling to obtain calf or porcine brains from the slaughterhouse, an effective, though rather expensive, alternative source of bovine BBB cells is now commercially available from Cambrex (bMVEC-B, Cat. AC-2509). These cells fit very well in the current protocol, exhibiting identical BBB properties as the currently described model, with only a few minor modifications. Briefly, the filters are also coated with fibronectin (30 min, 10 μg/ml, just as the flasks) before the astrocytes are seeded on the bottom. The bMVEC-B are also seeded at 30.000 cells/filter, but kept for 2 or 3 additional days on DMEM + S, as compared with the original protocol. In every tested setting, however, we were unable to obtain tight monolayers when the bMVEC-B were cultured on the recommended medium by the supplier, and so we suggest to use the DMEM + S as described earlier. Finally, also in these cells, the BBB properties were not induced with other sources of astrocytes (see Note 2).
   - 5.3. As another alternative source of readily available endothelial cells of relative constant composition to create a human BBB model, we also used pooled neonatal human dermal microvascular endothelial cells (Cambrex, HMVEC-d, Cat. CC-2516). Briefly, these cells should be cultured in uncoated flasks in the recommended media by the supplier (EGM-2 MV, Cat. CC-2516), for not more than six passages. Then, the cells should be passaged to astrocyte-containing filters just as described for the current model in EGM-2 MV instead of DMEM + S. After 5 days of co-culture with rat astrocytes and addition of

8-4-CPT-cAMP, RO-20-1724, and 0.1 μM dexamethasone, TEER increased to~85 Ω•cm². This effect on TEER was neither observed without astrocytes, nor with other sources of astrocytes (see Note 2). In these conditions, the human BBB model was also sensitive to DT (determined as described earlier) and lipopolysaccharide (LPS, as described in ref. [9]). We were, however, unable to find any indication of P-glycoprotein (P-gp) expression in these cells (determined as described in ref. [7]). The relatively low TEER across the BBB model and the possible absence of P-gp expression, however, limits the applicability of this constellation of human endothelial cells and rat astrocytes as a BBB cell culture model for the use for drug transport and effect screening purposes.

6. TEER experiments, but also drug transport experiments, are very sensitive to how one manipulates the filter inserts. Therefore, the experimental set-up should always include the proper control conditions to assess the influence of any manipulation to the cells. Care should be taken not to disturb the cells on the filter to much prior to the experiment; so it is for instance better not to change the medium on the filters to apply a solution of active compounds, but rather to apply the compounds in a small volume (as a rule we apply no more than 30 μl) of a concentrated stock solution to the existing medium. In addition, the influence of excipients (ethanol, DMSO, etc.) can be very strong, also on the expected effect of a stimulus, so these should be kept below 0.1% of the total volume on the filter insert. In addition, TEER and drug transport experiments can be conveniently combined, giving data from the effect of a drug on the BBB permeability in relation to the observed transport of the same drug across the filter [12]. For such experiments care should be taken to avoid contamination between the apical and basolateral compartment within and between filters. Washing the electrode in a series of six wells with clean medium between every TEER measurement was shown to be effective.
7. Although outside the scope of this paper, it is of relevance to state that next to the HRP-conjugates we have prepared similar conjugates to a number of other reporter labels and enzymes, including FITC and β-galactosidase, as well as CRM197-coated liposomes, and similar results were obtained (unpublished observations).
8. Finally and again outside the scope of this paper, it is of relevance to state that subsequently the CRM197-HRP conjugate was validated in guinea pigs as well and found to be specifically transported to and across the BBB *in vivo* [4].

Acknowledgments Part of the work described in this paper was supported by a grant from ZonMw-Dierproeven Begrensd (3170.0057). The authors would like to acknowledge Drs. Heleen Voorwinden, Dr. Corine Visser, and Drs. Arjen Brink for their excellent experimental contributions to the data presented in this paper.

References

1. de Boer AG, van der Sandt IC, Gaillard PJ. The role of drug transporters at the blood–brain barrier. Annu Rev Pharmacol Toxicol 2003;43:629–56.
2. de Boer AG, Gaillard PJ. Drug targeting to the brain. Annu Rev Pharmacol Toxicol 2007;47:9.1–9.33.
3. Gaillard PJ, Visser CC, de Boer AG. Targeted delivery across the blood–brain barrier. Expert Opin Drug Deliv 2005;2(2):299–309.
4. Gaillard PJ, Brink A, de Boer AG. Diphtheria-toxin receptor-targeted brain drug delivery. Int Congress Series 2005;1277(11):185–98.
5. Gaillard PJ, Voorwinden LH, Nielsen JL, Ivanov A, Atsumi R, Engman H, Ringbom C, de Boer AG, Breimer DD. Establishment and functional characterization of an in vitro model of the blood–brain barrier, comprising a co-culture of brain capillary endothelial cells and astrocytes. Eur J Pharm Sci 2001;12(3):215–22.

6. Franke H, Galla HJ, Beuckmann CT. An improved low-permeability in vitro-model of the blood–brain barrier: transport studies on retinoids, sucrose, haloperidol, caffeine and mannitol. Brain Res 1999;818(1):65–71.
7. Gaillard PJ, van der Sandt IC, Voorwinden LH, Vu D, Nielsen JL, de Boer AG, Breimer DD. Astrocytes increase the functional expression of P-glycoprotein in an in vitro model of the blood–brain barrier. Pharm Res 2000;17(10):1198–205.
8. van der Sandt IC, Gaillard PJ, Voorwinden HH, de Boer AG, Breimer DD. P-glycoprotein inhibition leads to enhanced disruptive effects by anti-microtubule cytostatics at the in vitro blood–brain barrier. Pharm Res 2001;18(5):587–92.
9. Gaillard PJ, de Boer AB, Breimer DD. Pharmacological investigations on lipopolysaccharide-induced permeability changes in the blood–brain barrier in vitro. Microvasc Res 2003;65(1):24–31.
10. Roux F, Couraud PO. Rat brain endothelial cell lines for the study of blood–brain barrier permeability and transport functions. Cell Mol Neurobiol 2005;25(1):41–58.
11. Weksler BB, Subileau EA, Perriere N, Charneau P, Holloway K, Leveque M, Tricoire-Leignel H, Nicotra A, Bourdoulous S, Turowski P, Male DK, Roux F, Greenwood J, Romero IA, Couraud PO. Blood–brain barrier-specific properties of a human adult brain endothelial cell line. FASEB J 2005;19(13):1872–4.
12. Gaillard PJ, de Boer AG. Relationship between permeability status of the blood–brain barrier and in vitro permeability coefficient of a drug. Eur J Pharm Sci 2000;12(2):95–102.

Chapter 9
Drug Delivery in Cancer Using Liposomes

Crispin R. Dass

Abstract There are various types of liposomes used for cancer therapy, but these can all be placed into three distinct categories based on the surface charge of vesicles: neutral, anionic and cationic. This chapter describes the more rigorous and easy methods used for liposome manufacture, with references, to aid the reader in preparing these formulations in-house.

Keywords Liposome; Cancer; Drug delivery; Formulation; Vesicle

1 Introduction

Liposomes generally have a large carrying capacity, but usually not large enough to ferry large molecules such as proteins. Hydrophilic drugs can be readily entrapped within the aqueous interior of the vesicles, while neutral and hydrophobic molecules may be carried within the hydrophobic bilayers of the vesicles. There are various types of liposome formulations available, some of which have arisen due to a marriage of ideas for specific types of liposomes [1]. The liposome field is more than 3 decades old now, with the commercialization of a handful of anti-cancer therapeutic agents, all non-biologicals. Such small molecule drugs that are used to treat certain types of neoplasms include the nucleic-acid-synthesis-interfering agents doxorubicin (available as Caelyx®) and daunorubicin (available as DaunoxomeZ®).

There are some excellent texts available [2, 3] on liposomes that serve as a good guide to the slightly initiated and most definitely for the more experienced. However, an easy guide to understanding liposome manufacture is generally hard to find, and the present concise chapter is intended to serve just this purpose. There are three methods that may be used for easy formulation of liposomes, and these have stood the test of time, at least for the past 12 years since the author has been using them. The most promising types of liposome used for anticancer purposes, cationic varieties, are discussed below.

From: *Methods in Molecular Biology, Vol. 437: Drug Delivery Systems*
Edited by: Kewal K. Jain © Humana Press, Totowa, NJ

2 Cationic Liposomes

Cationic liposomes (CLs), first brought to prominence in 1987 [4], have revolutionised the gene delivery arena, and are now being examined for other applications, such as delivery of small molecule drugs. Made from positively charged lipids such as DOTAP (dioleoyloxypropyl trimethylammonium chloride) and DC-cholesterol (DC-chol), this vesicle forms the major category of liposome used currently for drug and biomolecular delivery. Its current popularity is only due to its ability to deliver nucleic acids into cells both in vitro and in vivo for genetherapy of diseases such as cancer. They are also being evaluated clinically for this purpose for more than a decade now [5–9]. More recently, these liposomes have been found to be selectively delivered to tumour vascular endothelial cells [10]. Being positively charged, these liposomes are ideal for attachment to the slightly negatively charged cell membrane.

CLs are useful not only for delivering genetic constructs to the tumour vasculature, but also for causing an anti-vascular effect with small molecule cytotoxic agents. Kunstfeld and coworkers [11] demonstrated that paclitaxel encapsulated in CLs diminishes tumour angiogenesis and inhibits orthotopic melanoma growth in mice. In contrast, paclitaxel administered in its normal Cremophor EL vehicle, while showing an inhibitory effect in cell culture, was unable to significantly decrease angiogenesis and tumour growth in vivo.

Campbell and colleagues [12] found that CLs, stabilised with the addition of a 5 mol% of PEG, accumulated more in angiogenic vessels when CLs were used as opposed to electroneutral liposomes. PEG was used to increase circulation lifetime of the positively charged liposomes. Inclusion of PEGylated lipids in the vesicles has the added advantage of reducing aggregate formation [13], thus increasing both yield and injectability of complexes. Unmodified lipoplexes have a relatively shorter circulation half-life of less than 5 min [12]. Furthermore, when the percentage of cationic lipid was increased from 10 to 50 mol%, the accumulation in tumour vascular endothelial cells (VECs) increased 2-fold.

Thus, the use of cationic liposomes for cancer therapy cuts across different boundaries. There are limitations with these vesicles, but with additional research and development, better formulations will emerge. One significant achievement would be development of formulations that can ferry proteins in vivo.

3 Formulation Techniques

3.1 Ethanol Injection (Good for Cationic Liposomes)

The method is probably one of the easiest ones available. It was originally reported by Batzri and Korn in 1973 [14]. It involves the injection of a small volume of ethanolic solution of lipids into a large volume of water. The force of the injection ensures homogeneous mixing of the lipids, as does the immediate dilution of the ethanol in the large excess of water. This procedure can generate mainly small unilamellar vesicles (SUVs) with diameters around 25–50 nm. This process is probably the best for making anionic and cationic liposomes since at least some of the lipids dissolve readily in water. The charged lipid is dissolved in water while the hydrophobic one is dissolved in ethanol. The exact steps are as listed here using the example of a formulation prepared in-house:

1. Weigh out 12 mg of DDAB (dimethyl dioctadecyl ammonium bromide) into a clean glass test tube.
2. Dry down 4 mg of dioleoylphosphatidylethanolamine (DOPE) solution (in chloroform) using argon.
3. Dissolve the DDAB in 1 mL of absolute ethanol.
4. Dissolve the DOPE in 1 mL absolute ethanol. Use a 37°C bath to aid dissolution.
5. Prepare eight small glass test tubes with 0.75 mL nanopure water in each.
6. Add 0.125 mL of the ethanolic solution of DDAB to four glass test tubes.
7. Vortex one of the DDAB-containing tubes on high for 10 s.
8. Carefully, inject 0.125 mL of ethanolic DOPE into the vortexing tube in one swift motion with the tip of the pipettor below the surface of the vortexing liquid.
9. Vortex for further 10 s.
10. Extrude liposomes through a 220-nm polycarbonate filter 12 times through a poly(ether sulfone) (PES) membrane.
11. Transfer contents to storage vial and label appropriately.
12. Store liposome at 4°C.

3.2 Sonication (Good for Neutral Liposomes Used for Membrane Studies)

Lipids are mixed together and either sonicated using a probe or in a water bath. This procedure may be used for both charged and uncharged liposome manufacture, as sonication is a rather harsh method for rearranging the different lipids in a 3-D spherical conformation. The most important factor is to prevent heating of the sample during sonication, a phenomenon that could result in lipid degradation, leading to poor-quality liposomes. The exact steps are as listed as follows:

1. Dissolve lipids in a 20-mL glass vial and place on ice.
2. To create an ice bath for the sonication of lipids (to prevent heat-induced chemical degradation of lipids), fill a 50-mL Falcon tube three-fourths full of ice and pour in water to fill in the gaps between the pieces of ice.
3. Set probe sonicator to constant speed.
4. Immerse clean probe at least 1 cm into nanopure water and sonicate for 2 min to clean probe.
5. Now place probe into lipid solution and perform no more than 3 × 2-min bursts of sonication to prepare unilamellar vesicles (ULVs). Place vial on ice between sonication bursts. The opaque solution should change to an opalescent hue.

6. Centrifuge phospholipid vesicles (PLVs) to remove titanium bits that may have come off during sonication.
7. Store at 4°C.

3.3 Dried-Reconstituted Vesicles (Good for Sterically Stabilised Vesicles)

Lipids are mixed together and solvent is removed using freeze-drying. Then an aqueous solution is introduced and the lipid cake around the vessel wall is reconstituted. This method works best for manufacture of neutral liposomes, as the hydrophobic lipids readily dissolve in solvents such as chloroform and are deposited dry on the wall of the rotavapor vessel. Then the material to be encapsulated is dissolved in an aqueous solution and the dry film on the vessel wall is hydrated with this solution. The exact steps involved in the preparation of hydrogenated soy phosphatidylcholine (PHSPC)- and cholesterol-containing vesicles at a molar ratio of 60 to 40% are as listed here:

1. Prepare the following buffer: 10 mM Tris-HCl, 1 mM EDTA, 150 mM NaCl, pH = 7.4, osmolarity = 300 mOsm/kg.
2. Prepare the lipid film as follows: in a 500-mL flask with 0.401 g PHSPC and 0.13328 g cholesterol, add 7 mL chloroform. Drying is carried out at 40°C, on setting "4.5" of the rotavaporator, for 5 h, with the pump minimum pressure set at 66 mbar.
3. Drug agent to be encapsulated is added to 6 mL of the earlier-mentioned buffer.
4. The dried lipids are hydrated with the 6 mL of buffer with dissolved drug agent, and beads are added to aid homogeneous mixing. The mixing is done under nitrogen, at 40°C, on setting "4.5" of the Buchi Rotary Evaporator with water-cooled condensor coil, for 2 h.
5. Freeze and thaw (50°C) the preparation five times at 4 min for each cycle.
6. Sonicate preparation for 30 s.
7. Extrude preparation at 45°C, using 0.4-μm polycarbonate filter membranes at 200 psi. Change membrane if flow stops.
8. Sonicate formulation for 3 s.
9. Extrude through 200-nm filters at least six times.
10. Sonicate preparation for 1 min.
11. Extrude through 100-nm filters at least six times.
12. Separate lipid-encapsulated ones from free drug molecules using an appropriate column. In-house, for separation of free oligonucleotides from encapsulated ones, a Sepharose CL-2B, fractionation range 70,000–40,000,000, is used.
13. Concentrate liposome preparation if formulation is deemed to be too dilute using Vivaspin 20 mL Concentrator tubes with 100,000 MWCO PES.
14. Store the final preparation at 4°C

Glossary

DC-chol cholesteryl 3β-*N*-dimethyl aminoethyl carbamate hydrochloride
DDAB dimethyl dioctadecyl ammonium bromide
DOTAP dioleoyloxypropyl trimethylammonium chloride
PES polyether sulfone
PG phosphatidylglycerol
PS phosphatidylserine

References

1. Dass CR, Walker TL, Burton MA, Decruz EE. Enhanced anticancer therapy mediated by specialized liposomes. J Pharm Pharmacol. 1997;49(10):972–5.
2. Torchilin V and Weissig V (ed.), New RRC. Liposomes – a practical approach. Oxford University Press, Oxford, 1990, 1–251.
3. Torchilin VP, Weissig W. Liposomes – a practical approach, 2nd edn. Oxford University Press, Oxford, 2003.
4. Felgner PL, Gadek TR, Holm M, Roman R, Chan HW, Wenz M, Northrop JP, Ringold GM, Danielsen M. Lipofection: a highly efficient, lipid-mediated DNA-transfection procedure. Proc Natl Acad Sci USA. 1987;84:7413–7.
5. Caplen NJ, Alton EW, Middleton PG, Dorin JR, Stevenson BJ, Gao X, Durham SR, Jeffery PK, Hodson ME, Coutelle C, et al. Liposome-mediated CFTR gene transfer to the nasal epithelium of patients with cystic fibrosis. Nat Med. 1995;1(1):39–46.
6. Hyde SC, Southern KW, Gileadi U, Fitzjohn EM, Mofford KA, Waddell BE, Gooi HC, Goddard CA, Hannavy K, Smyth SE, Egan JJ, Sorgi FL, Huang L, Cuthbert AW, Evans MJ, Colledge WH, Higgins CF, Webb AK, Gill DR. Repeat administration of DNA/liposomes to the nasal epithelium of patients with cystic fibrosis. Gene Ther. 2000;7(13):1156–65.
7. Noone PG, Hohneker KW, Zhou Z, Johnson LG, Foy C, Gipson C, Jones K, Noah TL, Leigh MW, Schwartzbach C, Efthimiou J, Pearlman R, Boucher RC, Knowles MR. Safety and biological efficacy of a lipid-CFTR complex for gene transfer in the nasal epithelium of adult patients with cystic fibrosis. Mol Ther. 2000;1(1):105–14.
8. Ruiz FE, Clancy JP, Perricone MA, Bebok Z, Hong JS, Cheng SH, Meeker DP, Young KR, Schoumacher RA, Weatherly MR, Wing L, Morris JE, Sindel L, Rosenberg M, van Ginkel FW, McGhee JR, Kelly D, Lyrene RK, Sorscher EJ. A clinical inflammatory syndrome attributable to aerosolized lipid-DNA administration in cystic fibrosis. Hum Gene Ther. 2001; 12(7):751–61.
9. Stopeck AT, Jones A, Hersh EM, Thompson JA, Finucane DM, Gutheil JC, Gonzalez R. Phase II study of direct intralesional gene transfer of allovectin-7, an HLA-B7/beta2-microglobulin DNA-liposome complex, in patients with metastatic melanoma. Clin Cancer Res. 2001;7(8):2285–91.
10. Dass CR, Choong PF. Selective gene delivery for cancer therapy using cationic liposomes: in vivo proof of applicability. J Control Release. 2006;113(2):155–63.
11. Kunstfeld R, Wickenhauser G, Michaelis U, Teifel M, Umek W, Naujoks K, Wolff K, Petzelbauer P. Paclitaxel encapsulated in cationic liposomes diminishes tumor angiogenesis and melanoma growth in a "humanized" SCID mouse model. J Invest Dermatol. 2003;120:476–82.

12. Campbell RB, Fukumura D, Brown EB, Mazzola LM, Izumi Y, Jain RK, Torchilin VP, Munn LL. Cationic charge determines the distribution of liposomes between the vascular and extravascular compartments of tumors. Cancer Res. 2002;62:6831–6.
13. Meyer O, Kirpotin D, Hong K, Sternberg B, Park JW, Woodle MC, Papahadjopoulos D. Cationic liposomes coated with polyethylene glycol as carriers for oligonucleotides. J Biol Chem. 1998;273:15621–7.
14. Batzri S, Korn ED. Single bilayer liposomes prepared without sonication. Biochim Biophys Acta. 1973;298(4):1015–9.

Chapter 10
pH-Responsive Nanoparticles for Cancer Drug Delivery

Youqing Shen, Huadong Tang, Maciej Radosz, Edward Van Kirk, and William J. Murdoch

Abstract Solid tumors have an acidic extracellular environment and an altered pH gradient across their cell compartments. Nanoparticles responsive to the pH gradients are promising for cancer drug delivery. Such pH-responsive nanoparticles consist of a corona and a core, one or both of which respond to the external pH to change their soluble/insoluble or charge states. Nanoparticles whose coronas become positively charged or become soluble to make their targeting groups available for binding at the tumor extracellular pH have been developed for promoting cellular targeting and internalization. Nanoparticles whose cores become soluble or change their structures to release the carried drugs at the tumor extracellular pH or lysosomal pH have been developed for fast drug release into the extracellular fluid or cytosol. Such pH-responsive nanoparticles have therapeutic advantages over the conventional pH-insensitive counterparts.

Keywords Cancer drug delivery; Cytoplasmic drug delivery; Nanoparticles; pH-Responsive

Abbreviations

ATRP	Atom transfer radical polymerization
Cisplatin	*cis*-Diammineplatinum(II) dichloride
CMC	Critical micelle concentration
DOX	Doxorubicin
EPR	Enhanced permeability and retention effect
3LNPs	3-Layered pH-responsive nanoparticles
PA	Pullulan acetate
P(Asp)	Poly(aspartic acid)
PbAE	Poly(β-amino ester)
PBS	Phosphate-buffered saline
PCL	Polycaprolactone
PDEA	Poly[2-(*N,N*-diethylamino)ethyl methacrylate]

From: *Methods in Molecular Biology, Vol. 437: Drug Delivery Systems*
Edited by: Kewal K. Jain © Humana Press, Totowa, NJ

PDMA	Poly[2-(*N,N*-dimethylamino)ethyl methacrylate]
PEG	Poly(ethylene glycol)
PEO	Poly(ethylene oxide)
pH_e	Solid tumor extracellular pH
PHIM	pH-insensitive micelles
PHis	Poly(L-histidine)
PHSM	pH-sensitive micelles
pH_t	Polymer insoluble/soluble transition pH
PLGA	Poly(lactide-*co*-glycolide)
PLLA	Poly(L-lactic acid)
PNA	*N*-Phenyl-2-naphthylamine
PPO	Poly(propylene oxide)
RES	Reticuloendothelial system
SDM	Sulfadimethoxine

1 Introduction: Nanoparticles for Drug Delivery

Cancer has dethroned heart disease as the top killer among Americans under the age of 85 despite the significant progress in cancer detection and treatment in the past decades. It was estimated that there would be more than 1.4 million new cancer cases and about half a million cancer deaths in 2006 alone [1]. Chemotherapy is one of the several available arsenals to fight cancer, but it often has life-threatening side effects. For example, cisplatin, a widely used potent anticancer drug [2–4], has significant acute and chronic nephrotoxicity [5]. Common side effects of anticancer drugs include the decrease in the number of white blood cells, red blood cells, and platelets, nausea, vomiting, and hair loss.

An increased resistance to drug treatment is another serious challenge. Most of the cancer patients, although initially responsive, eventually develop and succumb to drug-resistant metastasis. For example, the success of typical postsurgical regimens for ovarian cancer, usually a platinum/taxane combination, is limited either by primary tumors being intrinsically refractory to treatment or by initially responsive tumors becoming refractory to treatment [6]. As a result, the first-line treatment of ovarian cancer yields about 30% pathologic remission and an overall response rate of 75%, but the disease usually recurs within 2 years of the initial treatment, ultimately causing death [7–9].

The intrinsic and acquired drug resistance mechanisms that mitigate the cytotoxic effects of anticancer drugs [6, 10–14] include the loss of surface receptors and transporters to slow drug influx, cell-membrane-associated multidrug resistance to remove drugs [12], specific drug metabolism or detoxification [15], intracellular drug sequestration [16], overexpression of Src tyrosine kinase [17] and splicing factor SPF45 [18], increased DNA-repair activity [19], altered expression of oncogenes and regulatory proteins [20, 21], and increased expression of antiapoptotic genes and mutations to resist apoptosis [14, 22]. For instance, melanoma cells are drug-resistant to a variety of chemotherapeutic drugs by exploiting their intrinsic resistance to apoptosis and by reprogramming their proliferation and survival pathways during melanoma progression [23]. Some mechanisms of drug resistance in cancer are shown in Fig. 10.1.

An important mechanism of multidrug resistance is that cancer cells overexpress ATP Binding Cassette (ABC) transporters in their plasma membranes [24], mainly the classical P-glycoprotein (PGP or MDR1, ABCB1), the multidrug resistance associated proteins (MRPs, in the ABCC subfamily), and the ABCG2 protein [25], all of which can very efficiently transport a variety of anticancer drugs out of the cells. [6, 10, 12, 13, 26–29] Therefore, as a consequence of the slowed drug entry but efficient drug removal by the P-gp pumps and the intracellular drug consumption (Eq. 1), the drug concentration in the cytoplasm is below the cell-killing threshold, resulting in a limited therapeutic efficacy.

$$\mathrm{d[D]/dt} = R_d + R_e + R_t - R_{p\text{-}gp} - \sum R_{i\ \text{drug resistance}} \tag{1}$$

where [D] is the cytosolic drug concentration, R_d, R_e, and R_t are the rates of drug entry by diffusion, endocytosis, and transport, respectively, $R_{\text{P-gp}}$ is the rate of drug

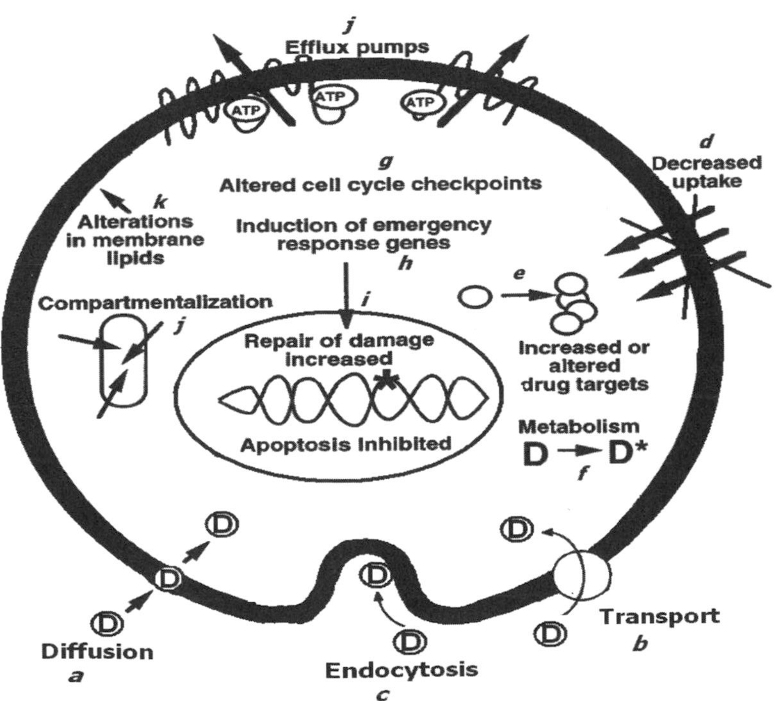

Fig. 10.1 Sketch of cancer cell's drug internalization (**a–c**) and forms of drug resistance (**d–k**). (modified from [12], with permission from Annual Reviews)

removal by the P-gp pumps, and R_i is the rate of drug consumption by other forms of drug resistance.

The drug toxicity to healthy tissues and the cell resistance to treatments described earlier pose a twofold challenge for drug delivery technology—to improve the delivery selectivity and to overcome the cell resistance—to simultaneously maximize the therapeutic efficacy and minimize the side effects.

One approach toward this goal is to exploit the enhanced permeability and retention effect (EPR) of cancerous tissues [30–36], which have unique pathophysiological characteristics, including the extensive angiogenesis and hence hypervasculature, defective vascular architecture, impaired lymphatic drainage/recovery system, and greatly increased production of permeability mediators. [30–36] The cutoff size for the permeation through the cancer blood capillaries was measured to be between 380 and 780nm [34–37]. Thus, nanosized drug carriers for example, polymer–drug conjugates [38–41], dendrimers [42–44], liposomes [45, 46], and polymer micelles or nanoparticles [47–54], can easily extravasate from the bloodstream and are trapped in tumor tissues, but not from the tight blood capillaries in healthy tissues [55], and thus preferentially deliver drugs to cancerous tissues. Furthermore, active tumor-targeting is also achieved by equipping the drug carriers with tumor-targeting groups such as folic acid [42, 56–73] and luteinizing-hormone-releasing hormone (LHRH) peptide [74–78]. These ligands can also induce receptor-mediated endocytosis for efficient cellular internalization of the drug carriers by cells overexpressing their receptors [79]. Therefore, by increasing bioavailability of drugs at sites of action, drugs in these carriers have shown enhanced efficacy against resistant tumors and fewer side effects.

Among the drug carriers mentioned earlier, polymeric nanoparticles are attracting much attention. [48, 51–53, 80–82] Their sizes can be easily optimized for penetrating through fine capillaries, crossing the fenestration into interstitial space, and efficient cellular uptake via endocytosis/phagocytosis. Furthermore, they can be equipped with a hydrophilic surface, for example, with a layer of poly(ethylene glycol) (PEG), to evade the recognition and subsequent uptake by the reticuloendothelial system (RES), and thus to have a circulation time that is long enough for passive accumulation in cancer tissues via the EPR effect [47, 50, 83–88].

Such stealthy nanoparticles with long circulation times can be fabricated from core-shell micelles formed by self-assembly of amphiphilic block copolymers (Fig. 10.2) [51–53, 89–92]. The hydrophobic inner core can carry drugs. The tight hydrophilic shell (e.g., PEG chains) prevents the protein adsorption and cellular adhesion, and thus protects the drug from degradation. The PEG chains also prevent the recognition by the RES [52, 87], which leads to an increased blood circulation time and enhanced drug accumulation in tumor tissues [52, 87]. Such nanoparticles have been used as drug carriers for cisplatin [93–98], doxorubicin [99–103], camptothecin [104–110], and paclitaxel [111–115]. These drugs were found to have a higher accumulation in tumors and lower toxicity. For example, in C26 tumor-bearing mice, the administration of doxorubicin (20mg/kg) resulted in toxic deaths, while the administration of doxorubicin in PEO-P(Asp) micelles permitted doses as high as 50mg/kg with no toxic deaths [116].

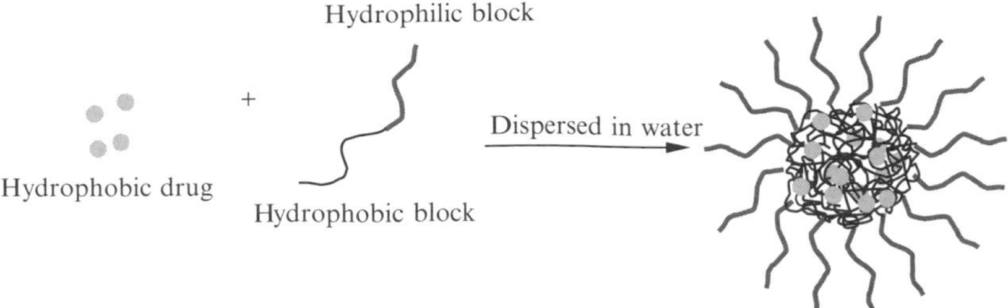

Fig. 10.2 Nanoparticles formed via polymer micelles by self-assembly of amphiphilic block copolymers

2 Rapid Drug Release: Key to Therapeutic Efficacy

Drugs become active only after they are liberated from their carriers [117, 118]. For example, doxorubicin (DOX) covalently bonded to the nanoparticle core of poly(lactide-*co*-glycolide) (PLGA) [119] or poly(aspartic acid)[116] showed low or even no anticancer activity because it could not be released on account of the absence of hydrolysable links between the drug and the polymer chains. By contrast, pH- [120, 121] or temperature [122–124]-induced rapid drug release led to a higher therapeutic efficacy. For example, Needham et al. formulated a DOX-containing liposome, the contents of which could be released rapidly in tens of seconds by a mild hyperthermic (39–40°C) induction. This new liposome, in combination with mild hyperthermia, was significantly more effective than free drug or temperature-insensitive liposomes in reducing tumor growth in a human squamous cell carcinoma xenograft line, producing 11 of 11 complete regressions lasting up to 60 days posttreatment [123]. Kirchmeier et al. found a significant correlation between the rate of nuclear accumulation of DOX and its in vitro cytotoxicity. The pH-sensitive immunoliposomes had faster release of DOX and consequently faster nuclear accumulations of the drug and higher cytotoxicities than did pH-insensitive liposomes [125].

The hydrophobic core of nanoparticles is mostly made of solid glassy polymers such as polycaprolactone, polylactide, and their random copolymers. Drugs are physically trapped and dispersed in the core. Except for the initial burst release period, the drug release from the solid nanoparticle cores tends to be a slow diffusion-controlled process [126]. Thus, nanoparticles responding to the acidic environments of tumor intercellular fluid or intracellular acidic compartments have been developed for fast drug release.

3 Tumor Extracellular pH-Responsive Nanoparticles for Drug Release

Tumors have a lower extracellular pH (pH_e) than do normal tissues, which is an intrinsic feature of the tumor phenotype [127–129]. This is resulted from the increased production of lactic acid [128–134] via anaerobic glycolysis due to their high metabolic rate but inadequate removal of lactic acid [128–134] and carbon dioxide [135], as well as other mechanisms [136]. Electrical and chemical probes showed that the mean pH of tumor extracellular fluid is typically about 7.06 with a range of 5.7–7.8 [137]. Low pH_e benefits tumor cells because it increases drug resistance by slowing the uptake of weakly basic drugs such as DOX and reducing their effects on tumors, promotes invasiveness [138], and induces vascular endothelial growth factor (VEGF) [139]. On the other hand, this acidic pH_e can be exploited as a drug-release trigger.

For example, Bae and coworkers reported pH-induced destabilizable poly (L-histidine) (PHis)-based micelles targeting the tumor acidic environment [140]. PHis has a pH-dependent water-solubility as a result of protonation of the unsaturated nitrogen of its imidazole ring at low pH [141]. Its soluble/insoluble transition pH (pH_t) is dependent on its molecular weight. PHis with a M_n more than 10 kDa becomes soluble at pH lower than 6 [141], but PHis with a M_n of 5 kDa becomes soluble at pH of about 6.5 [140]. Introducing water-soluble groups or polymers increases the pH_t [140, 142]. For instance, PHis5k-*block*-PEG2K (PHis-PEG) had a pH_t of about 7.0 [140]. This PHis-PEG formed micelles with diameters of about 100 nm. The critical micelle concentration (CMC) was dependent on the solution pH (Fig. 10.3). The micelles had low stability at pH lower than 7 due to the protonation of the hydrophobic PHis core (Fig. 10.4). When the solution pH was less than 5, no micelles could be detected.

As a result of this pH-dependent stability, the PHis-PEG micelles loaded with DOX exhibited a pH-dependent drug release. DOX was released faster at pH lower than 7.0 than at pH 7.4 and 8.0 (Fig. 10.5) [143]. It was hypothesized that PHis-PEG micelles were destabilized in a slightly acidic environment and hence released the DOX into the extracellular medium for enhanced drug permeation into the tumor cells due to high concentration gradients. The in vitro cytotoxicity was tested

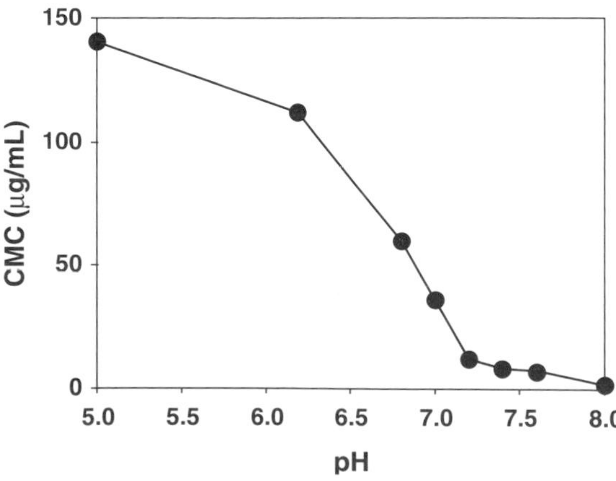

Fig. 10.3 The pH effect on the CMC of PHis5K–PEG2K micelles. (reproduced from [140], with permission from Elsevier)

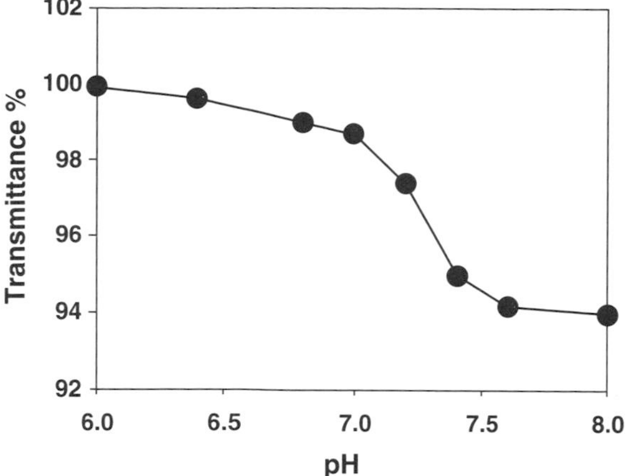

Fig. 10.4 The transmittance change of PHis5K-PEG2K micelles (0.1 g/l) with pH. (reproduced from [140], with permission from Elsevier)

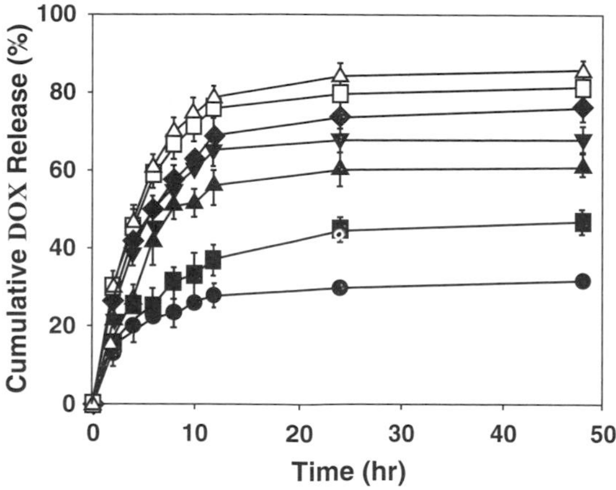

Fig. 10.5 pH-dependent cumulative DOX release from PHis-PEG micelles at pH 8.0 (●), 7.4 (■), 7.2 (▲), 7.0 (▼), 6.8 (♦), 6.2 (□), and 5.0 (Δ). (revised and reproduced from [143], with permission from Elsevier)

by incubating the DOX-loaded micelles with A2780 cells at pH 7.4 and 8.0. When cells were cultured with DOX/PHis-PEG at pH 6.8, a higher fraction of cells took up DOX, with a higher intracellular DOX concentration, and correspondingly lower cell viability, than when cultured at pH 7.4, as shown in Fig. 10.6 [144], which suggested that more DOX was available at the lower pH [144]. Also, compared with free DOX, the DOX-loaded micelles had a longer retention of DOX in blood, a higher DOX concentration in the cancer tissue, and more pronounced tumor

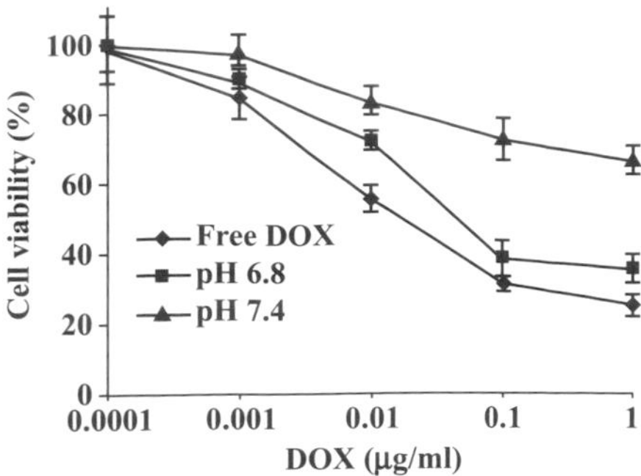

Fig. 10.6 Cytotoxicity of free DOX and DOX/PHis-PEG micelles at pH 7.4 and 6.8 after a 48-h incubation. (reproduced from [144], with permission from Taylor and Francis Ltd)

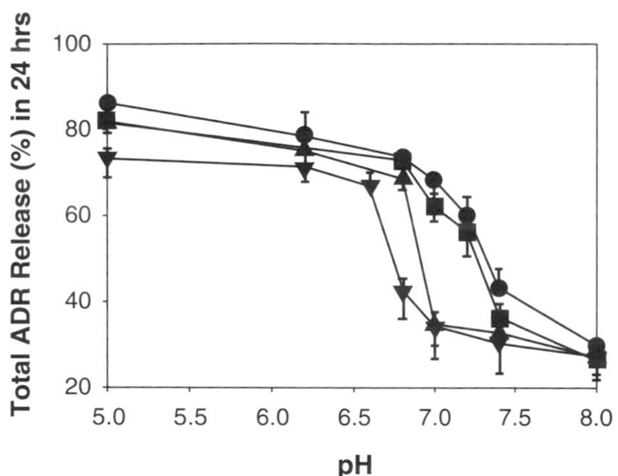

Fig. 10.7 pH-dependent 24-h-cumulative DOX release from the mixed micelles composed of PHis-PEG and PLLA-PEG with wt.% of PLLA-PEG at 0 (●), 10 (■), 25 (▲), and 40 (▼). (reproduced from [143], with permission from Elsevier)

suppression. It was presumed that the DOX release triggered by tumor extracellular pH (pH_e, 7.0) from the DOX/PHis-PEG accumulated in the tumor via the EPR effect contributed to the tumor inhibition.

As the data from Fig. 10.5 suggests, the PHis-PEG micelles are unstable and can release their contents at neutral pH. In order to deal with this problem, an amphiphilic block copolymer, PLLA-PEG, was added to make mixed micelles, in which the core was expected to contain poly(L-lactic acid) (PLLA) and PHis chains. The PLLA block in the core stabilized the micelles and hence suppressed the drug release at the near neutral pH. The optimum content of PLLA-PEG was found to be about 25–40% (Fig. 10.7). The DOX delivered from such mixed micelles showed low cytotoxicity at pH above 7.0 but high cytotoxicity at pH ~ 6.8 [143].

Another approach to tuning the pH-sensitivity proposed by Bae and coworkers is to introduce sulfadimethoxine (SDM; pK_a = 6.1; Scheme 10.1a) as a pH-sensitive motif [145, 146]. Sulfonamide is a weak acid because the proton of sulfonamide can be readily ionized to liberate a proton in basic solution [145, 146]. Thus, different from the amine-based PHis, SDM becomes hydrophobic by protonation at low pH. For example, pullulan acetate (PA) (Scheme 10.1b) conjugated with SDM moieties (PA/SDM) formed hydrogel nanoparticles that showed good stability at pH 7.4, but shrank and aggregated below pH 7.0, at which pH SDM became hydrophobic. As a result, the DOX release rate from the PA/SDM nanoparticles was low around the physiological pH but significantly enhanced below pH 6.8 (Fig. 10.8), which was consistent with the core structure changes observed using a fluorescent probe and a microviscosity probe [147].

The cytotoxicity of the DOX-loaded PA/SDM nanoparticles was evaluated using a breast-tumor cell line (MCF-7) to test the feasibility of the nanoparticles in targeting acidic tumor extracellular pH [148]. DOX-PA/SDM nanoparticles at pH 6.8 showed cytotoxicity similar to free DOX but higher cytotoxicity than the particles at pH 7.4. This pronounced cytotoxicity of the nanoparticles at low pH was partially attributed to the accelerated release of DOX triggered by pH changes. Another contributing factor, on the basis of the confocal laser microscopy characterization, was that at pH 6.8 and 6.4, SDM deionization caused the nanoparticle surfaces to become hydrophobic and thus aggressively bind to the cell membranes [148]. The cancer targeting and cellular uptake were promoted by introducing targeting groups such as vitamin H (biotin) [149].

Although the drug release from such pH-responsive nanoparticles triggered by the cancer pH_e can substantially reduce the systemic toxicity and enhance in vitro and in vivo anticancer activity, it does not solve the drug resistance problem. This is because cancer cells can lose the drug receptors to slow down drug uptake and

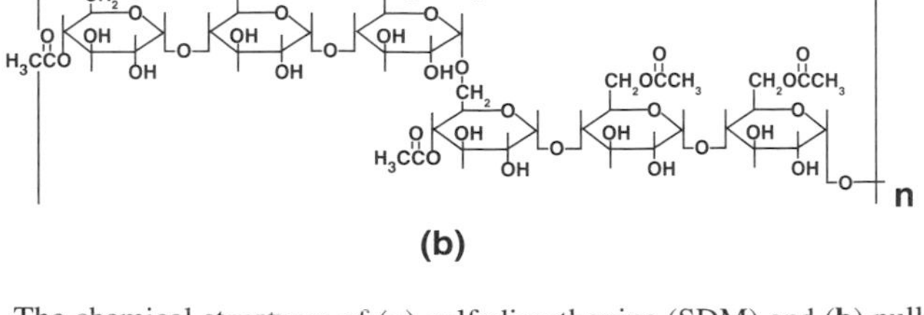

(a)

(b)

Scheme 10.1 The chemical structures of (**a**) sulfadimethoxine (SDM) and (**b**) pullulan acetate (PA). (modified from [146], with permission from Springer)

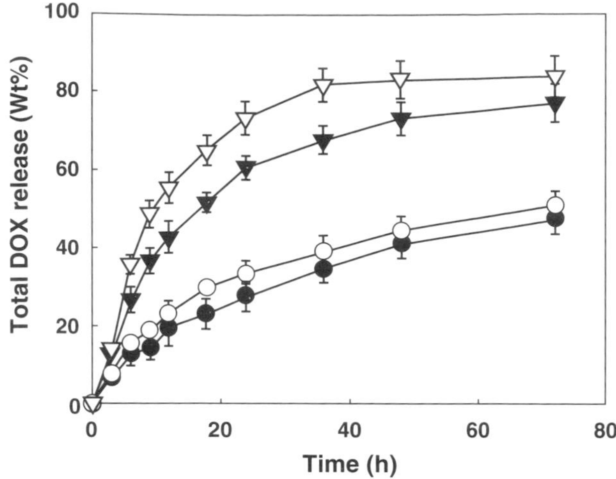

Fig. 10.8 DOX release from PA/SDM nanoparticles at pH 8.0 (•), 7.4 (○), 6.8 (▼), and 6.4 (∇). (The number of SDM groups per anhydroglucose unit of the pullulan acetate was 0.39.) (modified and reproduced from [146], with permission from Springer)

overexpress ATP Binding Cassette (ABC) transporters in their plasma membranes to efficiently transport cancer drugs out of the cells as they are entering the cell membranes [6, 10, 12, 13, 26–28], and thus reduce the cellular drug accumulation [29]. Therefore, the drug released in the extracellular fluid is difficult to reach the cytoplasm.

4 Tumor Extracellular pH-Responsive Nanoparticles for Cellular Targeting and Internalization

Figure 10.1 shows that cancer cells inhibit the drug uptake by losing the receptors/transporters and activating P-gp pumps that export the drugs associated with the plasma membrane. Hence, drugs released directly into the cytoplasm would circumvent this predicament [125, 150, 151]. Such an intracellular drug release requires the nanoparticles not only to preferentially accumulate in the cancer tissues but also to be fast internalized by the cells. However, the PEG chains that are needed for long circulation times, unfortunately, also slow down the nanoparticle cellular uptake as a result of steric repulsion [152, 153].

A common approach to increase the cellular uptake (internalization) is through receptor-mediated endocytosis by decorating nanoparticles with ligands that bind the receptors on tumor cells. The main challenge in this approach is to make these ligands selective for tumor cells and avoid healthy cells that may express the same receptors. To address this concern, Bae and coworkers designed a pH-sensitive multifunctional nanoparticle with a ligand that can bind its receptors only if exposed to an acidic tumor extracellular environment ($6.5 < pH < 7.0$) (Fig. 10.9) [154]. Specifically, the ligand biotin was anchored to the surface of the nanoparticles made from the mixed block copolymers PHis-PEG and PLLA-*b*-PEG-*b*-PHis-biotin. At neutral pH, the PHis was insoluble and collapsed on the hydrophobic core surface. Thus, the biotin was buried in the PEG corona and hence not available for binding. However, once the nanoparticle was exposed to the acidic extracellular fluid of solid tumors, the PHis became soluble, stretched, and exposed biotin for binding to the tumor receptors. The availability of the anchored biotin at low pH was confirmed using the avidin, a tetrameric protein with four biotin binding sites. The fast cellular uptake of the nanoparticles at acidic pH was confirmed by flow cytometry [154].

Another approach is to use polycation complexes, which can be electrostatically attracted and adsorbed to the negatively charged cell membranes and lead to electrostatically adsorptive endocytosis [155, 156]. For example, cationized polymer conjugates [157], proteins [158, 159], and nanoparticles [96, 160, 161] exhibit enhanced cellular internalizations when compared with their noncationized counterparts. Cationic charges, however, can cause severe serum inhibition and rapid clearance from the plasma compartment [162–164]; for instance, a cationized antibody had a 58-fold increase in the systemic clearance from the plasma compartment

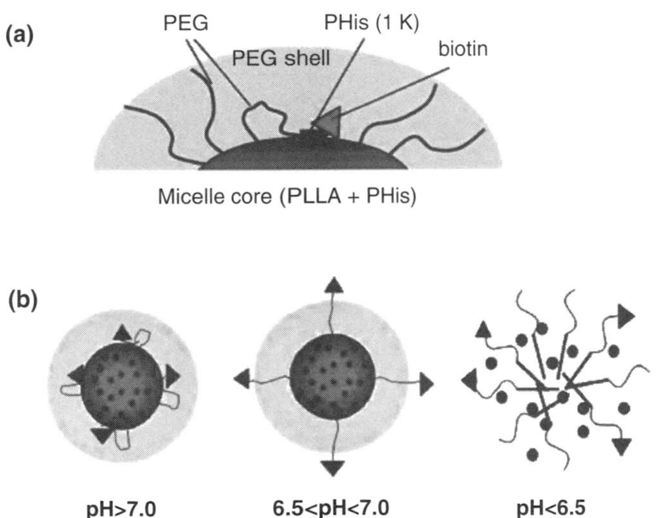

Fig. 10.9 pH-dependent biotin availability for binding. (**a**) Above pH 7.0, biotin anchored on the micelle core via pH-sensitive PHis is shielded by PEG shell of the micelle. (**b**) Biotin is exposed on the micelle surface at acidic conditions ($6.5 < pH < 7.0$) and can interact with cells, which facilitates biotin-receptor-mediated endocytosis. When the pH is further lowered ($pH < 6.5$), the micelle destabilizes, which enhances drug release. (modified and reproduced from [154], with permission from American Chemical Society)

and a 9-fold reduction in the mean residence time when compared with the native antibody [159].

To solve this problem, Shen and coworkers developed a three-layered pH-responsive nanoparticle (3LNP) that had a neutral corona at the physiological pH, but the corona became positively charged at the tumor extracellular pH, which resulted in rapid cellular internalization (Scheme 10.2) [165, 166].

Those nanoparticles (3LNPs) were fabricated via a pH-controlled hierarchical self-assembly of a tercopolymer brush (Schemes 10.2 and 10.3), which contained hydrophilic polycaprolactone (PCL) chains, water-soluble PEG chains, and pH-responsive poly[2-(*N,N*-diethylamino)ethyl methacrylate] (PDEA) chains. PDEA is a polybase that is soluble at low pH but insoluble at neutral pH [167–169]. The brush polymer was initially dispersed in a pH 5.0 solution where the PDEA chains were protonated and hence water-soluble. The hydrophobic PCL chains and drug molecules associated to form the hydrophobic core. The PEG and protonated PDEA chains formed a hydrophilic corona surrounding the core. After the solution pH was raised to 7.4, the PDEA chains were deprotonated and became hydrophobic, collapsing on the PCL core as a hydrophobic middle layer with only the PEG chains forming the hydrophilic corona (Scheme 10.3).

The 3LNPs had a pH-dependent zeta potential, shown in Fig. 10.10. They were neutral at pH 7.4 but gradually became positively charged at lower pH, indicating

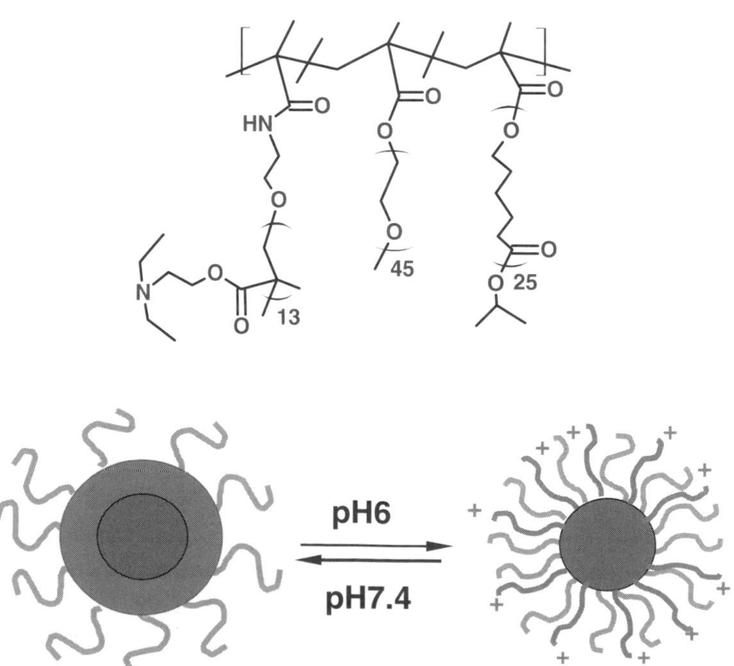

Scheme 10.2 The pH-responsive three-layered nanoparticles (3LNPs) [165, 166]

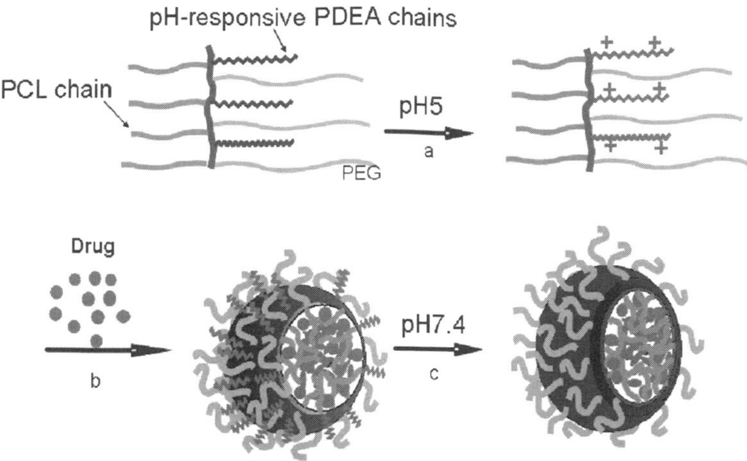

Scheme 10.3 Fabrication of drug-loaded pH-responsive three-layered onion-structured nanoparticles (3LNPs) via pH-controlled hierarchical self-assembly

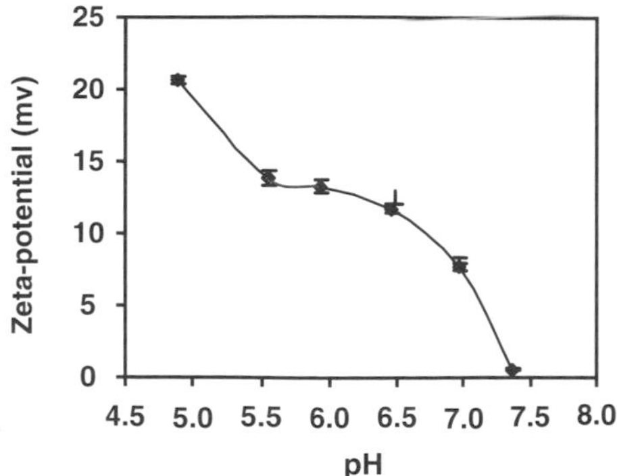

Fig. 10.10 Zeta potential of three-layered nanoparticles (3LNPs)

that as the solution pH decreased, the PDEA chains became positively charged and soluble. The effectiveness of the charged-PDEA-induced adsorption on the cell membrane and subsequent cellular uptake of the 3LNPs were estimated using confocal microscopy. A hydrophobic dye PKH26 was loaded in 3LNPs as the tracer. The cells were cultured with the PKH26/3LNPs at 4°C for 30 min at pH 7.4 and 6.0 respectively. The cells cultured with PKH26/3LNPs at pH 6.0 had bright red cell membranes, while those cultured at pH 7.4 were hardly visible (Fig. 10.11). Thus, we concluded that the 3LNPs were better attached to the cells at the low pH because of their positive charges.

The cellular uptake of these 3LNPs was observed using confocal scanning laser fluorescence microscopy. SKOV-3 ovarian cancer cells were cultured with

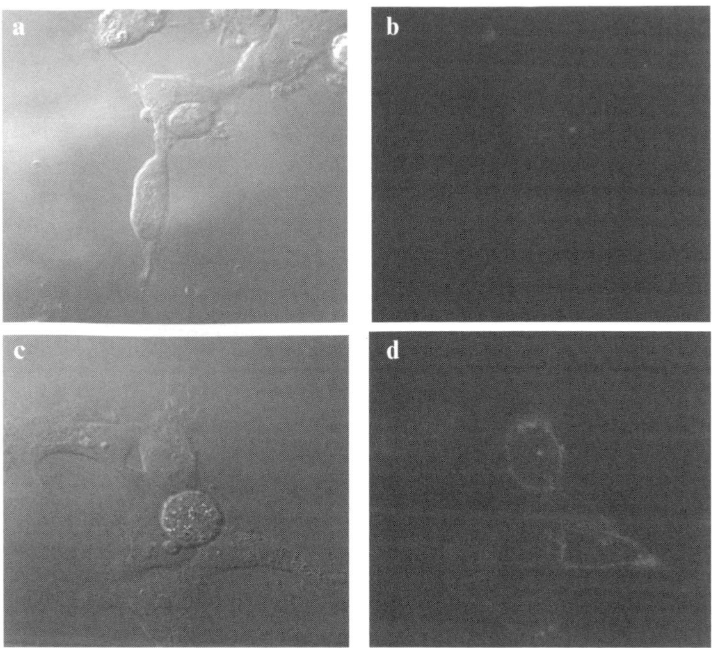

Fig. 10.11 Adsorption of 3LNPs with PKH26 on SKOV-3 cell membrane at pH 7.4 (**a**, **b**) and 6.0 (**c**, **d**) at 4°C observed with confocal scanning laser fluorescence microscopy. Differential interference contrast (a, c) and red fluorescence channel (b, d)

PKH26/3LNPs at pH 7.4 or 6.0 for 2.5 h. The cells cultured at pH 6.0 were found to have much more PKH26/3LNPs compared with those cultured at pH 7.4. The internalized 3LNPs were localized in lysosomes. The cellular internalization of PKH26/3LNPs was also quantitatively measured by flow cytometry in percent of the PKH26-positive cells (Fig. 10.12). Consistent with the confocal microscopy results, a significantly higher fraction of the cells took up PKH26/3LNPs at pH 6.0 when compared with that at pH 7.4. Thus, we concluded that, once localized in the acidic extracellular fluid of solid tumors, the 3LNPs would become positively charged and be efficiently internalized.

5 Lysosomal pH-Responsive Nanoparticles for Cytoplasmic Drug Delivery

Cytoplasmic drug release using liposomes [170–172], polymer-drug conjugates [173–176], and micellar nanoparticles [99, 100, 177–180] have been shown to bypass the P-gp-based multidrug resistance, probably because cells take up drugs in the carriers via endocytosis rather than diffusion through the cell membrane, where P-gp is located.

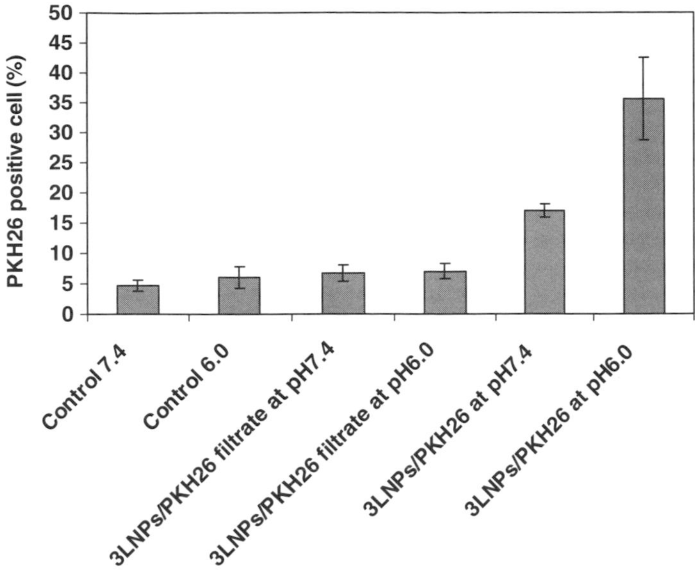

Fig. 10.12 PKH26-positive cells determined by flow cytometry of centrifuged PKH26 and nanoparticles encapsulating PKH26 at pH 7.4 and 6.0

As the internalized macromolecules [181–185] or nanoparticles [88, 186–188] progress through the endocytic pathway, they encounter compartments, namely, early endosomes, late endosomes, and lysosomes, of progressively increasing acidity. The pH of early endosomes is typically near 6 [184, 185, 189–193], with some at as low as 5.4 [192, 194, 195], that of late endosomes is near 5 [183, 193], and that of lysosomes is about 4–5. [189, 196, 197] Endosome acidification occurs rapidly after internalization, usually within minutes [189]. Drug-resistant cells often exhibit an altered pH gradient across different cell compartments to increase the drug-sequestering capacity of the compartments, particularly more acidic recycling endosomes, lysosomes, Golgi, and mitochondria, but more basic cytosol and nucleus, all aimed at resisting chemotherapeutic drugs [16, 198–201]. Thus, weakly basic anthracyclines and vinca alkaloids are sequestered away from the cytosol and nucleus into acidic cytoplasmic organelles, mainly in lysosomes [199, 202–206], and may be further extruded into the external medium [203, 207]. On the other hand, conventional nanoparticles, such as PCL-b-PEG nanoparticles, mainly stay in these acidic cytoplasmic organelles [208] and thus release their payload there. The released basic anticancer drugs such as DOX are then protonated and hence not able to escape from these compartments. Therefore, ideal nanoparticles should release the drug directly into the cytosol.

Bae and coworkers used PHis-PEG (75 wt.%)/PLLA-PEG-folate (25 wt.%) to make pH-sensitive micelles with folate-targeting groups (PHSM/f). The in vitro and in vivo anticancer activities of DOX-loaded PHSM/f were evaluated using

MCF-7 cells and their xenograph tumors, and compared with DOX-loaded pH-insensitive micelles made of PLLA-PEG with folate targeting groups (PHIM/f) [99]. The cellular localization of the nanoparticles was confirmed by confocal microscopy (Fig. 10.13). DOX delivered by PHSM/f was found uniformly distributed in the cytosol as well as in the nucleus, while DOX/PHIM/f was entrapped in endosome and multivesicular bodies. It was thus hypothesized that PHis, which is known to have an endosomal membrane-disruption activity induced by a "proton sponge" mechanism of its imidazole groups [209, 210], disrupted the compartment membrane and released DOX into the cytosol. As a result, DOX/PHSM/f showed much higher in vitro and in vivo anticancer activities toward DOX-resistant cells (Fig. 10.14).

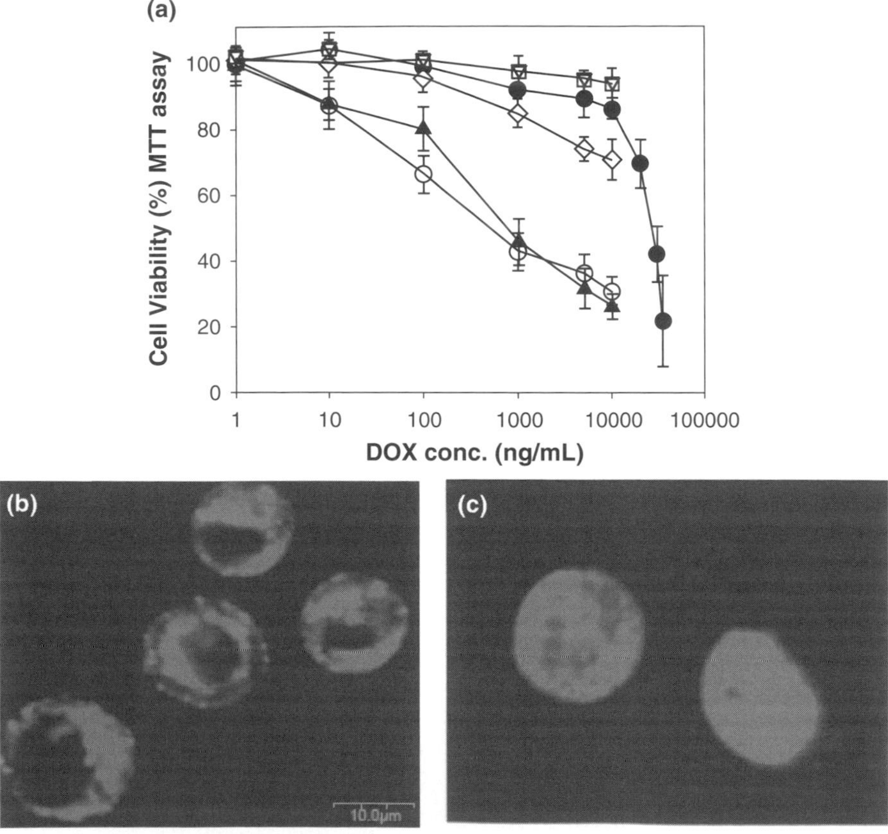

Fig. 10.13 **a** The cytotoxicity of DOX-loaded PHSM (□), PHSM/f (▲), PHIM (∇), PHIM/f (◊), and free DOX (•) against MCF-7/DOX[R] cells and free DOX against sensitive MCF-7 cells at pH 6.8 after a 48-h incubation (n = 7). Confocal microscopy images of MCF-7/DOX[R] cells treated with **b** DOX-loaded PHIM/f micelles at pH 6.8 and **c** DOC-loaded PHSM/f micelles at pH 6.8. The cells were incubated with the particles for 30 min. (reproduced from [99], with permission from Elsevier)

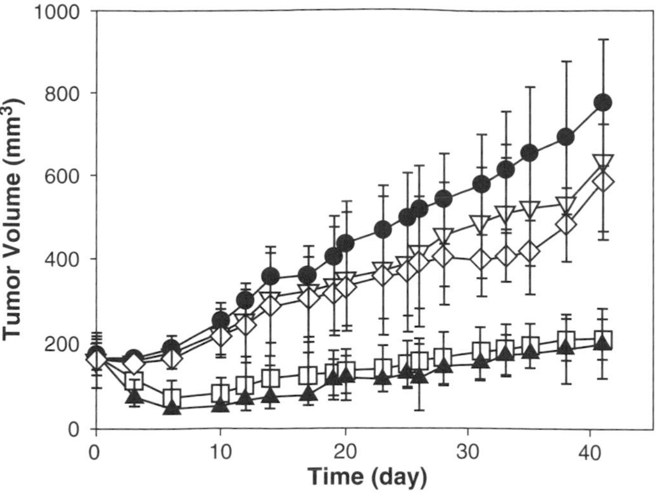

Fig. 10.14 Tumor growth inhibition of s.c. human breast MCF/DOX-resistant carcinoma xenografts in BLAB/c nude mice. Mice were injected with 10 mg/kg DOX equivalent dose of DOX-loaded PHSM (□), PHSM/f (▲), PHIM (∇), PHIM/f (◊), and DOX (●).Values are means ± SD. (reproduced from [99], with permission from Elsevier)

Amiji and coworkers used biodegradable and pH-sensitive poly(β-amino ester)s (PbAE) to make pH-responsive nanoparticles for tumor-targeted paclitaxel delivery [115, 211, 212]. PbAE is a biodegradable cationic polymer originally developed for gene delivery systems [213–219]. PbAE with various structures can be synthesized by condensation polymerization of diacrylates and amines [213, 218]. PbAE nanoparticles were prepared by solvent displacement method in the presence of an amphiphilic triblock copolymer containing poly(ethylene oxide) (PEO), poly(propylene oxide) (PPO), and Pluronic F-108. The nanoparticles were about 100–200 nm in diameter with positive surface charges (zeta potential of about +40 mV), and dissolved when the pH of the medium was less than 6.5. In vivo biodistribution data for these nanoparticles showed that despite their high surface positive charges [115], they had a prolonged circulating time, as evidenced by increased half-lives [212]. These nanoparticles efficiently delivered paclitaxel to the tumor sites via passive accumulation through the EPR effect, even more efficiently than the PEO-modified PCL nanoparticles (Figs. 10.15 and 10.16). It was assumed that the PbAE nanoparticles rapidly dissolved within the tumor low-pH environment.

We proposed PDEA-PEG micelles as lysosomal-pH-responsive fast drug-release nanoparticles for cytoplasmic drug delivery, as illustrated in Fig. 10.17 [97]. PDEA is soluble at acidic pH but insoluble at neutral pH [167–169]. Its block copolymer with PEG (PDEA-PEG) forms pH-responsive nanoparticles [168]. Once

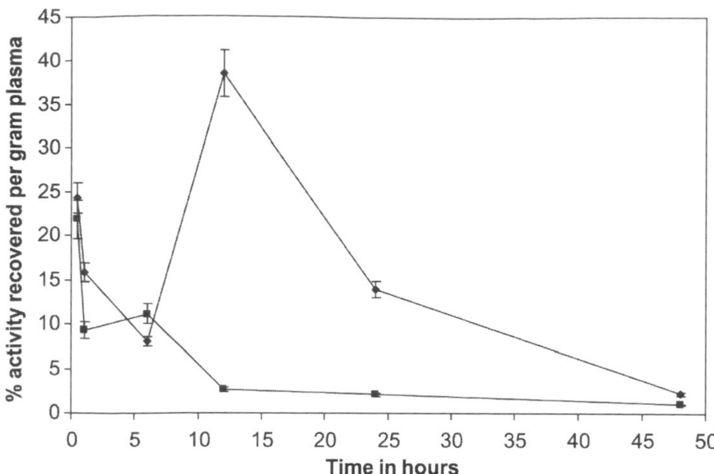

Fig. 10.15 Percentage of plasma activity as a function of time for indium-oxine-labeled poly(ethylene oxide)-modified poly(caprolactone) (PCL) nanoparticles (■) and poly(ethylene oxide)-modified poly(β-amino ester) (PbAE) nanoparticles(♦) in mice. (reproduced from [212], with permission from Springer)

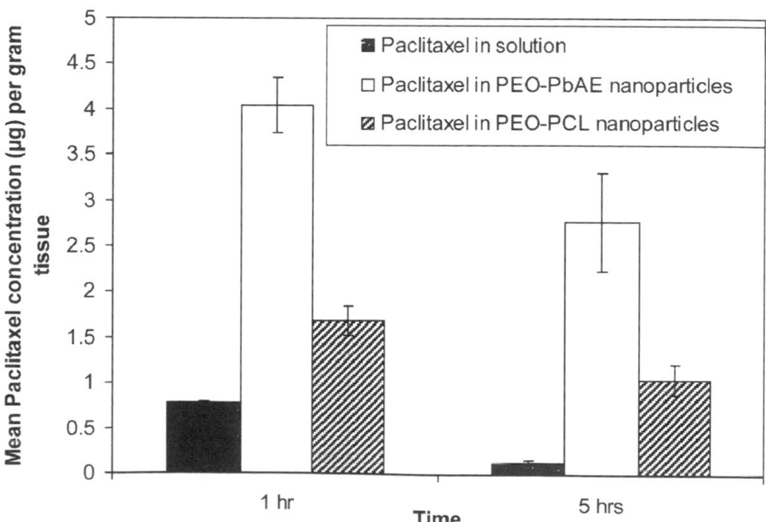

Fig. 10.16 Tumor paclitaxel concentrations upon intravenous administration to SKOV-3 human ovarian adenocarcinoma bearing nude mice. Paclitaxel was administered intravenously in aqueous solution, poly(ethylene oxide)-modified polycaprolactone (PCL) nanoparticles, or poly(ethylene oxide)-modified poly(β-amino ester) (PbAE) nanoparticles. (reproduced from [212], with permission from Springer)

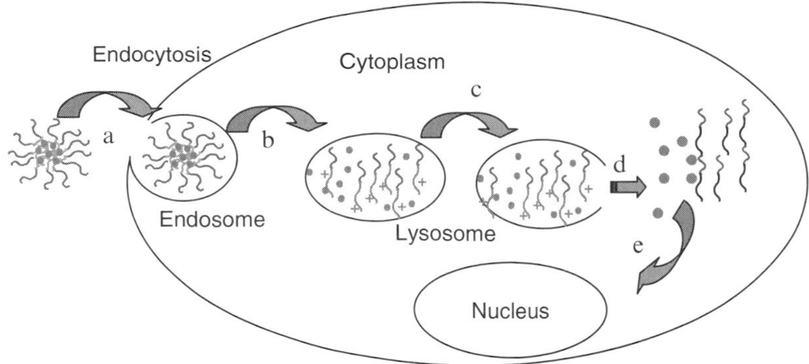

Fig. 10.17 Cytoplasmic drug delivery using lysosomal-pH-responsive fast-release nanoparticles: **a** the nanoparticle is internalized by endocytosis; **b** transferred to a lysosome; **c** the PDEA core is protonated at the lysosomal pH (4–5) and the nanoparticle dissolves, releasing the drug into the lysosome; **d** the continuous PDEA (poly[2-(*N,N*-diethylamino)ethyl methacrylate]) protonation causes an osmotic imbalance across the lysosome membrane, which finally ruptures the lysosome and hence releases the drug into cytoplasm

Scheme 10.4 Synthesis of the PDEA-PEG block copolymer

internalized by endocytosis and transferred into lysosomes, the nanoparticles would rapidly dissolve through protonation of their amine groups at the acidic lysosomal pH (<5.5). Continuous protonation of the amines would lead to an influx of electrolytes, and hence to osmotic swelling and lysosome rupture, reminiscent of that in gene delivery using polybases [220–223], and finally, to a rapid drug release into the cytoplasm. Such a rapid lysosomal release prevents drug degradation and makes all drug molecules available to surpass the intracellular drug resistance capacity.

The PDEA-PEG block copolymer was synthesized by a typical atom-transfer radical polymerization (ATRP) using PEG as the macroinitiator (Scheme 10.4). This block copolymer formed micelles with diameters less than 100 nm, which is ideal for endocytosis. The dissolution of the micelles at low pH was monitored using *N*-phenyl-2-naphthylamine (PNA) as a florescent probe which has a low fluorescent activity in hydrophilic environments. The PDEA-PEG micelle was loaded with PNA. Its fluorescence was monitored as a function of the solution pH. Figure 10.18 suggests that the micelles dissolved at a pH about 6.0, which implies that the nanoparticles can

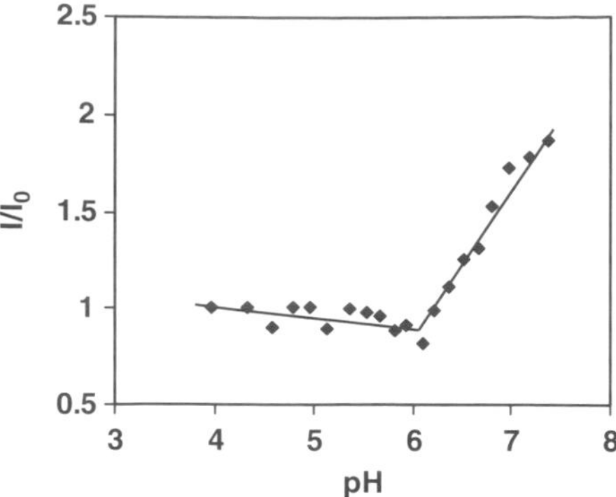

Fig. 10.18 The fluorescence intensity as a function of the solution pH for PNA-loaded PDEA-PEG micelles (relative to the fluorescence intensity at pH 4) [97]

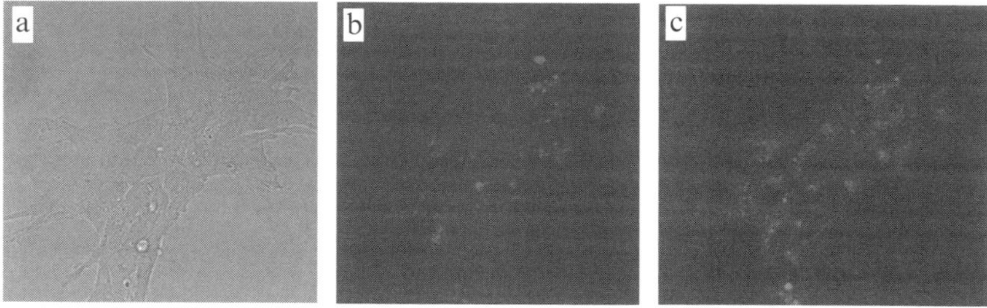

Fig. 10.19 Differential interference contrast (**a**) and confocal fluorescence scanning microscopy images of the cells obtained from the red channel (wavelength 550–620 nm) (**b**) and green channel (510–540 nm) (**c**). Cells were cultured with PHK-26-loaded PDEA-PEG nanoparticles for 90 min and then lysotracker [97]

dissolve in late lysosomes. The localization of the nanoparticles in lysosomes was observed using confocal microscopy. Figure 10.19 shows that the red fluorescent nanoparticles coincide with lysosomes marked with green fluorescence.

The cytotoxicity of the cisplatin-loaded PDEA-PEG nanoparticles was compared with the free cisplatin and cisplatin-loaded pH-insensitive PCL-PEG nanoparticles. Clearly, the cisplatin/PDEA-PEG had a significantly higher cytotoxicity than did the other two treatments (Fig. 10.20). Since PDEA-PEG micelles carried some positive charges (zeta potential ~ 9 mV) due to the protonation of the PDEA amines on the hydrophobic core surface, the effects of the surface charges on the micelles were also tested using PCL-block-poly[2-(*N,N*-dimethylamino) ethyl methacrylate] (PCL-PDMA) micelles. PDMA is water-soluble and carries cationic charges due to the protonation of its tertiary amines. The PCL-PDMA micelles had a zeta potential of +21 mV, much higher than those of the PDEA-PEG micelles. Figure 10.20 shows that cisplatin in these highly positive-charged PCL-PDMA nanoparticles had an inhibition efficiency of about 58%, still lower than cisplatin/PDEA-PEG

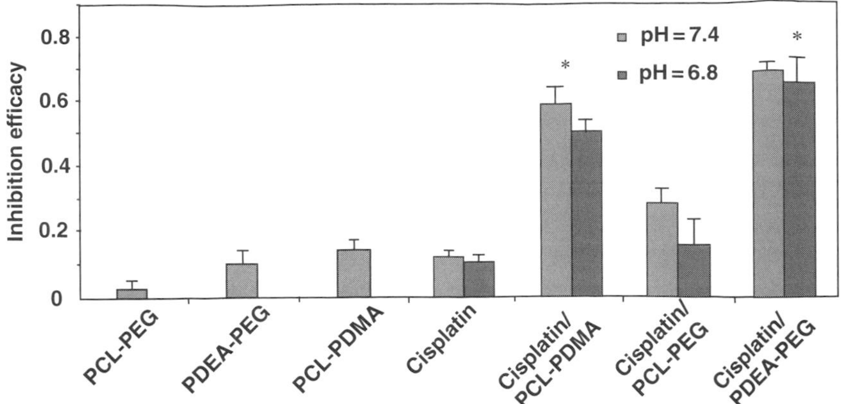

Fig. 10.20 Cytotoxicity of free and nanoparticle-encapsulated cisplatin to SKOV-3 adenocarcinoma cancer cells (2-h treatment) estimated with MTT Cell Proliferation Assay. Cisplatin dose, 0.25 μg/ml. Data represent mean ± SE ($n = 3$, $P < 0.05$) [97]

nanoparticles. This analysis indicates that cisplatin in the fast-release PDEA-PEG nanoparticles indeed had higher cytotoxicity than did that in the slow-release PCL-PEG nanoparticles. Different from the PHis-based nanoparticles, which showed higher cytotoxicity at lower pH (6.8) due to the particle destabilization at this pH, the drug in PDEA-PEG nanoparticles had slightly lower cytotoxicity at pH 6.8. This is because the PDEA-PEG nanoparticles were still stable at this pH and hence released little of cisplatin, while the cells were under stress at this low pH and only slowly took up the nanoparticles.

The in vivo antitumor activity of cisplatin encapsulated in the fast-release nanoparticles was tested using nude mice xenografted with intraperitoneal ovarian tumors simulating the advanced metastasis state of the disease in women. The morphometric analyses of histological sections of the mesentery/intestine tumors 6 h after the treatment were conducted to infer the acute effects of the treatment (Fig. 10.21). The tumors in the mice injected with PDEA-PEG nanoparticles were not different from those in the control group of mice injected with phosphate-buffered saline (PBS). The mice injected with PDEA-PEG had abundant blood vessels (labeled by an asterisk in the *right upper panel*) and very few pyknotic cells displaying highly condensed pyknotic nuclei, the most characteristic feature of apoptosis, which are considered to be dead cells. Therefore, PDEA-PEG nanoparticles alone had no in vivo anticancer activity even though they had some in vitro cytotoxicity. The tumors treated with free cisplatin or cisplatin/PDEA-PEG had significantly more pyknotic cells and fewer blood vessels, which is consistent with the suppressing neovascularization effect of cisplatin [224]. The tumors treated with cisplatin/PDEA-PEG had about twice as many pyknotic cells as those treated with free cisplatin. This suggests that cisplatin/PDEA-PEG is more efficient in inducing apoptosis, which is consistent with the in vitro results. The numbers of ovarian tumors were significantly reduced in mice treated twice (Fig. 10.22), but the most

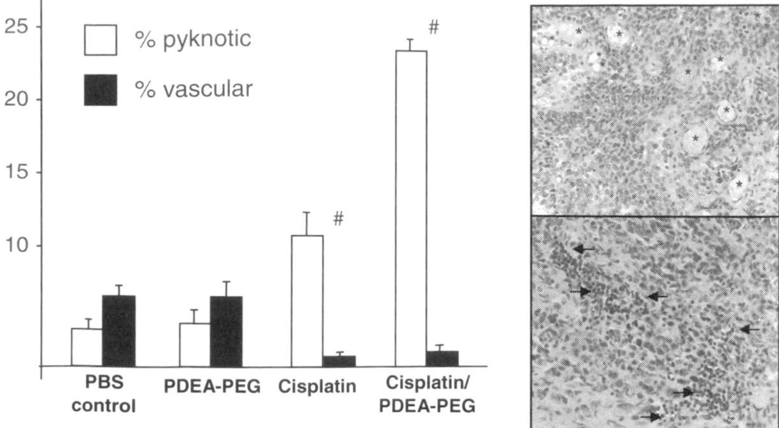

Fig. 10.21 Acute morphological responses of intraperitoneal ovarian tumors to PDEA-PEG, free cisplatin, and cisplatin/PDEA-PEG (6h after the injection). Representative histological sections are for tumor tissues from a vehicle control (*upper panel*: *asterisks* are placed in lumens of blood vessels) and from a cisplatin/PDEA-PEG-treated animal (*lower panel*: *arrows* indicate pyknotic/condensed cells). Data represent mean ± SE ($n = 4$, $P < 0.05$) [97]

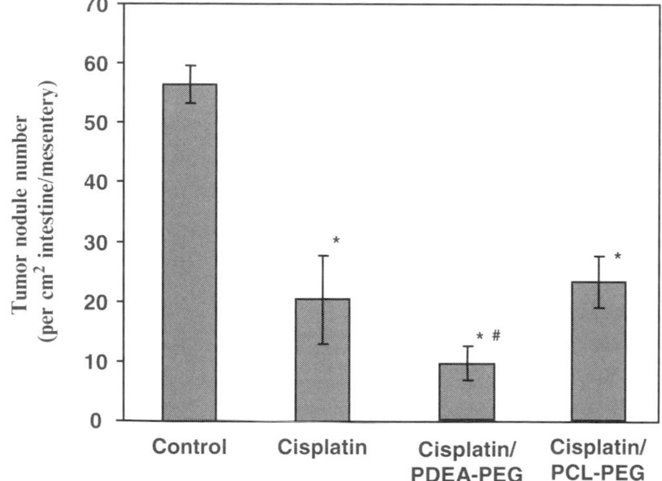

Fig. 10.22 The numbers of tumors on intestine/mesentery (per centimeter squared) of the nude mice. Cisplatin dose was 10mg/kg/treatment. Treated twice at the forth and fifth weeks after inoculation of SKOV-3 cells. Data represent mean ± SE ($n = 4$, $^*P < 0.05$ when compared with control group; $^\#P < 0.05$ when compared with free cisplatin and Cis/PCL-PEG) [97]

marked response occurred in the cisplatin/PDEA-PEG-treated group. There was almost no difference between the cisplatin/PCL-PEG (the slow-release nanoparticles) and cisplatin alone even though cisplatin/PCL-PEG nanoparticles showed in vitro cytotoxicity. To our knowledge, this was the first report demonstrating a substantial benefit of rapid cisplatin-release-nanoparticles for intraperitoneal ovarian cancer therapy.

6 Conclusion

Taking advantages of the altered pH gradients in tumor extracellular environments and in its intracellular compartments, pH-responsive nanoparticles have been designed to prove the concepts of specific cancer cell targeting, enhanced cellular internalization, and rapid drug release. Especially promising is the concept of intracellular drug delivery by the pH-responsive nanoparticles because it offers an efficient means of overcoming the multidrug resistance, one of the major causes for cancer treatment failures. Future directions are likely to be focused on multifunctional pH-responsive nanoparticles which combine diagnosis and treatments.

Acknowledgment The authors acknowledge the financial support from American Cancer Society (RSG-06–118–01-CDD), National Science Foundation (BES-0401982), and National Institutes of Health (NIH RR-016474).

References

1. American Cancer Society. Cancer facts and figures, 2006. Am Cancer Soc, 2006.
2. Rosenberg, B.; VanCamp, L.; Trosko, J. E.; Mansour, V. H. Platinum compounds: a new class of potent antitumor agents. Nature **1969**, 222, 385–386.
3. Takahara, P. M.; Rosenzweig, A. C.; Frederick, C. A.; Lippard, S. J. Crystal structure of double-stranded DNA containing the major adduct of the anticancer drug cisplatin. Nat. Rev. Drug Discov. **1995**, 377, 649–652.
4. Jamieson, E. R.; Lippard, S. J. Structure, recognition, and processing of cisplatin-DNA adducts. Chem. Rev. **1999**, 99, 2467–2498.
5. Pinzani, V.; Bressolle, F.; Hang, I. J.; Galtier, M.; Blayac, J. P.; Balmes, P. Cisplatin-induced renal toxicity and toxicity-modulating strategies: a review. Cancer Chemother. Pharmacol. **1994**, 35, 1–9.
6. Balch, C.; Huang, T. H.-M.; Brown, R.; Nephew, K. P. The epigenetics of ovarian cancer drug resistance and resensitization. Am. J. Obstet. Gynecol. **2004**, 191, 1552–1572.
7. American Cancer Society Inc. Cancer facts and figures, 2002. www.cancer.org.
8. Hodge, J. W.; Tsang, K.-Y.; Poole, D. J.; Schlom, J. Vaccine strategies for the therapy of ovarian cancer. Gynecol. Oncol. **2003**, 88, S97–S104.
9. Greenlee, R. T.; Hill-Harmon, M. B.; Murray, T.; Thun, M. Cancer statistics, 2001. CA Cancer J. Clin. **2001**, 51, 15–36.
10. Agarwal, R.; Kaye, S. B. Ovarian cancer: strategies for overcoming resistance to chemotherapy. Nat. Rev. Cancer **2003**, 3, 502–516.
11. Pastan, I.; Gottesman, M. M. Multidrug resistance. Annu. Rev. Med. **1991**, 42, 277–286.
12. Gottesman, M. M. Mechanisms of cancer drug resistance. Annu. Rev. Med. **2002**, 53, 615–627.
13. Vasey, P. A. Resistance to chemotherapy in advanced ovarian cancer: mechanisms and current strategies. Br. J. Cancer **2003**, 89, S23–S28.

14. Wang, G.; Reed, E.; Li, Q. Q. Molecular basis of cellular response to cisplatin chemotherapy in non-small cell lung cancer [Review]. Oncol. Rep. **2004**, 12, 955–965.
15. Michael, M.; Doherty, M. M. Tumoral drug metabolism: overview and its implications for cancer therapy J. Clin. Oncol. **2005**, 23, 205–229.
16. Duvvuri, M.; Krise, J. P. Intracellular drug sequestration events associated with the emergence of multidrug resistance: a mechanistic review. Front. Biosci. **2005**, 10, 1499–1509.
17. George, J. A.; Chen, T.; Taylor, C. C. Src tyrosine kinase and multidrug resistance protein-1 inhibitions act independently but cooperatively to restore paclitaxel sensitivity to paclitaxel-resistant ovarian cancer cells. Cancer Res. **2005**, 65, 10381–10388.
18. Perry III, W. L.; Shepard, R. L.; Sampath, J.; Yaden, B.; Chin, W. W.; Iversen, P. W.; Jin, S.; Lesoon, A.; O'Brien, K. A.; Peek, V. L.; Rolfe, M.; Shyjan, A.; Tighe, M.; Williamson, M.; Krishnan, V.; Moore, R. E.; Dantzig, A. H. Human splicing factor SPF45 (RBM17) confers broad multidrug resistance to anticancer drugs when overexpressed – a phenotype partially reversed by selective estrogen receptor modulators. Cancer Res. **2005**, 65, 6593–6600.
19. Parker, R. J.; Eastman, A.; Bostick-Bruton, F.; Reed, E. Acquired cisplatin resistance in human ovarian cancer cells is associated with enhanced repair of cisplatin-DNA lesions and reduced drug accumulation. J. Clin. Invest. **1991**, 87, 772–777.
20. Zhang, C.; Marme, A.; Wenger, T.; Gutwein, P.; Edler, L.; Rittgen, W.; Debatin, K.-M.; Altevogt, P.; Mattern, J.; Herr, I. Glucocorticoid-mediated inhibition of chemotherapy in ovarian carcinomas. Int. J. Oncol. **2006**, 28, 551–558.
21. Macleod, K.; Mullen, P.; Sewell, J.; Rabiasz, G.; Lawrie, S.; Miller, E.; Smyth, J. F.; Langdon, S. P. Altered ErbB receptor signaling and gene expression in cisplatin-resistant ovarian cancer. Cancer Res. **2005**, 65, 6789–6800.
22. Bergom, C.; Gao, C.; Newman, P. J. Mechanisms of PECAM-1-mediated cytoprotection and implications for cancer cell survival. Leuk Lymphoma **2005**, 46, 1409–1421.
23. Soengas, M. S.; Lowe, S. W. Apoptosis and melanoma chemoresistance. Oncogene **2003**, 22, 3138–3151.
24. Yakirevich, E.; Sabo, E.; Naroditsky, I.; Sova, Y.; Lavie, O.; Resnick, M. B. Multidrug resistance-related phenotype and apoptosis-related protein expression in ovarian serous carcinomas. Gynecol. Oncol. **2006**, 100, 152–159.
25. Austin, D. L.; Ross, D. D. Multidrug resistance mediated by the breast cancer resistance protein BCRP (ABCG2). Oncogene **2003**, 22, 7340–7358.
26. Bogman, K.; Peyer, A. K.; Torok, M.; Kusters, E.; Drewe, J. HMG-CoA reductase inhibitors and P-glycoprotein modulation. Br. J. Pharmacol. **2001**, 132, 1183–1192.
27. Glavinas, H.; Krajcsi, P.; Cserepes, J.; Sarkadi, B. The role of ABC transporters in drug resistance, metabolism, and toxicity. Curr. Drug Deliv. **2004**, 1, 27–42.
28. Ross, D. D.; Doyle, L. A. Mining our ABCs: Pharmacogenomic approach for evaluating transporter function in cancer drug resistance. Cancer Cell **2004**, 6, 105–107.
29. Keizer, H. G.; Schuurhuis, G. J.; Broxterman, H. J.; Lankelma, J.; Schoonen, W. G.; van Rijn, J.; Pinedo, H. M.; Joenje, H. Correlation of multidrug resistance with decreased drug accumulation, altered subcellular drug distribution, and increased P-glycoprotein expression in cultured SW-1573 human lung tumor cells. Cancer Res. **1989**, 49, 2988–2993.
30. Monsky, W. L.; Fukumura, D.; Gohongi, T.; Ancukiewcz, M.; Weich, H. A.; Torchilin, V. P.; Jain, R. K. Augmentation of transvascular transport of macromolecules and nanoparticles in tumors using vascular endothelial growth factor. Cancer Res. **1999**, 59, 4129–4135.
31. Maeda, H. The enhanced permeability and retention (EPR) effect in tumor vasculature: the key role of tumor-selective macromolecular drug targeting. Adv. Enzyme Regul. **2001**, 41, 189–207.
32. Maeda, H.; Seymour, L. W.; Miyamoto, Y. Conjugates of anticancer agents and polymers: advantages of macromolecular therapeutics in vivo. Bioconjug Chem. **1992**, 3, 351–362.
33. Jain, R. K. Delivery of molecular medicine to solid tumors: lessons from in vivo imaging of gene expression and function. J. Control. Release **2001**, 74, 7–25.
34. Hobbs, S. K.; Monsky, W. L.; Yuan, F.; Roberts, W. G.; Griffith, L.; Torchilin, V. P.; Jain, R. K. Regulation of transport pathways in tumor vessels: role of tumor type and microenvironment. Proc. Natl. Acad. Sci. USA **1998**, 95, 4607–4612.

35. Yuan, F.; Dellian, M.; Fukumura, D.; Leuning, M.; Berk, D. D.; Torchilin, V. P.; Jain, R. K. Vascular permeability in a human tumor xenograft: molecular size dependence and cutoff size. Cancer Res. **1995**, 55, 3752–3756.
36. Unezaki, S.; Maruyama, K.; Hosoda, J.-I.; Nagae, I.; Koyanagi, Y.; Nakata, M.; Ishida, O.; Iwatsuru, M.; Tsuchiya, S. Direct measurement of the extravasation of poly(ethylene glycol)-coated liposomes into solid tumor tissue by in vivo fluorescence microscopy. Int. J. Pharm. **1996**, 144, 11–17.
37. Yuan, F.; Lwunig, M.; Huang, S. K.; Berk, D. A.; Papahadjopoulos, D.; Jain, R. K. Microvascular permeability and interstitial penetration of sterically stabilized (stealth) liposomes in a human tumour xenograft. Cancer Res. **1994**, 54, 3352–3356.
38. Duncan, R. The dawning era of polymer therapeutics. Nat. Rev. **2003**, 2, 347–360.
39. Satchi-Fainaro, R.; Duncan, R.; Barnes, C. M. Polymer therapeutics for cancer: current status and future challenges. Adv. Polym. Sci. **2006**, 193, 1–65.
40. Thatte, S.; Datar, K.; Ottenbrite, R. M. Perspectives on: polymeric drugs and drug delivery systems. J. Bioact. Compat. Polym. **2005**, 20, 585–601.
41. Qiu, L. Y.; Bae, Y. H. Polymer architecture and drug delivery. Pharm. Res. **2006**, 23, 1–30.
42. Majoros, I. J.; Myc, A.; Thomas, T.; Mehta, C. B.; Baker, J. R. Jr. PAMAMdendrimer-based multifunctional conjugate for cancer therapy: synthesis, characterization, and functionality. Biomacromolecules **2006**, 7, 572–579.
43. Patri, A. K.; Kukowska-Latallo, J. F.; Baker, J. R. Targeted drug delivery with dendrimers: comparison of the release kinetics of covalently conjugated drug and non-covalent drug inclusion complex. Adv. Drug Deliv. Rev. **2005**, 57, 2203–2214.
44. Ambade, A. V.; Savariar, E. N.; Thayumanavan, S. Dendrimeric micelles for controlled drug release and targeted delivery. Mol. Pharm. **2005**, 2, 264–272.
45. Torchilin, V. P. Recent advances with liposomes as pharmaceutical carriers. Nature Rev. Drug Discov. **2005**, 4, 145–160.
46. Sapra, P.; Tyagi, P.; Allen, T. M. Ligand-targeted liposomes for cancer treatment. Curr. Drug Deliv. **2005**, 2, 369–381.
47. Torchilin, V. P. Targeted polymeric micelles for delivery of poorly soluble drugs. Cell. Mol. Life Sci. **2004**, 61, 2549–2559.
48. Kwon, G. S. Polymeric micelles for delivery of poorly water-soluble compounds. Crit. Rev. Ther. Drug Carrier Syst. **2003**, 20, 357–403.
49. Kataoka, K. H. A.; Nagasaki, Y. Block copolymer micelles for drug delivery: design, characterization and biological significance. Adv. Drug Deliv. Rev. **2001**, 47, 113–131.
50. Torchilin, V. P. Structure and design of polymeric surfactant-based drug delivery systems. J. Control. Release **2001**, 73, 137–172.
51. Labhasetwar, V.; Song, C.; Levy, R. J. Nanoparticle drug delivery system for restenosis. Adv. Drug Deliv. Rev. **1997**, 24, 63–85.
52. Brigger, I.; Dubernet, C.; Couvreur, P. Nanoparticles in cancer therapy and diagnosis. Adv. Drug Deliv. Rev. **2002**, 54, 631–651.
53. Panyam, J.; Labhasetwar, V. Biodegradable nanoparticles for drug and gene delivery to cells and tissue. Adv. Drug Deliv. Rev. **2003**, 55, 329–347.
54. Kim, Y.; Dalhaimer, P.; Christian, D. A.; Discher, D. E. Polymeric worm micelles as nanocarriers for drug delivery. Nanotechnology **2005**, 16, 484–491.
55. Seymour, L. W. Passive tumor targeting of soluble macromolecules and drug conjugates. Crit. Rev. Ther. Drug Carrier Syst. **1992**, 9, 135–187.
56. Lu, Y.; Low, P. S. Folate-mediated delivery of macromolecular anticancer therapeutic agents. Adv. Drug Deliv. Rev. **2002**, 54 675–693.
57. Lu, Y.; Low, P. S. Immunotherapy of folate receptor-expressing tumors: review of recent advances and future prospects. J. Control. Release **2003**, 91 17–29.
58. Gosselin, M. A.; Lee, R. J. Folate receptor-targeted liposomes as vectors for therapeutic agents. Biotechnol. Annu. Rev. **2002**, 8, 103–131.
59. Weitman, S. D.; Lark, R. H., Coney, L. R.; Fort, D. W.; Frasca, V.; Zurawski Jr., V. R.; Kamen, B. A. Distribution of the folate receptor GP38 in normal and malignant cell lines and tissues. Cancer Res. **1992**, 52, 3396–3401.

60. Ross, J. F.; Chaudhuri, P. K.; Ratnam, M. Differential regulation of folate receptor forms in normal and malignant tissues in vivo and in established cell lines. Physiologic and clinical Implications. Cancer **1994**, 73 2432–2443.
61. Li, P. Y.; Del Vecchio, S.; Fonti, R.; Carriero, M. V.; Potena, M. I.; Botti, G.; Miotti, S.; Lastoria, S.; Menard, S.; Colnaghi, M. I.; Salvatore, M. Local concentration of folate binding protein GP38 in sections of human ovarian carcinoma by in vitro quantitative autoradiography. J. Nucl. Med. **1996**, 37, 665–672.
62. Toffoli, G.; Cernigoi, C.; Russo, A.; Gallo, A.; Bagnoli, M.; Boiocchi, M. Overexpression of folate binding protein in ovarian cancers. Int. J. Cancer **1997**, 74, 193–198.
63. Corona, G.; Giannini, F.; Fabris, M.; Toffoli, G.; Boiocchi, M. Role of folate receptor and reduced folate carrier in the transport of 5-methyltetrahydrofolic acid in human ovarian carcinoma cells. Int. J. Cancer **1998**, 75, 125–133.
64. Atkinson, S. F.; Bettinger, T.; Seymour, L. W.; Behr, J.-P.; Ward, C. M. Conjugation of folate via gelonin carbohydrate residues retains ribosomal-inactivating properties of the toxin and permits targeting to folate receptor positive cells. J. Bio. Chem. **2001**, 276, 27930–27935.
65. Garin-Chesa, P.; Campbell, I.; Saigo, P. E.; Lewis Jr., J. L.; Old, L. J.; Rettig, W. J. Trophoblast and ovarian cancer antigen LK26. Sensitivity and specificity in immunopathology and molecular identification as a folate-binding protein. Am. J. Pathol. **1993**, 142, 557–567.
66. Wang, S.; Lee, R. J.; Cauchon, G.; Gorenstein, D. G.; Low, P. S. Delivery of antisense oligodeoxyribonucleotides against the human epidermal growth factor receptor into cultured KB cells with liposomes conjugated to folate via polyethylene glycol. Proc. Natl. Acad. Sci. USA **1995**, 92, 3318–3322.
67. Leamon, C. P.; Low, P. S. Folate-mediated targeting: from diagnostics to drug and gene delivery. Drug Discov. Today **2001**, 6, 44–51.
68. Leamon, C. P.; Reddy, J. A. Folate-targeted chemotherapy. Adv. Drug Deliv. Rev. **2004**, 56, 1127–1141.
69. Kennedy, M. D.; Jallad, K. N.; Lu, J.; Low, P. S.; Ben-Amotz, D. Evaluation of folate conjugate uptake and transport by the choroid plexus of mice. Pharm. Res. **2003**, 20, 714–719.
70. Sudimack, J.; Lee, R. J. Targeted drug delivery via the folate receptor. Adv. Drug Deliv. Rev. **2000**, 41, 147–162.
71. Hilgenbrink, A. R.; Low, P. S. Folate receptor-mediated drug targeting: from therapeutics to diagnostics. J. Pharm. Sci. **2005**, 94, 2135–2146.
72. Paranjpe, P. V.; Stein, S.; Sinko, P. J. Tumor-targeted and activated bioconjugates for improved camptothecin delivery. Anticancer Drugs **2005**, 16, 763–775.
73. Cavallaro, G.; Mariano, L.; Salmaso, S.; Caliceti, P.; Gaetano, G. Folate-mediated targeting of polymeric conjugates of gemcitabine. Int. J. Pharm. **2006**, 307, 258–269.
74. Bajo, A. M.; Schally, A. V.; Halmos, G.; Nagy, A. Targeted doxorubicin-containing luteinizing hormone-releasing hormone analogue AN-152 inhibits the growth of doxorubicin-resistant MX-1 human breast cancers. Clin. Cancer Res. **2003**, 9, 3742–3748.
75. Qi, L.; Nett, T. M.; Allen, M. C.; Sha, X.; Harrison, G. S.; Frederick, B. A.; Crawford, E. D.; Glode, L. M. Binding and cytotoxicity of conjugated and recombinant fusion proteins targeted to the gonadotropin-releasing hormone receptor. Cancer Res. **2004**, 64, 2090–2095.
76. Danila, D. C.; Schally, A. V.; Nagy, A.; Alexander, J. M. Selective induction of apoptosis by the cytotoxic analog AN-207 in cells expressing recombinant receptor for luteinizing hormone-releasing hormone. Proc. Natl. Acad. Sci. USA **1999**, 96, 669–673.
77. Dharap, S. S.; Qiu, B.; Williams, G. C.; Sinko, P.; Stein, S.; Minko, T. Molecular targeting of drug delivery systems to ovarian cancer by BH3 and LH-RH peptides. J. Control. Release **2003**, 91, 61–73.
78. Dharap, S. S.; Wang, Y.; Chandna, P.; Khandare, J. J.; Qiu, B.; Gunaseelan, S.; Sinko, P. J.; Stein, S.; Farmanfarmaian, A.; Minko, T. Tumor-specific targeting of an anticancer drug delivery system by LHRH peptide. Proc. Natl. Acad. Sci. USA **2005**, 102, 12962–12967.
79. Sheff, D. Endosomes as a route for drug delivery in the real world. Adv. Drug Deliv. Rev. **2004**, 56, 927–930.

80. van Vlerken, L. E.; Amiji, M. M. Multi-functional polymeric nanoparticles for tumour-targeted drug delivery. Expert Opin. Drug Deliv. 2006, 3, 205–216.
81. Kommareddy, S.; Tiwari, S. B.; Amiji, M. M. Long-circulating polymeric nanovectors for tumor-selective gene delivery. Technol. Cancer Res. Treat. 2005, 4, 615–625.
82. Hans, M. L.; Lowman, A. M. Biodegradable nanoparticles for drug delivery and targeting. Curr. Opin. Solid State Mater. Sci. 2002, 6, 319–327.
83. Gref, R. M. Y.; Peracchia, M. T.; Trubetskoy, V.; Torchilin, V.; Langer, R. Biodegradable long-circulating polymeric nanospheres. Science 1994, 263, 1600–1603.
84. Torchilin, V. P. Block copolymer micelles as a solution for drug delivery problems. Expert Opin. Ther. Patents 2005, 15, 63–75.
85. Torchilin, V. P.; Weissig, V. Polymeric micelles for the delivery of poorly soluble drugs. ACS Symp. Ser. 2000, 752, 297–313.
86. Bogdanov Jr., A.; Wright, S. C.; Marecos, E. M.; Bogdanova, A.; Martin, C.; Petherick, P.; Weissleder, R. A long-circulating copolymer in "passive targeting" to solid tumors. J. Drug Target 1997, 4, 321–330.
87. Moghimi, S. M.; Hunter, A. C.; Murray, J. C. Longcirculating and target-specific nanoparticles: theory to practice. Pharmacol. Rev. 2001, 53, 283–318.
88. Kaul, G.; Amiji, M. Long-circulating poly(ethylene glycol)-modified gelatin nanoparticles for intracellular delivery. Pharm. Res. 2002, 19, 1061–1067.
89. Kwon, G. S.; Kataoka, K. Block copolymer micelles as long-circulating drug vehicles. Adv. Drug Deliv. Rev. 1995, 16 295–309.
90. Savic, R.; Eisenberg, A.; Maysinger, D. Block copolymer micelles as delivery vehicles of hydrophobic drugs: micelle-cell interactions. J. Drug Target. 2006, 14, 343–355.
91. Torchilin, V. P. Recent approaches to intracellular delivery of drugs and DNA and organelle targeting. Ann. Rev. Biomed. Eng. 2006, 8, 343–375.
92. Yokoyama, M. Polymeric micelle drug carriers for tumor targeting. ACS Symp. Ser. 2006, 923, 27–39.
93. Yokoyama, M.; Okano, T.; Sakurai, Y.; Suwa, S.; Kataoka, K. Introduction of cisplatin into polymeric micelle. J. Control. Release 1996, 39, 351–356.
94. Cabral, H.; Nishiyama, N.; Okazaki, S.; Koyama, H.; Kataoka, K. Preparation and biological properties of dichloro(1,2-diaminocyclohexane)platinum(II) (DACHPt)-loaded polymeric micelles. J. Control. Release 2005, 101, 223–232.
95. Nishiyama, N.; Okazaki, S.; Cabral, H.; Miyamoto, M.; Kato, Y.; Sugiyama, Y.; Nishio, K.; Matsumura, Y.; Kataoka, K. Novel cisplatin-incorporated polymeric micelles can eradicate solid tumors in mice. Cancer Res. 2003, 63, 8977–8983.
96. Xu, P.; Van Kirk, E. A.; Li, S.; Murdoch, W. J.; Ren, J.; Hussain, M. D.; Radosz, M.; Shen, Y. Highly stable core-surface-crosslinked nanoparticles as cisplatin carriers for cancer chemotherapy. Colloids Surf. B: Biointerface 2006, 48, 50–57.
97. Xu, P.; Van Kirk, E. A.; Murdoch, W. J.; Zhan, Y.; Isaak, D. D.; Radosz, M.; Shen, Y. Anticancer efficacies of cisplatin-releasing pH-responsive nanoparticles. Biomacromolecules 2006, 7, 829–835.
98. Uchino, H.; Matsumura, Y.; Negishi, T.; Koizumi, F.; Hayashi, T.; Honda, T.; Nishiyama, N.; Kataoka, K.; Naito, S.; Kakizoe, T. Cisplatin-incorporating polymeric micelles (NC-6004) can reduce nephrotoxicity and neurotoxicity of cisplatin in rats. Br. J. Cancer 2005, 93, 678–687.
99. Lee, E. S.; Na, K.; Bae, Y. H. Doxorubicin loaded pH-sensitive polymeric micelles for reversal of resistant MCF-7 tumor. J. Control. Release **2005**, 103, 405–418.
100. Aouali, N.; Morjani, H.; Trussardi, A.; Soma, E.; Giroux, B.; Manfait, M. Enhanced cytotoxicity and nuclear accumulation of doxorubicin-loaded nanospheres in human breast cancer MCF7 cells expressing MRP1. Int. J. Oncol. 2003, 23, 1195–1201.
101. Kataoka, K.; Matsumoto, T.; Yokoyama, M.; Okano, T.; Sakurai, Y.; Fukushima, S.; Okamoto, K.; Kwon, G. S. Doxorubicin-loaded poly(ethylene glycol)-poly([beta]-benzyl-aspartate) copolymer micelles: their pharmaceutical characteristics and biological significance. J. Control. Release 2000, 64, 143–153.

102. Kwon, G.; Naito, M.; Yokoyama, M.; Okano, T.; Sakurai, Y.; Kataoka, K. Block copolymer micelles for drug delivery: loading and release of doxorubicin. J. Control. Release 1997, 48 195–201.
103. Bennis, S.; Chapey, C.; Couvreur, P.; Robert, J. Enhanced cytotoxicity of doxorubicin encapsulated in polyhexylcyanoacrylate nanospheres against multi-drug-resistant tumour cells in culture. Eur. J. Cancer 1994, 30A, 89–93.
104. Kawano, K.; Watanabe, M.; Yamamoto, T.; Yokoyama, M.; Opanasopit, P.; Okano, T.; Maitani, Y. Enhanced antitumor effect of camptothecin loaded in long-circulating polymeric micelles. J. Control. Release 2006, 112, 329–332.
105. Opanasopit, P.; Yokoyama, M.; Watanabe, M.; Kawano, K.; Maitani, Y.; Okano, T. Influence of serum and albumins from different species on stability of camptothecin-loaded micelles. J. Control. Release 2005, 104, 313–321.
106. Onishi, H.; Machida, Y. Macromolecular and nanotechnological modification of camptothecin and its analogs to improve the efficacy. Curr. Drug Discov. Technol. 2005, 2, 169–183.
107. Zhang, L.; Hu, Y.; Jiang, X.; Yang, C.; Lu, W.; Yang, Y. H. Camptothecin derivative-loaded poly(caprolactone-co-lactide)-b-PEG-b-poly(caprolactone-co-lactide) nanoparticles and their biodistribution in mice. J. Control. Release 2004, 96, 135–148.
108. Tyner, K. M.; Schiffman, S. R.; Giannelis, E. P. Nanobiohybrids as delivery vehicles for camptothecin. J. Control. Release 2004, 95, 501–514.
109. Miura, H.; Onishi, H.; Sasatsu, M.; Machida, Y. Antitumor characteristics of methoxypolyethylene glycol-poly(dl-lactic acid) nanoparticles containing camptothecin. J. Control. Release 2004, 97, 101–113.
110. Williams, J.; Lansdown, R.; Sweitzer, R.; Romanowski, M.; LaBell, R.; Ramaswami, R.; Unger, E. Nanoparticle drug delivery system for intravenous delivery of topoisomerase inhibitors. J. Control. Release 2003, 91, 167–172.
111. Koziara, J. M.; Whisman, T. R.; Tseng, M. T.; Mumper, R. J. In-vivo efficacy of novel paclitaxel nanoparticles in paclitaxel-resistant human colorectal tumors. J. Control. Release 2006, 112, 312–319.
112. Xu, Z.; Gu, W.; Huang, J.; Hong, S.; Zhou, Z.; Yang, Y.; Yan, Z.; Li, Y. In vitro and in vivo evaluation of actively targetable nanoparticles for paclitaxel delivery. Int. J. Pharm. 2005, 288, 361–368.
113. Soga, O.; van Nostrum, C. F.; Fens, M.; Rijcken, C. J. F.; Schiffelers, R. M.; Storm, G.; Hennink, W. E. Thermosensitive and biodegradable polymeric micelles for paclitaxel delivery. J. Control. Release 2005, 103, 341–353.
114. Park, E. K.; Kim, S. Y.; Lee, S. B.; Lee, Y. M. Folate-conjugated methoxy poly(ethylene glycol)/poly(vepsiln-caprolactone) amphiphilic block copolymeric micelles for tumor-targeted drug delivery. J. Control. Release 2005, 109, 158–168.
115. Potineni, A.; Lynn, D. M.; Langer, R.; Amiji, M. M. Poly(ethylene oxide)-modified poly(beta-amino ester) nanoparticles as a pH-sensitive biodegradable system for paclitaxel delivery. J. Control. Release 2003, 86, 223–234.
116. Yokoyama, M.; Fukushima, S.; Uehara, R.; Okamoto, K.; Kataoka, K.; Sakurai, Y.; Okano, T. Characterization of physical entrapment and chemical conjugation of adriamycin in polymeric micelles and their design for in vivo delivery to a solid tumor. J. Control. Release 1998, 50, 79–92.
117. Putnam, D.; Kope ek, J. Polymers with anticancer activity. Adv. Polym. Sci. 1995, 122, 55–123.
118. Kope ek, J.; Kopeckova, P.; Minko, T. L. Z. R.; Peterson, C. M. Water soluble polymers in tumor targeted delivery. J. Control. Release 2001, 74, 147–158.
119. Yoo, H. S.; Lee, K. H.; Oh, J. E.; Park, T. G. In vitro and in vivo anti-tumor activities of nanoparticles based on doxorubicin-PLGA conjugates. J. Control. Release 2000, 68, 419–431.
120. Simoes, S.; Moreira, J. N.; Fonseca, C.; Duzgunes, N.; Pedroso de Lima, M. C. On the formulation of pH-sensitive liposomes with long circulation times. Adv. Drug Deliv. Rev. 2004, 56, 947–965.

121. Maeda, M.; Kumano, A.; Tirrell, D. A. H+-induced release of contents of phosphatidylcholine vesicles bearing surface-bound polyelectrolyte chains. J. Am. Chem. Soc. 1988, 110, 7455–7459.
122. Chen, Q.; Tong, S.; Dewhirst, M. W.; Yuan, F. Targeting tumor microvessels using doxorubicin encapsulated in a novel thermosensitive liposome. Mol. Cancer Ther. 2004, 3, 1311–1317.
123. Needham, D.; Anyarambhatla, G.; Kong, G.; Dewhirst, M. W. A new temperature-sensitive liposome for use with mild hyperthermia: characterization and testing in a human tumor xenograft model. Cancer Res. 2000, 60, 1197–1201.
124. Zou, Y.; Yamagishi, M.; Horikoshi, I.; Ueno, M.; Gu, X.; Perez-Soler, R. Enhanced therapeutic effect against liver W256 carcinoma with temperature-sensitive liposomal adriamycin administered into the hepatic artery. Cancer Res. 1993, 53, 3046–3051.
125. Kirchmeier, M. J.; Shida, T.; Chevrette, J.; Allen, T. M. Correlations between the rate of intracellular release of endocytosed liposomal doxorubicin and cytotoxicity as determined by a new assay. J. Liposome Res. 2001, 11, 15–29.
126. Zhang, Z.; Feng, S. S. In vitro investigation on poly(lactide)-Tween 80 copolymer nanoparticles fabricated by dialysis method for chemotherapy. Biomacromolecules 2006, 7, 1139–1146.
127. Thistlethwaite, A. J.; Leeper, D. B.; Moylan, D. J. 3rd; Nerlinger, R. E. pH distribution in human tumors. Int. J. Rad. Oncol. Biol. Phys. 1985, 11, 1647–52.
128. Kallinowski, F.; Vaupel, P. pH distributions in spontaneous and isotransplanted rat tumors. Br. J. Cancer 1988, 58, 314–321.
129. Martin, G. R.; Jain, R. K. Noninvasive measurement of interstital pH profiles in normal and neoplastic tissue using fluorescence ratio imaging microscopy. Cancer Res. 1994, 54, 5670–5674.
130. Stubbs, M.; McSheehy, P. M. J.; Griffiths, J. R.; Bashford, C. L. Causes and consequences of tumor acidity and implications for treatment. Mol. Med. Today 2000, 6, 15–19.
131. Yamagata, M.; Hasuda, K.; Stamato, T.; Tannock, I. F. The contribution of lactic acid to acidification of tumors: studies of variant cells lacking lactate dehydrogenase. Br. J. Cancer 1998, 77, 1726–1731.
132. Parkins, C. S.; Stratford, M. R. L.; Dennis, M. F.; Stubbs, M.; Chaplin, D. J. The relationship between extracellular lactate and tumor pH in a murine tumor model of ischemia-reperfusion. Br. J. Cancer 1997, 75, 319–323.
133. Dellian, M.; Helmlinger, G.; Yuan, F.; Jain, R. K. Fluorescence ratio imaging of interstitial pH in solid tumours: effect of glucose on spatial and temporal gradients. Br. J. Cancer 1996, 74, 1206–1215.
134. Leeper, D. B.; Engin, K.; Thistlethwaite, A. J.; Hitchon, H. D.; Dover, J. D.; Li, D.-J.; Tupchong, L. Human tumor extracellular pH as a function of blood glucose concentration. Int. J. Rad. Oncol. Biol. Phys. 1994, 28, 935–943.
135. Newell, K.; Franchi, A.; Pouyssegur, J.; Tannock, I. Studies with glycolysis-deficient cells suggest that production of lactic acid is not the only cause of tumor acidity. Proc. Natl. Acad. Sci. USA 1993, 90, 1127–1131.
136. De Milito, A.; Fais, S. Tumor acidity, chemoresistance and proton pump inhibitors. Future Oncol. 2005, 1, 779–786.
137. Engin, K.; Leeper, D. B.; Cater, J. R.; Thistlethwaite, A. J.; Tupchong, L.; McFarlane, J. D. Extracellular pH distribution in human tumours. Int. J. Hyperthermia 1995, 11, 211–216.
138. Martinez-Zaguilan, R.; Seftor, E. A.; Seftor, R. E. B.; Chu, Y.-W.; Gillies, R. J.; Hendrix, M. J. C. Acidic pH enhances the invasive behavior of human melanoma cells. Clin. Exp. Metastasis 1996, 14, 176–186.
139. Xu, L.; Fukumura, D.; Jain, R. K. Acidic extracellular pH induces vascular endothelial growth factor (VEGF) in human glioblastoma cells via ERK1/2 MAPK signaling pathway. Mechanism of low ph-induced VEGF. J. Biol. Chem. 2002, 277, 11368–11374.
140. Lee, E. S.; Shin, H. J.; Na, K.; Bae, Y. H. Poly(l-histidine)-PEG block copolymer micelles and pH-induced destabilization. J. Control. Release 2003, 90, 363–374.

141. Pack, D. W.; Putnam, D.; Langer, R. Design of imidazolecontaining endosomolytic biopolymers for gene delivery. Biotech. Bioeng. **2000**, 67 217–223.
142. Asayama, S.; Kawakami, H.; Nagaoka, S. Design of a poly(L-histidine)-carbohydrate conjugate for a new pH-sensitive drug carrier. Polym. Adv. Technol. **2004**, 15, 439–444.
143. Lee, E. S.; Na, K.; Bae, Y. H. Polymeric micelle for tumor pH and folate-mediated targeting. J. Control. Release **2003**, 91, 103–113.
144. Gao, Z. G.; Lee, D. H.; Kim, D. I.; Bae, Y. H. Doxorubicin loaded pH-sensitive micelle targeting acidic extracellular pH of human ovarian A2780 tumor in mice. J. Drug Target. **2005**, 13, 391–397.
145. Kang, S. I.; Na, K.; Bae, Y. H. Sulfonamide-containing polymers: a new class of pH-sensitive polymers and gels. Macromol. Symp. **2001**, 172, 149–156.
146. Na, K.; Bae, Y. H. Self-assembled hydrogel nanoparticles responsive to tumor extracellular pH from pullulan derivative/sulfonamide conjugate: characterization, aggregation, and adriamycin release in vitro. Pharm. Res. **2002**, 19, 681–688.
147. Na, K.; Lee, K. H.; Bae, Y. H. pH-sensitivity and pH-dependent interior structural change of self-assembled hydrogel nanoparticles of pullulan acetate/oligo-sulfonamide conjugate. J. Control. Release **2004**, 97, 513–525.
148. Na, K.; Lee, E. S.; Bae, Y. H. Adriamycin loaded pullulan acetate/sulfonamide conjugate nanoparticles responding to tumor pH: pH-dependent cell interaction, internalization and cytotoxicity in vitro. J. Control. Release **2003**, 87, 3–13.
149. Na, K.; Lee, T. B.; Park, K.-H.; Shin, E.-K.; Lee, Y.-B.; Choi, H.-K. Self-assembled nanoparticles of hydrophobically-modified polysaccharide bearing vitamin H as a targeted anti-cancer drug delivery system. Eur. J. Pharm. Sci. **2003**, 18, 165–173.
150. Goren, D.; Horowitz, A. T.; Tzemach, D.; Tarshish, M.; Zalipsky, S.; Gabizon, A. Nuclear delivery of doxorubicin via folate-targeted liposomes with bypass of multidrug-resistance efflux pump. Clin. Cancer Res. **2000**, 6, 1949–1957.
151. Larsen, A. K.; Escargueil, A. E.; Skladanowski, A. Resistance mechanisms associated with altered intracellular distribution of anticancer agents. Pharmacol. Ther. **2000**, 85, 217–229.
152. De Jaeghere, F.; Allemann, E.; Feijen, J. K. T.; Doelker, E.; Gurny, R. Cellular uptake of PEO surface-modified nanoparticles: evaluation of nanoparticles made of PLA:PEO diblock and triblock copolymers. J. Drug Target. **2000**, 8, 143–153.
153. Vittaz, M.; Bazile, D.; Spenlehauer, G.; Verrecchia, T.; Veillard, M.; Puisieux, F.; Labarre, D. Effect of PEO surface density on long-circulating PLA-PEO nanoparticles which are very low complement activators. Biomaterials **1996**, 17, 1575–1581.
154. Lee, E. S.; Na, K.; Bae, Y. H. Super pH-sensitive multifunctional polymeric micelle. Nano Lett. **2005**, 5, 325–329.
155. Kabanov, A. V.; Felgner, P. L.; Seymour, L. W. (eds) Self-assembling complexes for gene delivery. Wiley, Chichester, UK, 1998.
156. Blau, S.; Jubeh, T. T.; Haupt, S. M.; Rubinstein, A. Drug targeting by surface cationization. Crit. Rev. Ther. Drug Carrier Syst. **2000**, 17, 425–465.
157. Hamblin, M. R.; Rajadhyaksha, M.; Momma, T.; Soukos, N. S.; Hasan, T. In vivo fluorescence imaging of the transport of charged chlorine6 conjugates in a rat orthotopic prostate tumor. Br. J. Cancer **1999**, 81, 261–268.
158. Pardridge, W. M.; Bickel, U.; Buciak, J.; Yang, J.; Diagne, A.; Aepinus, C. Cationization of a monoclonal antibody to the human immunodeficiency virus REV protein enhances cellular uptake but does not impair antigen binding of the antibody. Immun. Lett. **1994**, 42, 191–195.
159. Pardridge, W. M.; Kang, Y.-S.; Yang, J.; Buciak, J. L. Enhanced cellular uptake and in vivo biodistribution of a monoclonal antibody following cationization. J. Pharm. Sci. **1995**, 84, 943–948.
160. Labhasetwar, V.; Song, C.; Humphrey, W.; Shebuski, R.; Levy, R. J. Arterial uptake of biodegradable nanoparticles: effect of surface modifications. J. Pharm. Sci. **1998**, 87, 1229–1234.

161. Nam, Y. S.; Kang, H. S.; Park, J. Y.; Park, T. G.; Han, S.-H.; Chang, I.-S. New micelle-like polymer aggregates made from PEI-PLGA diblock copolymers: micellar characteristics and cellular uptake. Biomaterials **2003**, 24, 2053–2059.
162. Ma, S.-F.; Nishikawa, M.; Katsumi, H.; Yamashita, F.; Hashida, M. Cationic charge-dependent hepatic delivery of amidated serum albumin. J. Control. Release **2005**, 102, 583–594.
163. Ma, S.-F.; Nishikawa, M.; Katsumi, H.; Yamashita, F.; Hashida, M. Liver targeting of catalase by cationization for prevention of acute liver failure in mice. J. Control. Release **2006**, 110, 273–282.
164. Lee, H. J.; Pardridge, W. M. Monoclonal antibody radiopharmaceuticals: cationization, pegylation, radiometal chelation, pharmacokinetics, and tumor imaging. Bioconjug. Chem. **2003**, 14, 546–553.
165. Zhan, Y.; Van Kirk, E.; Xu, P.; Murdoch, W. J.; Radosz, M.; Shen, Y. pH-responsive nanoaprticles for fast cytoplasmic drug delivery. Bioconjuate Chem. **2006** (submitted).
166. Zhan, Y.; Van Kirk, E.; Xu, P.; Murdoch, W. J.; Radosz, M.; Shen, Y. pH-responsive three layer onion-structured nanoparticles for drug delivery. Polym. Mater. Sci. Eng. Prepr. **2006**, 94, 139–140.
167. Cai, Y.; Armes, S. P. Synthesis of well-defined Y-shaped zwitterionic block copolymers via atom-transfer radical polymerization. Macromolecules **2005**, 38, 271–279.
168. Lee, A. S.; Buetuen, V.; Vamvakaki, M.; Armes, S. P.; Pople, J. A.; Gast, A. P. Structure of pH-dependent block copolymer micelles: Charge and ionic strength dependence. Macromolecules **2002**, 35, 8540–8551.
169. Butun, V.; Armes, S. P.; Billingham, N. C. Synthesis and aqueous solution properties of near-monodisperse tertiary amine methacrylate homopolymers and diblock copolymers. Polymer **2001**, 42, 5993–6008.
170. Pan, X.; Lee, R. J. Tumour-selective drug delivery via folate receptor-targeted liposomes. Expert Opin. Drug Deliv. **2004**, 1, 7–17.
171. Mamot, C.; Drummond, D. C.; Hong, K.; Kirpotin, D. B.; Park, J. W. Liposome-based approaches to overcome anticancer drug resistance. Drug Resist. Updat. **2003**, 6, 271–279.
172. Sadava, D.; Coleman, A.; Kane, S. E. Liposomal daunorubicin overcomes drug resistance in human breast, ovarian and lung carcinoma cells. J. Liposome Res. **2002**, 12, 301–309.
173. Nori, A.; Kopecek, J. Intracellular targeting of polymer-bound drugs for cancer chemotherapy. Adv. Drug Deliv. Rev. **2005**, 57, 609–636.
174. Duncan, R.; Vicent, M. J.; Greco, F.; Nicholson, R. I. Polymer-drug conjugates: towards a novel approach for the treatment of endocrine-related cancer. Endocr Relat Cancer **2005**, 12 (Suppl. 1), S189–S199.
175. Nan, A.; Ghandehari, H.; Hebert, C.; Siavash, H.; Nikitakis, N.; Reynolds, M.; Sauk, J. J. Water-soluble polymers for targeted drug delivery to human squamous carcinoma of head and neck. J. Drug Target. **2005**, 13(3), 189–197.
176. Hamann, P. R.; Hinman, L. M.; Beyer, C. F.; Greenberger, L. M.; Lin, C.; Lindh, D.; Menendez, A. T.; Wallace, R.; Durr, F. E.; Upeslacis, J. An anti-MUC1 antibody-calicheamicin conjugate for treatment of solid tumors. Choice of linker and overcoming drug resistance. Bioconjug. Chem. **2005**, 16, 346–353.
177. Laurand, A.; Laroche-Clary, A.; Larrue, A.; Huet, S.; Soma, E.; Bonnet, J.; Robert, J. Quantification of the expression of multidrug resistance-related genes in human tumour cell lines grown with free doxorubicin or doxorubicin encapsulated in polyisohexylcyanoacrylate nanospheres. Anticancer Res. **2004**, 24, 3781–3788.
178. Rapoport, N. Combined cancer therapy by micellar-encapsulated drug and ultrasound. Int. J. Pharm. **2004**, 277, 155–162.
179. Cuenca, A. G.; Jiang, H.; Hochwald, S. N.; Delano, M.; Cance, W. G.; Grobmyer, S. R. Emerging implications of nanotechnology on cancer diagnostics and therapeutics. Cancer **2006**, 107, 459–466.
180. Vauthier, C.; Dubernet, C.; Chauvierre, C.; Brigger, I.; Couvreur, P. Drug delivery to resistant tumors: the potential of poly(alkyl cyanoacrylate) nanoparticles. J. Control. Release **2003**, 93, 151–160.

181. Merion, M.; Schlesinger, P.; Brooks, R. M.; Moerhing, J. M.; Moerhing, T. J.; Sly. W. S. Defective acidification of endosomes in Chinese hamster ovary cell mutants "cross-resistant" to toxins and viruses. Proc. Natl. Acad. Sci. USA **1983**, 80, 5315–5319.
182. Kielian, M. C.; Marsh, M.; Helenius, A. Kinetics of endosome acidification detected by mutant and wild-type Semliki Forest virus. EMBO J. **1986**, 5, 3103–3109.
183. Schmid, S. L.; Fuchs, R.; Male, P.; Mellman, I. Two distinct subpopulations of endosomes involved in membrane recycling and transport to lysosomes. Cell **1988**, 52, 73–83.
184. Murphy, R. F.; Powers, S.; Cantor, C. R. Endosome pH measured in single cells by dual fluorescence flow cytometry: rapid acidification of insulin to pH 6. J. Cell Biol. **1984**, 98, 1757–1762.
185. Yamashiro, D. J.; Maxfield, F. R. Kinetics of endosome acidification in mutant and wild-type Chinese hamster ovary cells. J. Cell Biol. **1987**, 105, 2713–2721.
186. Shukla, R.; Bansal, V.; Chaudhary, M.; Basu, A.; Bhonde, R. R.; Sastry, M. Biocompatibility of gold nanoparticles and their endocytotic fate inside the cellular compartment: a microscopic overview. Langmuir **2005**, 21, 10644–10654.
187. Gao, H.; Shi, W.; Freund, L. B. Mechanics of receptor-mediated endocytosis. Proc. Natl. Acad. Sci. USA **2005**, 102, 9469–9474.
188. Garcia-Garcia, E.; Andrieux, K.; Gil, S.; Kim, H. R.; Doan, T. L.; Desmaele, D.; d'Angelo, J.; Taran, F.; Georgin, D.; Couvreur, P. A methodology to study intracellular distribution of nanoparticles in brain endothelial cells. Int. J. Pharm. **2005**, 298, 310–314.
189. Steinman, R. M.; Mellman, I. S.; Muller, W. A.; Cohn, Z. A. Endocytosis and the recycling of plasma membrane. J. Cell Biol. **1983**, 96, 1–27.
190. Sipe, D. M.; Murphy, R. F. High-resolution kinetics of transferrin acidification in BALB/c 3T3 cells: exposure to pH 6 followed by temperature-sensitive alkalinization during recycling. Proc. Natl. Acad. Sci. USA **1987**, 84, 7119–7123.
191. Cain, C. C.; Sipe, D. M.; Murphy, R. F. Regulation of endocytic pH by the sodium-potassium ATPase in living cells. Proc. Natl. Acad. Sci. USA **1989**, 86, 544–548.
192. Killisch, I.; Steinlein, P.; Römisch, K.; Hollinshead, R.; Beug, H.; Griffiths, G. Characterization of early and late endocytic compartments of the transferrin cycle: transferrin receptor antibody blocks erythroid differentiation by trapping the receptor in the early endosome. J. Cell Sci. **1992**, 103, 211–232.
193. Schmid, S.; Fuchs, R.; Kielian, M.; Helenius, A.; Mellman, I. Acidification of endosome subpopulations in wild-type Chinese hamster ovary cells and temperature-sensitive acidification-defective mutants. J. Cell Bio. **1989**, 108, 1291–1300.
194. Sipe, D. M.; Jesurum, A.; Murphy, R. F. Absence of sodium-potassium ATPase regulation of endosomal acidification in K562 erythroleukemia cells. Analysis via inhibition of transferrin recycling by low temperatures. J. Bio. Chem. **1991**, 266, 3469–3474.
195. Rybak, S. L.; Murphy, R. F. Primary cell cultures from murine kidney and heart differ in endosomal pH. J. Cell. Physiol. **1998**, 176, 216–222.
196. Reijngoud, D. J.; Tager, J. M. The permeability properties of the lysosomal membrane. Biochim. Biophys. Acta. **1977**, 472, 419–449.
197. Barret, A.; Heath, M. Lysosomes: a laboratory handbook, 2nd edn. North-Holland, New York, 1977.
198. Miraglia, E.; Viarisio, D.; Riganti, C.; Costamagna, C.; Ghigo, D.; Bosia, A. Na^+/H^+ exchanger activity is increased in doxorubicin-resistant human colon cancer cells and its modulation modifies the sensitivity of the cells to doxorubicin. Int. J. Cancer **2005**, 115, 924–929.
199. Gong, Y.; Duvvuri, M.; Krise, J. P. Separate roles for the Golgi apparatus and lysosomes in the sequestration of drugs in the multidrug-resistant human leukemic cell line HL-60. J. Bio. Chem. **2003**, 278, 50234–50239.
200. Laochariyakul, P.; Ponglikitmongkol, M.; Mankhetkorn, S. Functional study of intracellular P-gp- and MRP1-mediated pumping of free cytosolic pirarubicin into acidic organelles in intrinsic resistant SiHa cells. Can. J. Physiol. Pharmacol. **2003**, 81, 790–799.

201. Arancia, G.; Calcabrini, A.; Meschini, S.; Molinari, A. Intracellular distribution of anthracyclines in drug resistant cells. Cytotechnology **1998**, 27, 95–111.
202. Ouar, Z.; Lacave, R.; Bens, M.; Vandewalle, A. Mechanisms of altered sequestration and efflux of chemotherapeutic drugs by multidrug-resistant cells. Cell Biol. Toxicol. **1999**, 15 91–100.
203. Schindler, M.; Grabski, S.; Hoff, E.; Simon, S. M. Defective pH regulation of acidic compartments in human breast cancer cells (MCF-7) is normalized in adriamycin-resistant cells (MCF-7adr). Biochemistry **1996**, 35, 2811–2817.
204. Weisburg, J. H.; Roepe, P. D.; Dzekunov, S.; Scheinberg, D. A. Intracellular pH and multidrug resistance regulate complement-mediated cytotoxicity of nucleated human cells. J. Biol. Chem. **1999**, 274, 10877–10888.
205. Altan, N.; Chen, Y.; Schindler, M.; Simon, S. M. Defective acidification in human breast tumor cells and implications for chemotherapy. J. Exp. Med. **1998**, 187, 1583–1598.
206. Burrow, S. M.; Phoenix, D. A.; Wainwright, M.; Tobin, M. J. Intracellular localisation studies of doxorubicin and Victoria Blue BO in EMT6-S and EMT6-R cells using confocal microscopy. Cytotechnology **2002**, 39, 15–25.
207. Simon, S. M. Role of organelle pH in tumor cell biology and drug resistance. Drug Discov. Today **1999**, 4, 32–38.
208. Savic, R.; Luo, L.; Eisenberg, A.; Maysinger, D. Micellar nanocontainers distribute to defined cytoplasmic organelles. Science **2003**, 300, 615–618.
209. Pichon, C.; Goncalves, C.; Midoux, P. Histidine-rich peptides and polymers for nucleic acids delivery. Adv. Drug Deliv. Rev. **2001**, 53 75–94.
210. James, M. A.; Claire, G.; Antoine, K.; Gilles, P.; Dominique, A.; Marie-Helene, M.-B.; Burkhard, B. Enhanced membrane disruption and antibiotic action against pathogenic bacteria by designed histidine-rich peptides at acidic pH. Antimicrob. Agents Chemother. **2006**, 50, 3305–3311.
211. Shenoy, D.; Little, S.; Langer, R.; Amiji, M. Poly(ethylene oxide)-modified poly(b-amino ester) nanoparticles as a pH-sensitive system for tumor-targeted delivery of hydrophobic drugs. 1. In vitro evaluations. Mol. Pharm. **2005**, 2, 357–366.
212. Shenoy, D.; Little, S.; Langer, R.; Amiji, M. Poly(ethylene oxide)-modified poly(β-amino ester) nanoparticles as a pH-sensitive system for tumor-targeted delivery of hydrophobic drugs: Part 2. In vivo distribution and tumor localization studies. Pharm. Res. **2005**, 22, 2107–2114.
213. Anderson, D. G.; Akinc, A.; Hossain, N.; Langer, R. Structure/property studies of polymeric gene delivery using a library of poly(beta-amino esters). Mol. Ther. **2005**, 11, 426–434.
214. Little, S. R.; Lynn, D. M.; Puram, S. V.; Langer, R. Formulation and characterization of poly(β-amino ester) microparticles for genetic vaccine delivery. J. Control. Release **2005**, 107, 449–462.
215. Lynn, D. M.; Anderson, D. G.; Akinc, A.; Langer, R. Degradable poly(β-amino ester)s for gene delivery. Polym. Gene Deliv. **2005**, 227–241.
216. Little, S. R.; Lynn, D. M.; Ge, Q.; Anderson, D. G.; Puram, S. V.; Chen, J.; Eisen, H. N.; Langer, R. Poly(β-amino ester)-containing microparticles enhance the activity of nonviral genetic vaccines. Proc. Natl. Acad. Sci. USA **2004**, 101, 9534–9539.
217. Akinc, A.; Anderson, D. G.; Lynn, D. M.; Langer, R. Synthesis of poly(β-amino ester)s optimized for highly effective gene delivery. Bioconjug. Chem. **2003**, 14, 979–988.
218. Akinc, A.; Lynn, D. M.; Anderson, D. G.; Langer, R. Parallel synthesis and biophysical characterization of a degradable polymer library for gene delivery. J. Am. Chem. Soc. **2003**, 125, 5316–5323.
219. Lynn, D. M.; Langer, R. Degradable poly(beta-amino esters): synthesis, characterization, and self-assembly with plasmid DNA. J. Am. Chem. Soc. **2000**, 122 10761–10768.
220. Van de Wetering, P.; Moret, E. E.; Schuurmans-Nieuwenbroek, N. M. E.; van Steenbergen, M. J.; Hennink, W. E. Structure-activity relationships of water-soluble cationic methacrylate/methacrylamide polymers for nonviral gene delivery. Bioconjug. Chem. **1999**, 10, 589–597.

221. Cherng, J. Y.; Van de Wetering, P.; Talsma, H.; Crommelin, D. J. A.; Hennink, W. E. Effect of size and serum proteins on transfection efficiency of poly(2-(dimethylamino)-ethyl methacrylate)-plasmid nanoparticles. Pharmacol. Res. **1996**, 13, 1038–1042.
222. Richardson, S.; Ferruti, S.; Duncan, R. Poly(amidoamine)s as potential endosomolytic polymers: evaluation in vitro and body distribution in normal and tumor bearing animals. J. Drug Target. **1996**, 6, 391–397.
223. Demeneix, B. A.; Behr, J.-P. The proton sponge: a trick the viruses did not exploit. In: Felgner, P. L.; Heller, M. J.; Lehn, P.; Behr, J. P.; Szoka Jr., F. C. (eds) Artificial self-assembling systems for gene delivery. American Chemical Society, Washington, DC, 1996; pp. 146–151.
224. Yoshikawa, A.; Saura, R.; Matsubara, T.; Mizuno, K. A mechanism of cisplatin action: antineoplastic effect through inhibition of neovascularization. Kobe J. Med. Sci. **1997**, 43, 109.

Chapter 11
Extended-Release Oral Drug Delivery Technologies: Monolithic Matrix Systems

Sandip B. Tiwari and Ali R. Rajabi-Siahboomi

Abstract Oral drug delivery is the largest and the oldest segment of the total drug delivery market. It is the fastest growing and most preferred route for drug administration. Use of hydrophilic matrices for oral extended release of drugs is a common practice in the pharmaceutical industry. This chapter presents different polymer choices for fabrication of monolithic hydrophilic matrices and discusses formulation and manufacturing variables affecting the design and performance of the extended-release product by using selected practical examples.

Keywords Hydrophilic matrix; Monolithic; Drug delivery; Extended release; Formulation; Cellulose ethers; Hypromellose (hydroxypropyl methylcellulose, HPMC)

1 Introduction

Oral administration of drugs has been the most common and preferred route for delivery of most therapeutic agents. It remains the preferred route of administration investigated in the discovery and development of new drug candidates and formulations. The popularity of the oral route is attributed to patient acceptance, ease of administration, accurate dosing, cost-effective manufacturing methods, and generally improved shelf-life of the product. For many drugs and therapeutic indications, conventional multiple dosing of immediate release formulations provides satisfactory clinical performance with an appropriate balance of efficacy and safety. The rationale for development of an extended-release formulation of a drug is to enhance its therapeutic benefits, minimizing its side effects while improving the management of the diseased condition. Table 11.1 lists some of the advantages offered by extended-release dosage forms [1–3]. Besides its clinical advantages, an innovative extended-release formulation provides an opportunity for a pharmaceutical company to manage its product life-cycle. The dearth of new chemical entities is forcing many pharmaceutical companies to reformulate an existing conventional formulation to an extended-release product as a strategy of life-cycle management and retaining

From: *Methods in Molecular Biology, Vol. 437: Drug Delivery Systems*
Edited by: Kewal K. Jain © Humana Press, Totowa, NJ

Table 11.1 Advantages and limitations of a drug formulated into an extended release (ER) dosage form

Clinical advantages
Reduction in frequency of drug administration
Improved patient compliance
Reduction in drug level fluctuation in blood
Reduction in total drug usage when compared with conventional therapy
Reduction in drug accumulation with chronic therapy
Reduction in drug toxicity (local/systemic)
Stabilization of medical condition (because of more uniform drug levels)
Improvement in bioavailability of some drugs because of spatial control
Economical to the health care providers and the patient
Commercial/industrial advantages
Illustration of innovative/technological leadership
Product life-cycle extension
Product differentiation
Market expansion
Patent extension
Potential limitations
Delay in onset of drug action
Possibility of dose dumping in the case of a poor formulation strategy
Increased potential for first pass metabolism
Greater dependence on GI residence time of dosage form
Possibility of less accurate dose adjustment in some cases
Cost per unit dose is higher when compared with conventional doses
Not all drugs are suitable for formulating into ER dosage form

market share. Moreover, the enactment of the Hatch-Waxman Act – 1984 (Drug Price Competition and Patent Term Restoration Act) has led to a sudden surge of extended-release formulations being introduced into the market by generic manufacturers. In fact, the last decade has witnessed the highest number of new drug applications (NDA) and abbreviated new drug applications (ANDAs) filed with FDA for extended-release formulations [4]. The first commercial oral extended-release formulation was the pellet-filled capsule (Spansules®) which was introduced in the 1950 by Smith, Kline and French [5]. Spansule capsules were formulated by coating a drug onto nonpareil sugar beads and further coating with glyceryl stearate and wax. Since then, a number of strategies have been developed to obtain extended release of a drug in the body. These vary from simple matrix tablets or pellets to more technologically sophisticated extended-release systems which have been introduced into the marketplace [6, 7]. Successful commercialization of an extended-release dosage form is usually challenging and involves consideration of many factors such as the physicochemical properties of the drug [nature and form of the drug, Biopharmaceutical Classification System (BCS) class, dose and stability of the drug in the gastrointestinal (GI) tract], physiological factors (route of administration, site and mode of absorption, metabolism and elimination) and manufacturing variables (choice of excipients, equipment and manufacturing methods). This chapter will mainly focus on monolithic hydrophilic matrix systems (tablets) as a common strategy used in the industry to achieve extended release of drugs.

Various polymer choices for fabrication of monolithic matrices as well as formulation and manufacturing variables affecting the design and performance of the extended-release product are discussed here using selected practical examples. The technology of extended-release dosage forms, the theoretical basis for their formulation, and their clinical performance have been extensively discussed and reported in the literature [2, 8–15]. Our aim is not to duplicate this effort but rather to focus on the practical perspective of the formulation design and manufacture of hydrophilic matrices and to provide general guidelines. Such practical aspect usually involves generalizations for which there are occasional exceptions.

1.1 Definition of Terminologies

The United States Pharmacopoeia (USP) defines the modified-release (MR) dosage form as "the one for which the drug release characteristics of time course and/or location are chosen to accomplish therapeutic or convenience objectives not offered by conventional dosage forms such as solutions, ointments, or promptly dissolving dosage forms" [16]. One class of MR dosage form is an extended-release (ER) dosage form and is defined as the one that allows at least a 2-fold reduction in dosing frequency or significant increase in patient compliance or therapeutic performance when compared with that presented as a conventional dosage form (a solution or a prompt drug-releasing dosage form). The terms "controlled release (CR)", "prolonged release", "sustained or slow release (SR)" and "long-acting (LA)" have been used synonymously with "extended release". The commercial branded products in this category are often designated by suffixes such as CR, CD (controlled delivery), ER, LA, PD (programmed or prolonged delivery), Retard, SA (slow-acting), SR, TD (timed delivery), TR (timed release), XL and XR (extended release). Nearly all of the currently marketed monolithic (*mono* meaning single, *lith* is stone or block of material taken to mean a single unit or a tablet) oral ER dosage forms fall into one of the following two technologies:

1. Hydrophilic, hydrophobic or inert matrix systems: These consist of a rate-controlling polymer matrix through which the drug is dissolved or dispersed.
2. Reservoir (coated) systems where drug-containing core is enclosed within a polymer coatings. Depending on the polymer used, two types of reservoir systems are considered
   (a) Simple diffusion/erosion systems where a drug-containing core is enclosed within hydrophilic and/or water-insoluble polymer coatings. Drug release is

achieved by diffusion of the drug through the coating or after the erosion of the polymer coating.

(b) Osmotic systems where the drug core is contained within a semi-permeable polymer membrane with a mechanical/laser drilled hole for drug delivery. Drug release is achieved by osmotic pressure generated within the tablet core.

2 Extended-Release Oral Drug Delivery: Monolithic Hydrophilic Matrices

A matrix tablet is the simplest and the most cost-effective method to fabricate an extended-release dosage form. The majority of commercially available matrix formulations are in the form of tablets and their manufacture is similar to conventional tablet formulations consisting of granulation, blending, compression and coating steps. In its simplest form, a typical ER matrix formulation consists of a drug, release retardant polymer (hydrophilic or hydrophobic or both), one or more excipients (as filler or binder), flow aid (glidant) and a lubricant. Other functional ingredients such as buffering agents, stabilizers, solubilizers and surfactants may also be included to improve or optimize the release and/or stability performance of the formulation system.

2.1 Hydrophilic Matrices

Hydrophilic matrices are the most commonly used oral extended-release systems because of their ability to provide desired release profiles for a wide range of drugs, robust formulation, cost-effective manufacture, and broad regulatory acceptance of the polymers. Table 11.2 shows a list of hydrophilic polymers commonly used for fabrication of matrices [17, 18]. Hydrophobic materials are also used either alone (hydrophobic matrix systems) or in conjugation with hydrophilic matrix systems (hydrophilic-hydrophobic matrix systems) and are also listed in Table 11.2. Cellulose ethers, in particular hypromellose (hydroxypropyl methylcellulose, HPMC), have been the polymers of choice for the formulation of hydrophilic matrix systems.

For illustration of basic formulation principles, matrices of cellulose ethers, HPMC in particular, are discussed here. Nevertheless, the fundamentals for the design and performance of the most hydrophilic matrices remain the same.

2.2 Cellulose Ethers in Hydrophilic Matrices

Chemically, HPMC is mixed-alkyl hydroxyalkyl cellulose ether containing methoxyl and hydroxypropoxyl groups. A general structure of cellulose ether polymers is shown in Fig. 11.1, where the R group can be a single or a combination

Table 11.2 Polymers commonly studied for fabrication of extended release monolithic matrices [17, 18]

Hydrophilic polymers
- Cellulosic
  - Methylcellulose
  - Hypromellose (Hydroxypropylmethylcellulose, HPMC)
  - Hydroxypropylcellulose (HPC)
  - Hydroxyethylcellulose (HEC)
  - Sodium carboxymethylcellulose (Na-CMC)
- Noncellulosic: gums/polysaccharides
  - Sodium alginate
  - Xanthan gum
  - Carrageenan
  - Ceratonia (locust bean gum)
  - Chitosan
  - Guar gum
  - Pectin
  - Cross-linked high amylose starch
- Noncellulosic: others
  - Polyethylene oxide
  - Homopolymers and copolymers of acrylic acid

Water-insoluble and hydrophobic polymers
- Ethylcellulose
- Hypromellose acetate succinate
- Cellulose acetate
- Cellulose acetate propionate
- Methycrylic acid copolymers
- Poly(vinyl acetate)

Fatty acids/alcohols/waxes
- Bees' wax
- Carnauba wax
- Candelilla wax
- Paraffin waxes
- Cetyl alcohol
- Stearyl alcohol
- Glyceryl behenate
- Glyceryl monooleate, monosterate, palmitostearate
- Hydrogenated vegetable oil
  - Hydrogenated palm oil
  - Hydrogenated cottonseed oil
  - Hydrogenated castor oil
  - Hydrogenated soybean oil

of substituents. Type and distribution of the substituent groups affect the physicochemical properties such as rate and extent of hydration, surface activity, biodegradation and mechanical plasticity of the polymers. These properties plus molecular

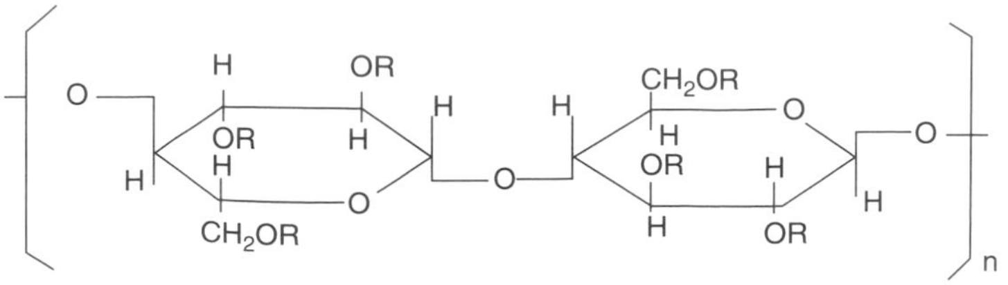

Fig. 11.1 General structure of cellulose with three possible substitution sites indicated by R (not all are substituted) (hydroxyl groups) on each D-anhydroglucose monomer. Hydroxypropyl methylcellulose (HPMC) contains methoxyl (CH_3–O) and hydroxypropoxyl ($CH_3CHOHCH_2$–O) substituents

weight distribution of cellulose ethers make them versatile for use in ER formulation of a wide range of drugs with different solubilities and doses. Moreover, they are non-ionic water-soluble polymers, and hence the possibility of ionic interaction or complexation with other formulation components is greatly reduced and their matrices exhibit pH-independent drug release profile. Aqueous solutions of HPMC are stable over a wide pH range (pH 3–11) and are resistant to enzymatic degradation. HPMC is available commercially from Dow Chemical Company under the trade name of Methocel™ [19]. Methocel is available in four different chemistries (A, E, F and K) depending on the degree of hydroxypropoxyl and methoxyl group substitutions. Methocel E (hypromellose 2910 USP) and K (hypromellose 2208, USP) chemistries are most widely used in extended-release formulations and are distributed worldwide by Colorcon, Inc. The USP classification code is based on the substitution type with the first two digits representing the mean % methoxyl substitution and the last two digits representing the mean % hydroxypropoxyl substitution [20]. The chemical substitution specification of these cellulose ethers are summarized in Table 11.3. Water-soluble cellulose ethers are also graded based on viscosity (in cPs) of a 2% (w/v) aqueous solution at 20°C, as shown in Table 11.3 [19].

HPMC is highly hydrophilic and hence hydrates rapidly when in contact with water. On the other hand, since the hydroxypropyl group is hydrophilic and methoxyl group is hydrophobic, the ratio of hydroxypropyl to methoxyl content affects the extent of polymer interaction with water. This property will in turn influence water mobility in a hydrated gel layer and drug release [21, 22]. Methocel grades for extended-release formulations include E50LV, K100LV, K4M CR, K15M CR, K100M CR, E4M CR and E10M CR. The viscosity of a 2% aqueous solution of these polymers ranges from 50 to 100,000 cPs at 20°C. Similar grades of HPMC are also available from other suppliers such as ShinEtsu, Japan [23] and Aqualon division of Hercules Inc., USA [24].

Other non-ionic cellulose ethers which have been studied in the formulation of hydrophilic matrices include high viscosity grades of hydroxypropylcellulose (HPC) and hydroxyethylcellulose (HEC) [24]. The ionic cellulose ether, sodium carboxymethylcellulose (Na CMC), with low or medium viscosity grades has

Table 11.3 Pharmaceutical grades of Methocel cellulose ethers

Dow product	USP hypromellose	% methoxyl substitution	% hydroxypropoxyl substitution	Viscosity grades (cPs)
Methocel E	2910	28–30	7–12	3, 5, 6, 15, 50, 4000, 10000
Methocel K	2208	19–24	7–12	3, 100, 4000, 15000, 100000

USP United States Pharmacopoeia

also been studied in combination with other non-ionic polymers [25]. A Na-CMC matrix does not fully hydrate to form a gel when placed in a media with low pH (e.g. pH 1.2) and it may disintegrate. Non-cellulosic hydrophilic polymers used for fabrication of matrices include water soluble/swellable polysaccharides (xanthan gum and sodium alginate), polymers of acrylic acid (e.g. Carbopol®) and poly(ethylene oxide) (POLYOX™) [15, 26–29]. Polymers of acrylic acid are synthetic high molecular weight polymers that are cross-linked with either allyl sucrose or allyl ethers of pentaerythritol [26]. Because these polymers are cross-linked, they are not water soluble but they swell on hydration and form a gel layer. As discussed earlier, HPMC swelling is because of the hydration of the polymer, leading to relaxation of polymer chains and subsequent entanglement of these polymer chains (cross-linking) to form a viscous gel. In case of acrylic acid polymers, surface gel formation is not because of the entanglement of the polymer chains (as the polymers are already cross-linked) but because of the formation of the discrete microgels made up of many polymer particles [26].

Poly(ethylene oxide) is also a non-ionic water-soluble resin, available in a variety of molecular weight grades ranging from 100,000 to 7,000,000 Daltons. The common grades of PEO which are used for extended-release applications include POLYOX WSR-205 NF, WSR-1105 NF, WSR N-12K NF, WSR N-60K NF, WSR-301 NF, WSR-303 NF and WSR Coagulant NF [27]. They are the fastest hydrating water-soluble polymers amongst the hydrophilic polymers, which makes PEO products a suitable choice for applications where slower initial drug release is required [27].

2.3 Drug Release from Hydrophilic Matrices

The mechanism of drug release from hydrophilic matrix tablets after ingestion is complex but it is based on diffusion of the drug through, and erosion of, the outer hydrated polymer on the surface of the matrix. Typically, when the matrix tablet is exposed to an aqueous solution or gastrointestinal fluids, the surface of the tablet is wetted and the polymer hydrates to form a gelly-like structure around the matrix, which is referred to as the "gel layer". This process is also termed as the glassy to rubbery state transition of the (surface layer) polymer. This leads to relaxation and swelling of the matrix which also contributes to the mechanism of drug release. The core of the tablet remains essentially dry at this stage. In the case of a highly soluble drug, this phenomenon may lead to an initial burst release due to the presence of the drug on the surface of the matrix tablet. The gel layer (rubbery state) grows with time as more water permeates into the core of the matrix, thereby increasing the thickness of the gel layer and providing a diffusion barrier to drug release [21]. Simultaneously, as the outer layer becomes fully hydrated, the polymer chains become completely relaxed and can no longer maintain the integrity of the gel layer, thereby leading to disentanglement and erosion of the surface of the matrix. Water continues to penetrate towards the core of the tablet, through the gel layer, until it has been completely eroded. Soluble drugs can be released by a combination of diffusion and erosion mechanisms whereas erosion is the predominant mechanism for insoluble drugs [30]. For successful extended release of drugs, it is essential that polymer hydration and surface gel layer formation are quick so as to prevent immediate tablet disintegration and premature drug release. For this reason, polymers for hydrophilic matrices are usually supplied in small particle size (such as Methocel CR grades) to ensure rapid hydration and consistent formation of the gel layer on the surface of the tablet.

A large number of mathematical models have been developed to describe drug release profiles from matrices [31–36]. The simple and more widely used model is the one derived by Korsmeyer et al. [37] and is as follows:

$$\mathrm{M}_t / \mathrm{M}_\alpha = k\, t^n \quad (1)$$

where $\mathrm{M}_t/\mathrm{M}_\alpha$ is the fraction of drug release, k is the diffusion rate constant, t is the release time and n is the release exponent indicative of the mechanism of drug release. The equation was modified by Ford et al. [38] to account for any lag time or initial burst release of the drug

$$\mathrm{M}_t / \mathrm{M}_\alpha = k\, (t - l)^n \quad (2)$$

where l is the lag time. It is clear from both equations that when the exponent n takes a value of 1.0, the drug release rate is independent of time. This case corresponds to zero-order release kinetics (also termed as case II transport). Here, the polymer relaxation and erosion [39] are the rate-controlling steps. When $n = 0.5$, Fickian diffusion is the rate-controlling step (case I transport). Values of n between 0.5 and 1 indicate the contribution of both the diffusion process as well as polymer relaxation in controlling the release kinetics (non-Fickian, anomalous or first-order release). It should be noted that the two extreme values of $n = 0.5$ and 1 are only valid for slab geometry. For cylindrical tablets, these values range from $0.45 < n < 0.89$ for Fickian, anomalous or case II transport respectively [33].

2.4 Formulation of Hydrophilic Matrices

Typical formulation of a hydrophilic matrix consists of drug, polymer and excipients. These components can be compressed into tablets directly or after granulation by dry, wet or hot melt method depending on the nature of the drug, excipients and the preference for process in a particular pharmaceutical company. The various formulation and manufacturing considerations in the design of hydrophilic matrices are listed in Table 11.4. The development of hydrophilic matrices has largely been empirical. There is no universal recipe/methodology for designing an ER matrix formulation. One can formulate an ER matrix product with different hydrophilic and/or hydrophobic polymers using various manufacturing principles and processes. A metformin hydrochloride (HCl) extended-release tablet (Glucophage® XR, Bristol Myers Squibb) is a good example of a use of polymer combinations to achieve a desired release profile. The formulation consists of a dual hydrophilic

Table 11.4 Formulation and manufacturing considerations in the design of hydrophilic matrices for extended release of drugs

Material/process	Parameter for consideration
Formulation components	
Drug	Solubility and permeability, pK_a, dose, stability, particle size
Polymer	Particle size, type, level
Excipient	
Filler	Level/type (solubility)
Other excipients	
Lubricants	Level/type (stearates, non-stearates, fatty acids/oils)
Others	Release rate modifiers, stabilizers, solubilizer, surfactant, buffering agents
Manufacturing aspects	
Manufacturing method	
Direct compression	Particle size of polymer/drug, flow aid
Dry granulation	Slugging/roller compaction
Wet granulation	
Solvent	Aqueous/non-aqueous
Binders	Water-soluble/insoluble, enteric polymers, fatty acids/waxes
Process	Low shear
	High shear
	Fluidized bed/foam granulation
Characteristic of dosage form	
Physical properties	Hardness, size, shape, volume and friability
Presence of coating	
Functional	Water-soluble/insoluble polymers, enteric polymers, fatty acids/waxes
Non-functional	Elegance/aesthetics

polymer matrix system where the drug is combined with an ionic release-controlling polymer (sodium carboxymethylcellulose) to form an "inner" phase, which is then incorporated as discrete particles into an "external" phase of a second non-ionic polymer, HPMC [40, 41]. There are many other extended-release formulations of metformin HCl approved by US FDA [42]. These formulations range from simple monolithic hydrophilic matrix systems of a single polymer to combination of hydrophilic polymers with or without water-insoluble polymers (including enteric polymers) and hydrophobic matrices [43, 44]. Although these formulations vary in their design and compositions, they all achieve similar extended-release profiles when tested *in vitro* and *in vivo* (bioequivalent). In the following sections, some selected fundamental formulation parameters and manufacturing considerations for HPMC matrices are discussed as a general guideline.

2.5 Key Formulation Considerations

2.5.1 Drug Properties

Drug solubility and dose are the most important factors to consider in the design of ER matrices. In general, extended-release formulation of extreme drug solubilities coupled with a high dose is challenging. Drugs with very low solubility (e.g. < 0.01 mg/mL) may dissolve slowly and have slow diffusion through the gel layer of a hydrophilic matrix. Therefore, the main mechanism of release would be through erosion of the surface of the hydrated matrix. In these cases, the control over matrix erosion to achieve consistent extended release throughout the GI tract is critical. For drugs with very high water solubility, the drug dissolves within the gel layer (even with small amounts of free water) and diffuses out into the media. Therefore, it is important to control the factors that affect drug diffusivity (e.g. pH, gel strength and availability of free water) within the gel layer and parameters that ensure integrity of the gel layer after the drug has been dissolved and released from the gel layer. Drug solubility, therefore, is an important factor determining the mechanism of drug release from HPMC hydrophilic matrices, influencing the choice of polymer viscosity, chemistry and choice of excipients. Use of an appropriate viscosity grade will enable a formulation scientist to design matrices based on diffusion, diffusion and erosion or erosion only mechanisms. For water-soluble drugs, high viscosity grades of HPMC (Methocel K4M CR, K15M CR or K100M CR) tend to generate consistent diffusion-controlled systems (n approaching ~ 0.45). For drugs with poor water solubility, low viscosity grades of HPMC (Methocel K100LV CR and E50LV) are recommended where erosion is the predominant release mechanism ($n \sim 0.9$). Depending on drug solubility, it may be necessary to blend polymers of different viscosities to obtain an intermediate viscosity grade of HPMC and achieve desired release kinetics. It should be noted that as drug diffusion is dependent on its molecular weight, chemistry and other excipients within the gel layer, drug release too is dependant on these properties [45].

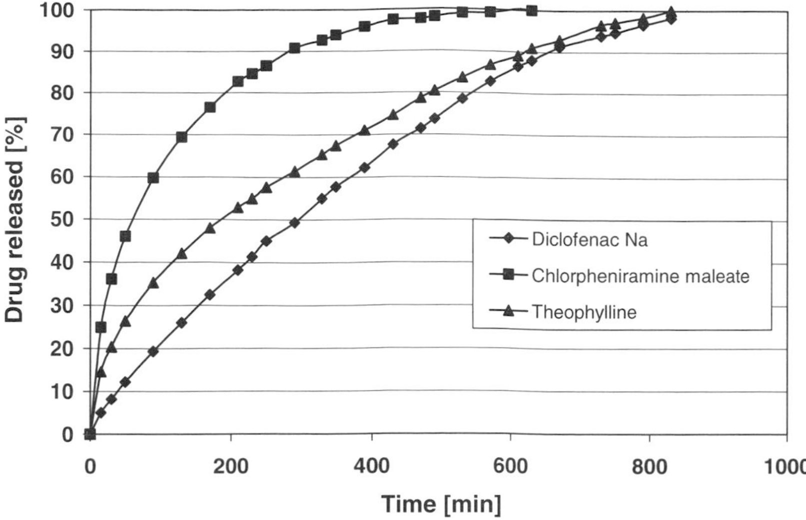

Fig. 11.2 Influence of drug solubility on their release profiles from a Methocel K4M CR hydrophilic matrix formulation consisting of 30% (w/w) polymer and microcrystalline cellulose as filler

Figure 11.2 shows the influence of drug solubility on release profiles for chlorpheniramine maleate, diclofenac sodium and theophylline from a Methocel K4M CR matrix formulation, keeping all other matrix composition and properties constant [46]. As aqueous solubility of the drug decreased, drug release rate also decreased.

For poorly soluble drugs, particle size of the drug has a major influence on its release profile [47–49]. A decrease in particle size of the drug causes increase in solubility and hence faster drug release rate.

Hydrophilic matrix formulation of high-dose drugs (approximating ~1.0 g) is challenging because of the overall dosage weight limitations versus the quantity of the polymer required to achieve desired release profiles. It has been reported that very large tablets that are formulated to be swallowed whole (e.g. ER and delayed-release formulations) lead to poor patient compliance and therefore reduced market acceptability [50].

2.5.2 Polymer Considerations

Polymer level and viscosity grade are the major drug release controlling factors in HPMC hydrophilic matrices. Depending on dosage size and desired release rate, the typical use level can vary from ~20% to 50% (w/w) [19]. For drugs with high water solubility, there is a threshold level of polymer for achieving extended release, and further increase in polymer level may not decrease the drug release rate. However, for obtaining a robust formulation with consistent performance and insensitivity to minor variations in raw materials or manufacturing processes, a usage level of ≥30% (w/w) has been recommended [51, 52].

Particle size of the polymer is another important factor. The finer the particle size, the faster the rate of hydration of the polymer and hence better the control of drug release [53]. Coarser polymer particles used in a direct compression formulation have been reported to result in faster drug release than finer particles [54]. The coarser the particle size, the slower the hydration rate and gel layer formation. The way to circumvent this problem is the use of fine particle size grades of the polymer. For example, Methocel K Premium CR grades have more than 90% of particles below 149 μm or 100 mesh.

The methoxyl to hydroxypropoxyl substitution ratio of HPMC polymer also influences drug release which generally follows Methocel E (hypromellose 2910) > K (hypromellose 2208) [30]. Matrices formulated with high viscosity grades of HPMC form gel layers with higher gel strengths [55], which results in slower diffusion and erosion rates and hence slower drug release.

2.5.3 Presence of Other Excipients

Fillers

Soluble (e.g. lactose), insoluble (e.g. microcrystalline cellulose, dicalcium phosphate) and/or partially soluble (e.g. partially pregelinized starch) fillers are generally used in hydrophilic matrices to enhance pharmacotechnical properties of tablets (improve compressibility, flow and mechanical strength) or to modify the drug release profile. The inclusion of fillers affects the dissolution performance of a matrix by a “dilution effect” on the polymer. The magnitude of the effect on the performance of matrices is dependant on the drug, the polymer level and the level of excipient itself. The presence of water-soluble fillers in high concentrations in the matrix leads to faster and greater water uptake by the matrix, resulting in weaker gel strength, higher erosion of the gel layer and therefore faster drug release. Insoluble but weakly swellable fillers such as microcrystalline cellulose remain within the gel structure and generally result in decreased release rate [11]. The presence of partially pregelatinzed starch such as Starch 1500® in HPMC matrices has been reported to decrease the drug release rate [56]. For a highly soluble or sparingly soluble drug, the rank order of release rate was as follows: lactose > microcrystalline cellulose > partially pregelatinzed starch [56].

Release Modifiers and Stabilizers

As discussed previously, HPMC is a non-ionic polymer and hence the polymer hydration and gel formation of its matrix is essentially independent of pH of a typical dissolution media used. However, when drugs with pH-dependent aqueous solubility (weak acids or bases) are formulated in HPMC matrices, they may exhibit pH-dependent drug release. Formulating ER matrices of such drugs may lead to lower drug release due to exposure of the dosage form to increasing pH media of the GI tract (from pH 1.2 to 7) [57]. Formulating pH-independent ER matrices for such drugs would not only ensure adequate release throughout the physiological pH, but also lower intra- and inter-patient variability [58, 59]. Development of such pH-independent matrices for weakly basic drugs has been shown with the incorporation of acidic excipients (weak acids or salts of strong acids) that lower the micro-environmental pH within the gel layer and thus maintain high local solubility of the drug independent of the external release media [60–66]. Two types of acidic excipients have been used. The first category is "small molecules" or non-polymeric pH modifiers such as adipic, citric, malic, succinic, tartaric, ascorbic or fumaric acid and salts of strong acids such as L-cysteine hydrochloride and glycine hydrochloride. The second category is "large molecules" or polymeric pH modifiers such as sodium alginate, Carbopol and enteric polymers. The extent of micro-environmental control of pH is dependant on the ionization constant and solubility of the release modifier. In general, the higher the pK_a of the acid, the higher the micro-environmental pH. In addition to control of the micro-environmental pH, the polymeric pH modifiers may also alter the gel strength and erosion rate of the matrix and therefore the release rate of the drug [60, 64]. The combination of these two opposing effects could also contribute to pH-independent release profiles.

Similar to basic drugs, development of pH-independent ER matrices for weakly acidic drugs is possible with incorporation of non-polymeric bases/salts of strong bases and polymeric pH modifiers [67–70]. The examples of basic excipients are sodium, potassium or magnesium salts of (bi)carbonate, phosphate or hydroxide, magnesium oxide, 2-amino-2-methyl-1,3-propanediol (AMPD) and Eudragit® E100.

The effectiveness of this approach often depends on the properties of the drug and the release-modifying agent as well as the ratio of the drug to release-modifying excipient. In matrix systems, a small molecule pH modifier (such as tartaric acid or citric acid) that is water soluble can leach out of viscous gel layer fairly quickly, resulting in a limited change of pH in the gel layer over an extended duration of the drug release. Thus, it is important to design a system that retains the release-modifying agent in a delivery device suitable for the extended period of release. Polymeric pH modifiers are a better choice in such situations as they have higher molecular weights and provide longer residence times in the matrix. However, the magnitude of pH modulation provided by the polymeric pH modifiers is not expected to be comparable to that provided by non-polymeric acids, and in some cases it might be necessary to include an additional "small molecule" pH modifier in the matrix formulation.

pH modifiers are also used for improving the stability of active pharmaceutical ingredients in the matrix composition. Bupropion hydrochloride, for example, is an antidepressant drug that undergoes degradation in an alkaline environment. To formulate an acceptable ER solid dosage form, the use of weak acids or salts of strong acids as stabilizers in the formulation (tartaric acid, citric acid, ascorbic acid, L-cysteine hydrochloride and glycine hydrochloride) has been suggested in the literature [71]. These stabilizers provide an acidic environment surrounding the active drug that prevents its decomposition [71].

Effect of Salts and Electrolytes

In general, as the concentration of ions in a polymer solution increases, polymer hydration or solubility decreases [72]. The amount of water available to hydrate the polymer is reduced because more water molecules are required to keep the ions in solution. Moreover, the types of ions in solution affect polymer hydration to varying degrees. The susceptibility of cellulose ethers to ionic effects follows the lyotropic series of the ions (chloride < tartarate < phosphates and potassium < sodium) [73]. Changes in the hydration state of a polymer in solution are manifested primarily by changes in solution viscosity and turbidity or cloud point [55]. At low ionic strengths, the polymer hydration is unaffected, but higher ionic strengths may lead to a loss of gel integrity of the matrix. The extent of this influence depends on the polymer type and lyotropic series of the ions. The effect of electrolytes or salts is important only in cases where high concentrations of salts or electrolytes are present as tablet components or as constituents of dissolution media. *In-vivo* conditions, however, have fairly low ionic strength (ionic strength of gastrointestinal fluids, $\mu = 0.01–0.15$) to affect the polymer hydration and have significant impact on release rate [74].

2.5.4 Method of Manufacture

Hydrophilic matrix tablets are manufactured using traditional tablet manufacturing methods of direct compression (DC), wet granulation or dry granulation (roller compaction or slugging) depending on formulation properties or on manufacturer's preference.

HPMC polymers generally have very good compressibility and results in tablets with high mechanical strength [75]. It has been reported that high molecular weight grades of HPMC may undergo less plastic flow than the low molecular weight grades and thus require higher pressures to deform [76]. In a matrix formulation, the inclusion of DC excipients and other ingredients may render the formulation for direct compression with acceptable mechanical properties of the tablets.

Aqueous wet granulation is generally achieved using a spray system to avoid formation of a lumpy mass [77]. Addition of a binder may not be necessary as HPMC itself has excellent binder properties when hydrated. Over-granulation or high concentration of a binder beyond the optimal level could adversely affect the compressibility of the granules. To reduce the formation of a lumpy mass during granulation and improve process efficiency, a novel foam granulation technology has recently been introduced [78]. In this method, using a simple foam apparatus, air is incorporated into a solution of conventional water-soluble polymeric binder such as a low viscosity grade HPMC to generate foam. Application of such foam for granulation results in an increased surface area and volume of polymeric binder and therefore improves the distribution of water/binder system throughout the powder bed.

The effect of compression force on drug release from hydrophilic matrices is minimal when tablets are made with sufficient strength and optimum levels of polymers are used [48]. One could relate variation in compression forces to a change in the porosity of the tablets. However, as the porosity of the hydrated matrix is independent of the initial porosity, the compression force is expected to have little influence on drug release rate [79]. Once a sufficient tablet hardness suitable for processing and handling is achieved, tablet hardness would have little further effect on drug release profile. To ensure consistent porosity and avoid entrapment of air within the dry tablet core, a pre-compression step may have to be considered in the manufacture of matrices.

Compression speed has been reported to adversely affect the tensile strength of the tablets and lower compression speed has been suggested for obtaining a product with better mechanical quality [80–82]. A robust formulation, which is insensitive to changes in the manufacturing processes such as over-granulation effect or variable tablet hardness, may be obtained by reducing the amount of intragranular HPMC and replacing it as an extragranular component [83].

2.5.5 Characteristic of Dosage Form

Variation in tablet shape and size may cause changes in surface area available for drug release and hence influences drug release profiles from HPMC matrices. A constant surface area to volume ratio (S/V) of different size and shape tablets for a HPMC formulation would lead to similar drug release profiles [84]. The size of the tablet may also dictate the polymer level requirement. Smaller tablets have been reported to require higher polymer content because of their higher surface area to volume ratio and thus shorter diffusion pathways [85]. Siepmann et al. have reported a model which calculates the size and shape of hydrophilic matrices required to achieve a particular release rate [85]. Variation in the aspect ratio (radius/height) of the tablets leading to an optimal shape can be calculated to achieve a desired release profile. One technology proposed for modifying the matrix surface area to volume ratio was by physical restriction of the swelling of hydrophilic matrix by partially coating the matrix with insoluble polymers or multi-layered tablets (Geomatrix® technology) [86, 87]. The other technology which demonstrates the advantages of modulated S/V is Dome Matrix® technology [88]. The unique shape of this Matrix technology was designed to allow possible addition (stacking) of several of these systems to produce different geometric designs.

2.5.6 Presence of Coating

Application of film coatings to tablet formulations is a common practice in the pharmaceutical industry. Tablets are coated for a variety of reasons such as improving the stability of the formulation, taste masking, enhancing the aesthetic appearance, identification and branding, improving the packaging process or modifying drug release profile. Coating of hydrophilic matrices with water-soluble polymers such as Opadry® or low-viscosity HPMC generally does not alter drug release profiles [89, 90]. Coating with water-insoluble polymers such as ethylcellulose with or without permeability modifiers (e.g., low viscosity grades of HPMC or Opadry) may be used for modulating the drug release profile from HPMC matrices [91, 92].

2.5.7 Modulation of Drug Release from HPMC Hydrophilic Matrices

Drug release profiles from HPMC hydrophilic matrices are generally first-order for highly soluble drugs or zero-order for practically insoluble drugs, with the release exponent, n (Eq. 1) ranging from 0.5 to 0.8. In most cases, the choice of polymer, filler type and their levels would determine the release kinetics of drugs. Various other strategies have been investigated to further modulate drug release from these matrices, including use of other polymers, e.g. ethylcellulose [92], enteric polymers [93], hydrophobic materials [91, 92], other ionic or non-ionic hydrophilic polymers [25, 45], polysaccharides [29, 94, 95] (Table 11.2); restriction of the swelling characteristics of the HPMC matrices [86]; use of compression coating with hydrophilic polymers [96] or insoluble film coating [92] and multi-layer tablets and dosage shape [85].

As shown in this chapter, formulating HPMC matrices can be a complex process; critical factors being drug solubility, dosage level, rate-controlling polymer and excipient choice. In order to help the pharmaceutical scientists with a starting formula for hydrophilic matrix tablets, Colorcon, Inc. has developed a predictive formulation service called HyperStart® [97]. This system is based on mathematical models and relationships, validated with extensive experimental data. Use of the HyperStart formulation service may help to simplify the formulation and development process and reduce the time to market.

2.6 Case Studies: Formulation of Hydrophilic Matrices

In this section, practical examples demonstrating the formulation and process development of extended-release matrices are presented. Verapamil hydrochloride (HCl) (soluble), metformin HCl (freely soluble), carbamazepine (practically insoluble) and venlafaxine HCl (freely soluble) extended-release matrix formulations are used to illustrate the different formulation and manufacturing principles (direct compression and wet granulation).

2.6.1 Verapamil Hydrochloride Extended-Release Matrices (240 mg ER tablets) [98]

Verapamil HCl is a high-dose, soluble drug (1 in 20 parts of water), and extended-release tablet formulations of Verapamil HCl have been studied and marketed. Isoptin® SR (Abbott Laboratories) utilizes hydrophilic polymers as the release-controlling agent [99], while Covera HS® (Pfizer Inc.) utilizes an osmotic pump tablet formulation [100] based on a controlled onset extended release platform. The objective of this study was to develop an 8-h extended release, HPMC matrix formulation of Verapamil HCl 240 mg with a release profile that meets the requirements of "Verapamil HCl Extended Release Tablets" (method III, USP 28/NF 23). The formulation composition and procedure proposed by the HyperStart service for manufacturing of the tablets are depicted in Table 11.5. This is an example of wet granulation method to overcome the poor compressibility and flow of a high-dose, high-solubility drug. The dissolution results show that the formulation met the release profile requirements set by USP 28/NF 23 (Fig. 11.3.).

2.6.2 Carbamazepine Extended-Release Matrices (200 mg tablets) [101]

Although carbamazepine presents formulation challenges due to its inherent poor compressibility, high dose and low water solubility (1 in 10,000), extended-release

Table 11.5 Hydrophilic matrix formulation for verapamil HCl (240 mg)

Ingredient	% w/w	Quantity per tablet (mg)	Quantity per 10,000 tablets[a] (g)
Verapamil HCl	47.80	240.00	2,400.00
HPMC (Methocel K100 LV CR)	29.90	150.00	1,500.00
HPMC (Methocel E5 LV)	0.40	2.00	20.00
Lactose (Fast Flo® NF)	20.90	105.00	1,050.00
Colloidal silicon dioxide (Cab-O-Sil® M-5)	0.50	2.50	25.00
Magnesium stearate NF	0.50	2.50	25.00
Purified water	q.s.	q.s.	q.s.
Total	*100.00*	*502.00*	*5,020.00*

q.s. quantity sufficient, [a]Calculated quantity for better understanding to reader

Verapamil HCl and spray-dried lactose were blended in a Hobart mixer for 5 min. This blend was then granulated with 2% (w/w) aqueous solution of Methocel E5. The wet mass was dried in an oven at 40°C for 10 h. The resulting granules were passed through a 16-mesh screen (1.18 mm) and blended with Methocel K100LV and colloidal silicon dioxide in a twin shell blender (Patterson Kelley, USA) for 10 min. Magnesium stearate was added to the mixture and blended for an additional 3 min. The blend was then compressed on an instrumented rotary tablet press using 11-mm standard concave tooling.

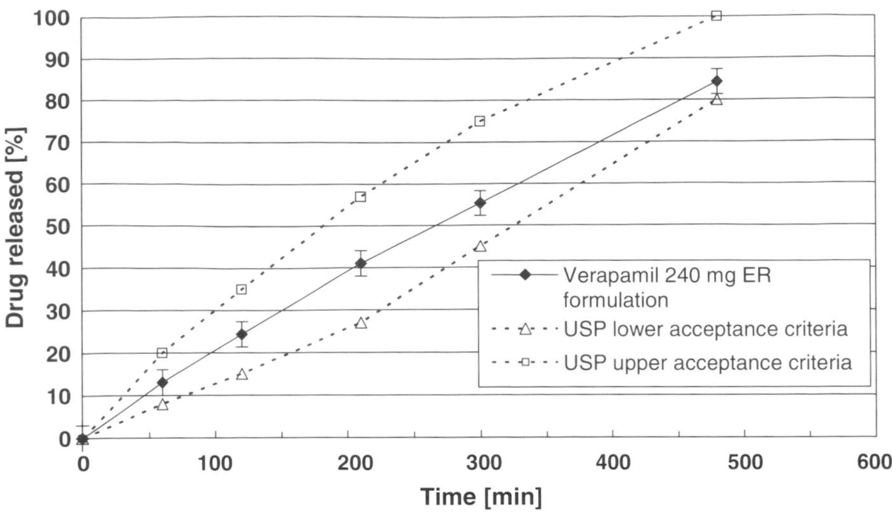

Fig. 11.3 Drug release profile of verapamil HCl (240 mg) extended release matrix tablets in simulated gastric (0–2 h) and intestinal fluid (2–8 h) without enzymes (900 mL), using USP apparatus II at 50 rpm

tablet formulations of carbamazepine have been studied and marketed [102]. This case study is an example of a 24-h extended release of Carbamazepine (200 mg) using a Methocel matrix formulation proposed by HyperStart (Table 11.6) with a drug release profile within the USP acceptance criteria. The formulation composition and procedure for manufacturing tablets are shown in Table 11.6. A wet granulation method has been used to improve compressibility and flow properties of the formulation. Moreover, a surfactant has been included in the formulation to improve wettability and solubility of carbamazepine within the formulation. Figure 11.4 shows the dissolution profile for this formulation, which meets the USP requirements.

2.6.3 Metformin Hydrochloride Extended-Release Matrices (500 mg tablets) [103]

Metformin HCl presents formulation challenges due to its poor inherent compressibility, high dose and high water solubility (4 in 1 part at 25°C). The objective of this case study was to develop a 12-h slow release HPMC matrix formulation of Metformin HCl 500 mg with a release profile similar to the marketed brand product Glucophage® XR (Bristol Myers Squibb). The formulation composition and procedure for manufacturing tablets are shown in Table 11.7. The dissolution results show that the use of Methocel matrices alone resulted in an extended drug release formulation of a highly water-soluble drug (Fig. 11.5).

Table 11.6 Hydrophilic matrix formulation for carbamazepine (200 mg)

Ingredient	% w/w	Quantity per tablet (mg)	Quantity per 10,000 tablets[a] (g)
Carbamazepine	57.14	200.00	2,000.00
HPMC (Methocel K100LV CR)	30.00	105.00	1,050.00
Microcrystalline cellulose (Avicel® PH102)	10.95	38.32	383.20
HPMC (Methocel E3LV)[b]	0.16	0.56	5.60
Sodium lauryl sulphate[c]	0.50	1.75	17.50
Fumed silica (Aerosil® 200)	1.00	3.50	35.00
Magnesium stearate	0.25	0.87	8.70
Purified water	q.s.	q.s.	q.s.
Total	*100.00*	*350.00*	*3,500.00*

[a]Calculated quantity for better understanding to reader, [b]Methocel E3LV was used as a wet granulation binder, [c]Sodium lauryl sulphate (SLS), a surfactant, was used within the binder solution to improve carbamazepine solubility.

Carbamazepine and 50% of Methocel K100LV CR were loaded in a fluid bed granulator (Glatt GPCG 1) and granulated using an aqueous solution of Methocel E3 LV containing sodium lauryl sulfate with a spray rate of 18 g/min and atomizing air pressure of 1.5 bar. The resulting granules were screened using a mesh (no. 25), loaded to Turbula mixer, and the remaining quantity of Methocel K100LV CR, fumed silica and microcrystalline cellulose was added. The mixture was blended for 5 min. Lubricant (magnesium stearate) was added and blending was performed for an additional 1 min. The blend was then compressed on an instrumented rotary tablet press using 9-mm standard concave tooling.

Note: Processing conditions in fluid bed granulation could vary depending on the quantity of total composition and type of fluid bed machine utilized

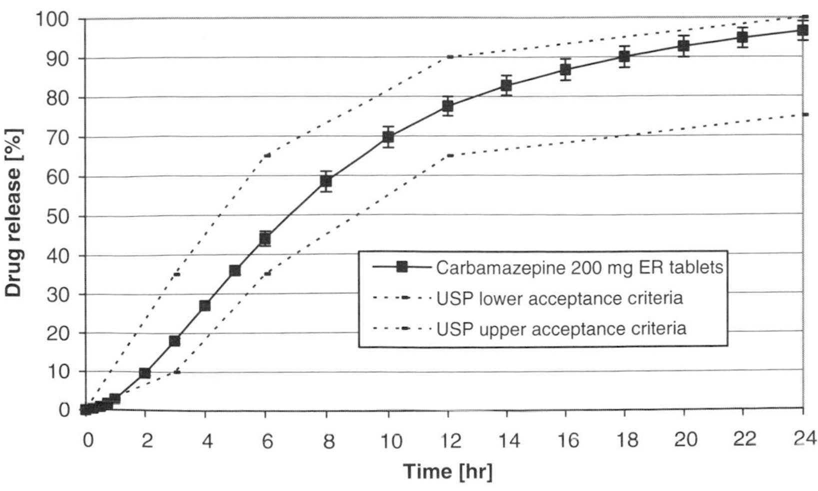

Fig. 11.4 Drug release profile of carbamazepine (200 mg) extended release matrix tablets in water (900 mL) using USP apparatus I at 100 rpm

Table 11.7 Metformin HCl extended release matrix formulation (500 mg tablets)

Ingredient	% w/w	Quantity per tablet (mg)	Quantity per 10,000 tablets[a] (g)
Metformin	50.00	500.00	5,000.00
HPMC (Methocel K100M CR)	30.00	300.00	3,000.00
Microcrystalline cellulose (Avicel PH102)	19.00	190.00	1,900.00
Fumed silica (Aerosil 200)	0.50	5.00	50.00
Magnesium stearate	0.50	5.00	50.00
Total	*100.00*	*1,000.00*	*10,000.00*

[a]Calculated quantity for better understanding to reader

Microcrystalline cellulose and fumed silica were screened using a mesh (no. 50). All ingredients except magnesium stearate were mixed in Turbula mixer for 5 min. Magnesium stearate was added and the mixture was blended for an additional 2 min. The blend was compressed on an instrumented rotary tablet press using 7.0 × 18.0 mm caplet tooling.

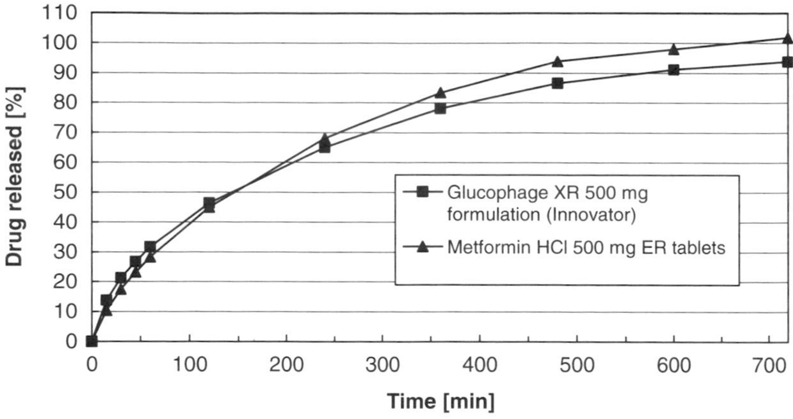

Fig. 11.5 Drug release profile of metformin HCl (500 mg) extended release matrix tablets in water (1000 mL) using USP apparatus II at 100 rpm

2.6.4 Venlafaxine Hydrochloride Extended-Release Matrices (37.5 mg tablets) [92]

Highly water-soluble drugs formulated with HPMC matrices may be characterized with an initial burst effect. The purpose of this study was to modulate drug release of a highly water-soluble active, venlafaxine HCl, from HPMC matrices without an initial burst effect. In this case study, aqueous ethylcellulose dispersion (Surelease®) was used to coat the matrix to suppress the initial burst. The formulation composition and procedure for manufacturing tablets are shown in Table 11.8. The dissolution results show that using Surelease as an insoluble coating resulted in an extended drug release formulation of venlafaxine HCl without the typical initial burst effect (Fig. 11.6).

Table 11.8 Venlafaxine HCl matrix formulation (37.5 mg tablets)

Ingredient	% w/w	Quantity per tablet (mg)	Quantity per 10,000 tablets[a] (g)
Venlafaxine HCl	12.50	37.50	375.00
Starch 1500	25.00	75.00	750.00
HPMC (Methocel K15M CR)	30.00	90.00	900.00
Microcrystalline cellulose (Avicel PH102)	31.50	94.50	945.00
Magnesium stearate	0.50	1.50	15.00
Fumed silica (Aerosil 200)	0.50	1.50	15.00
Total	*100.00*	*300.00*	*3,000.00*
Coating of formulation			
Aqueous dispersion of ethylcellulose (Surelease)	q.s.	q.s.	

[a]Calculated quantity for better understanding to reader

Venlafaxine HCl, Methocel K15M CR and microcrystalline cellulose were blended for 10 min. Lubricant (magnesium stearate) and glidant (fumed silica) were added and blending was performed for an additional 5 min. The blend was compressed on an instrumented rotary tablet press using 10 mm standard concave tooling. Tablets were then coated with a diluted (15% w/v solids) aqueous dispersion of ethylcellulose (Surelease) to a 4% weight gain using a side vented coating machine (O'Hara Lab coat-I).

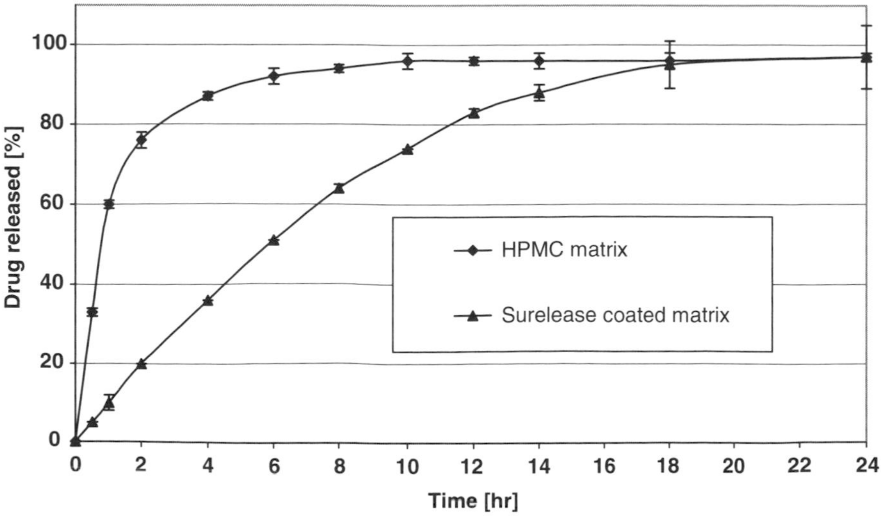

Fig. 11.6 Drug release profile of venlafaxine HCl (37.5 mg) extended release matrix tablets in water (900 mL) using USP apparatus II at 100 rpm

3 Notes

1. Drug solubility and dose are the most important factors to consider in the formulation of HPMC ER matrices. Use of an appropriate viscosity grade will enable a formulation scientist to design matrices based on diffusion, erosion or diffusion and erosion mechanisms. For water-soluble drugs, high viscosity grades of HPMC (Methocel K4M CR, K15M CR or K100M CR) tend to generate consistent diffusion-controlled systems (n approaching ~0.45). For drugs with poor water solubility, low viscosity grades of HPMC (Methocel K100LV CR and E50LV) are recommended where erosion is the predominant release mechanism ($n \sim$ 0.9). Depending on drug solubility, it may be necessary to blend polymers of different viscosities to obtain intermediate viscosity grades of HPMC and achieve desired release kinetics.
2. Polymer level is also the major drug release rate controlling factor in HPMC matrices. Depending on dosage form, size and desired release rate, the typical use level can vary from ~20 to 50% (w/w) [19]. For obtaining a robust formulation with consistent performance and which is insensitive to minor variations in raw materials or manufacturing processes, usage level of ≥30% (w/w) is generally recommended.
3. Particle size of the HPMC is another important factor. The finer the particle size, the faster the rate of hydration of the polymer and hence better the control of dug release [53]. In ER hydrophilic matrices, it is generally recommended to use fine particle size grades of the polymer (e.g. Methocel K Premium CR grades have more than 90% of particles below 149 μm or no. 100 mesh).
4. Hydrophilic HPMC matrices are manufactured using traditional manufacturing methods such as direct compression (DC), wet granulation or dry granulation (roller compaction or slugging). The choice of the method depends on the formulation properties or on the manufacturer's preference, or both.
5. The effect of compression force on drug release from hydrophilic matrices is minimal when tablets have enough hardness (to withstand handling) and optimum levels of polymers are used [48]. To ensure consistent quality of the tablets, a pre-compression step may have to be considered in the manufacture of hydrophilic matrices.
6. In a wet granulation process, inclusion of a portion of the HPMC as inter-granular and a portion as extra-granular may be beneficial.
7. Smaller tablets have been reported to require higher polymer contents because of their higher surface area to volume ratio and thus shorter diffusion pathways.
8. Coating of hydrophilic matrices with water-soluble polymers such as Opadry or low-viscosity HPMC generally does not alter drug release profiles. Coating with water-insoluble polymers such as ethylcellulose may be used for modulating the drug release profile from HPMC matrices.
9. Further modification and fine tuning of drug release from HPMC matrices may be achieved by the use of other non-ionic/ionic polymers, water-insoluble polymers, polysaccharides or hydrophobic excipients.
10. In order to help pharmaceutical scientists with a starting formula for hydrophilic matrix tablets predictive mathematical models such as HyperStart® has been developed [97]. Use of this service will simplify the development process and reduce the time to market.

Methocel™ and Polyox™ are trademarks of the Dow Chemical Company. Carbopol® and Eudragit® are registered trademarks of Lubrizol Advanced Materials, Inc., and Roehm GMBH Ltd respectively. Surelease®, Starch 1500®, Opadry® and HyperStart® are registered trademarks of BPSI Holdings LLC. Fast Flo®, Cab-O-Sil® and Avicel® are registered trademarks of Foremost Farms USA, Cabot Corporation, and FMC Corporation respectively. Geomatrix® is a registered trademark of SkyePharma PLC.

Acknowledgements The authors thank Kurt Fegely, Dr. Marina Levina, Dr. Abhijit Gothoskar and Viena Diaz (all from Colorcon, Inc.) for their experimental contributions to the data presented in this chapter.

References

1. Lordi, N.G. (1987) Sustained release dosage forms. In: Lachman, L., Lieberman, H.A. and Kanig, J.L. (eds.) The theory and practice of industrial pharmacy, 3rd edn (Indian edn). Varghese Publishing House, Bombay, pp. 430–456.
2. Chiao, C.S.L. and Robinson, J.R. (1995) Sustained-release drug delivery systems. In: Gennaro, A.R. (ed.) Remington: the science and practice of pharmacy, vol. II. Mack Publishing Company, Easton, PA, pp. 1660–1675.
3. Kydonieus, A.F. (1890) Fundamental concepts of controlled release. In: Kydonieus, A.F. (ed.) Controlled release technologies: methods, theory, and applications, vol. I. CRC, Boca Raton, Florida, pp. 1–19.
4. U.S. Food and Drug Administration, Center for Drug Evaluation and Research, Drug Approval Reports. http://www.accessdata.fda.gov/scripts/cder/drugsatfda/index.cfm?fuseaction=Reports.ReportsMenu.
5. GlaxoSmithKline. Our Histroy [see 1952, online]. http://www.gsk.com/about/histroy-noflash.htm.
6. Sastry, S.V., Nyshadham, J.R. and Fix, J.A. (2000) Recent technological advances in oral drug delivery – a review. Pharm Sci Technol Today, 3, 138–145.
7. Rathbone, M.J., Hadgraft, J. and Roberts, M.S. (eds.) (2003) Modified release drug delivery technology. Marcel Dekker, New York.
8. Buri, P., and Doelker, E. (1980) [Formulation of sustained-release tablets. II. Hydrophilic matrices]. Pharm Acta Helv, 55, 189–197.
9. Conte, U., Colombo, P., Caramella, C. and La Manna, A. (1979) Sustained release nitrofurantoin tablets by direct compression. Farmaco [Prat], 34, 306–316.
10. Conte, U., Colombo, P., Gazzaniga, A., Sangalli, M.E. and La Manna, A. (1988) Swelling-activated drug delivery systems. Biomaterials, 9, 489–493.
11. Lee, B.-J., Ryu, S.-G. and Cui, J.-H. (1999) Formulation and release characteristics of hydroxypropyl methylcellulose matrix tablet containing melatonin. Drug Dev Ind Pharm, 25, 493–501.
12. Salomen, J.L. and Doelker, E. (1980) [Formulation of sustained release tablets. I. Inert matrices]. Pharm Acta Helv, 55, 174–182.
13. Wilding, I.R., Davis, S.S., Melia, C.D., Hardy, J.G., Evans, D.F., Short, A.H. and Sparrow, R.A. (1989) Gastrointestinal transit of Sinemet CR in healthy volunteers. Neurology, 39, 53–58.
14. Wilding, I.R., Hardy, J.G., Davis, S.S., Melia, C.D., Evans, D.F., Short, A.H., Sparrow, R.A. and Yeh, K.C. (1991) Characterisation of the *in vivo* behaviour of a controlled-release formulation of levodopa (Sinemet CR). Clin Neuropharmacol, 14, 305–321.
15. Zhu, Y., Shah, N.H., Malick, A.W., Infeld, M.H. and McGinity, J.W. (2006) Controlled release of a poorly water-soluble drug from hot-melt extrudates containing acrylic polymers. Drug Dev Ind Pharm, 32, 569–583.
16. USP 29-NF 24. General Chapters: <1090> *In vivo* bioequivalence guidance – general guidelines. http://www.uspnf.com/uspnf/login.
17. Inactive ingredient search for approved drug products. http://www.accessdata.fda.gov/scripts/cder/iig/index.cfm.
18. Rowe, R.C., Sheskey, P.J. and Owen, S.C. (2006) Handbook of pharmaceutical excipients, 5th edn. The Pharmaceutical Press, London, UK.
19. Dow excipients, Methocel™ products. http://www.dow.com/dowexcipients/products/methocel.htm.
20. USP monographs: hypromellose. http://www.uspnf.com/uspnf/login.
21. Rajabi-Siahboomi, A.R., Bowtell, R.W., Mansfield, P., Davies, M.C. and Melia, C.D. (1996) Structure and behavior in hydrophilic matrix sustained release dosage forms: 4. Studies of water mobility and diffusion coefficients in the gel layer of HPMC tablets using NMR imaging. Pharm Res, 13, 376–380.

22. McCrystal, C.B., Ford, J.L. and Rajabi-Siahboomi, A.R. (1999) Water distribution studies within cellulose ethers using differential scanning calorimetry. 1. Effect of polymer molecular weight and drug addition. J Pharm Sci, 88, 792–796.
23. Metolose®. http://www.metolose.jp/e/pharmaceutical/metolose.shtml.
24. Aqualon pharmaceutical excipients. http://www.herc.com/aqualon/pharm/index.html.
25. Dabbagh, M.A., Ford, J.L., Rubinstein, M.H., Hogan, J.E. and Rajabi-Siahboomi, A.R. (1999) Release of propranolol hydrochloride from matrix tablets containing sodium carboxymethylcellulose and hydroxypropylmethylcellulose. Pharm Dev Technol, 4, 313–324.
26. Carbopol brand polymers. http://www.carbopol.com/.
27. Dow excipients, POLYOX products. http://www.dow.com/dowexcipients/products/polyox.htm.
28. Timmins, P., Delargy, A.M. and Howard, J.R. (1997) Optimization and characterization of a pH independant extended release hydrophilic matrix tablet. Pharm Dev Technol, 2, 25–31.
29. Kelly, M.L., Tobyn, M.J. and Staniforth, J.N. (2000) Tablet and capsule hydrophilic matrices based on heterodisperse polysaccarides having porosity-independant *in vitro* release profiles. Pharm Dev Technol, 5, 59–66.
30. Rajabi-Siahboomi, A.R. and Jordan, M.P. (2000) Slow release HPMC matrix systems. Eur Pharm Rev, 5, 21–23.
31. Siepmann, J., Kranz, H., Bodmeier, R. and Peppas, N.A. (1999) HPMC-matrices for controlled drug delivery: a new model combining diffusion, swelling, and dissolution mechanisms and predicting the release kinetics. Pharm Res, 16, 1748–1756.
32. Siepmann, J. and Peppas, N.A. (2000) Hydrophilic matrices for controlled drug delivery: an improved mathematical model to predict the resulting drug release kinetics (the "sequential layer" model). Pharm Res, 17, 1290–1298.
33. Siepmann, J. and Peppas, N.A. (2001) Modeling of drug release from delivery systems based on hydroxypropyl methylcellulose (HPMC). Adv Drug Deliv Rev, 48, 139–157.
34. Siepmann, J. and Peppas, N.A. (2001) Mathematical modeling of controlled drug delivery. Adv Drug Deliv Rev, 48, 137–138.
35. Siepmann, J., Podual, K., Sriwongjanya, M., Peppas, N.A. and Bodmeier, R. (1999) A new model describing the swelling and drug release kinetics from hydroxypropyl methylcellulose tablets. J Pharm Sci, 88, 65–72.
36. Siepmann, J., Streubel, A. and Peppas, N.A. (2002) Understanding and predicting drug delivery from hydrophilic matrix tablets using the "sequential layer" model. Pharm Res, 19, 306–314.
37. Korsmeyer, R.W., Gurny, R., Doelker, E., Buri, P. and Peppas, N.A. (1983) Mechanisms of potassium chloride release from compressed, hydrophilic, polymeric matrices: effect of entrapped air. J Pharm Sci, 72, 1189–1191.
38. Ford, J.L., Mitchell, K., Rowe, P., Armstrong, D.J., Elliott, P.N.C., Rostron, C. and Hogan, J.E. (1991) Mathematical modelling of drug release from hydroxypropylmethylcellulose matrices: effect of temperature. Int J Pharm, 71, 95–104.
39. Bajwa, G.S., Hoebler, K., Sammon, C., Timmins, P. and Melia, C.D. (2006) Microstructural imaging of early gel layer formation in HPMC matrices. J Pharm Sci, 95, 2145–2157.
40. Timmins, P., Dennis, A.B. and Vyas, K.A. (2002) Biphasic controlled release delivery system for high solubility pharmaceuticals and method. US Pat. 6,475,521.
41. Timmins, P., Dennis, A.B. and Vyas, K.A. (2003) Method of use of a biphasic controlled release delivery system for high solubility pharmaceuticals and method. US Pat. 6,660,300.
42. Electronic orange book query. http://www.fda.gov/cder/ob/docs/queryai.htm.
43. International patent applications. http://www.wipo.int/pctdb/en/.
44. Lohray, B.B. and Tiwari, S.B. (2005) A controlled release delivery system for metformin. WO/2005/123134.
45. Rao, K.V., Devi, P.K. and Buri, P. (1990) Influence of molecular size and water solubility of the solute on its release from swelling and erosion controlled polymeric matrices. J Control Release, 12, 133–141.

46. Influence of drug solubility on release profiles of drugs from Methocel K4M CR matrices [Internal report]. Colorcon Inc., West Point, PA.
47. Hogan, J.E. (1989) Hydroxypropylmethylcellulose sustained release technology. Drug Dev Ind Pharm, 15, 975–999.
48. Velasco, M.V., Ford, J.L., Rowe, P. and Rajabi-Siahboomi, A.R. (1999) Influence of drug: hydroxypropylmethylcellulose ratio, drug and polymer particle size and compression force on the release of diclofenac sodium from HPMC tablets. J Control Release, 57, 75–85.
49. Mitchell, K., Ford, J.L., Armstrong, D.J., Elliott, P.N., Hogan, J.E. and Rostron, C. (1993) The influence of drugs on the properties of gels and swelling characteristics of matrices containing methylcellulose or hydroxypropylmethylcellulose. Int J Pharm, 100, 165–173.
50. Brand protection: tablet shape. http://www.colorcon.com/best/c06_tablet_shape.html.
51. Ford, J.L., Rubinstein, M.H., Changela, A. and Hogan, J.E. (1985) The influence of pH on the dissolution of promethazine hydrochloride from hydroxypropyl methylcellulose controlled release tablets. J Pharm Pharmacol, 37, 115P.
52. Levina, M., Gothoskar, A. and Rajabi-Siahboomi, A. (2006) Application of a modelling system in the formulation of extended release hydrophilic matrices. Pharm Technol Eur, 18, 20–26.
53. Alderman, D.A. (1984) A review of cellulose ethers in hydrophilic matrices for oral controlled-release dosage forms. Int J Pharm Tech Prod Manuf, 5, 1–9.
54. Shah, N., Railkar, A.S., Watnanee, P., Zeng, F.-U., Chen, A., Infeld, M.H. and Malick, A.W. (1996) Effects of processing techniques in controlling the release rate and mechanical strength of hydroxypropyl methylcellulose based hydrogel matrices. Eur J Pharm Biopharm, 42, 183–187.
55. Sarkar, N. (1979) Thermal gelation properties of methyl and hydroxypropyl methylcellulose. J Appl Polym Sci, 24, 1073–1087.
56. Levina, M. and Rajabi-Siahboomi, A. (2004) The influence of excipients on drug release from hydroxypropyl methylcellulose matrices. J Pharm Sci, 93, 2746–2754.
57. Horter, D. and Dressman, J.D. (1997) Influence of physicochemical properties on dissolution of drugs in the gastrointestinal tract. Adv Drug Deliv Rev, 25, 3–14.
58. Vashi, V.I. and Meyer, M.C. (1988) Effect of pH on the *in vitro* dissolution and *in vivo* absorption of controlled release theophylline in dogs. J Pharm Sci, 77, 760–774.
59. Kohri, N., Miyata, N., Takahashi, M., Endo, H., Iseki, K., Miyazaki, K., Takechi, S. and Nomura, A. (1992) Evaluation of pH-independant sustained release granules of dipyridamole by using gastric-acidity controlled rabbits and human subjects. Int J Pharm, 81, 49–58.
60. Tatavarti, A.S. and Hoag, S.W. (2006) Microenvironmental pH modulation based release enhancement of a weakly basic drug from hydrophilic matrices. J Pharm Sci, 95, 1459–1468.
61. Siepe, S., Herrmann, W., Borchert, H.H., Lueckel, B., Kramer, A., Ries, A. and Gurny, R. (2006) Microenvironmental pH and microviscosity inside pH-controlled matrix tablets: an EPR imaging study. J Control Release, 112, 72–78.
62. Tatavarti, A.S., Mehta, K.A., Augsburger, L.L. and Hoag, S.W. (2004) Influence of methacrylic and acrylic acid polymers on the release performance of weakly basic drugs from sustained release hydrophilic matrices. J Pharm Sci, 93, 2319–2331.
63. Siepe, S., Lueckel, B., Kramer, A., Ries, A. and Gurny, R. (2006) Strategies for the design of hydrophilic matrix tablets with controlled microenvironmental pH. Int J Pharm, 316, 14–20.
64. Streubel, A., Siepmann, J., Dashevsky, A. and Bodmeier, R. (2000) pH-independent release of a weakly basic drug from water-insoluble and -soluble matrix tablets. J Control Release, 67, 101–110.
65. Gabr, K. (1992) Effect of organic acids on release patterns of weakly basic drugs from inert sustained release matrix tablets. Eur J Pharm Biopharm, 38, 199–202.
66. Varma, M.V.S., Kaushal, A.M. and Garg, S. (2005) Influence of micro-environmental pH on the gel layer behavior and release of a weakly basic drug from various hydrophilic matrices. J Control Release, 103, 499–510.

67. O'Connor, K.M. and Corrigan, O.I. (2002) Effect of a basic organic excipient on the dissolution of diclofenac salts. J Pharm Sci, 91, 2271–2281.
68. Rao, V.M., Engh, K. and Qiu, Y. (2003) Design of pH-independent controlled release matrix tablets for acidic drugs. Int J Pharm, 252, 81–86.
69. Michelucci, J.J., Sherman, D.M. and DeNeale, R.J. (1990) Sustained release etodolac. US Pat. 4,966,768.
70. Riis, T., Bauer-Brandl, A., Wagner, T. and Kranz, H. (2007) pH-independent drug release of an extremely poorly soluble weakly acidic drug from multiparticulate extended release formulations. Eur J Pharm Biopharm, 65, 78–84.
71. Ruff, M.D., Kalidindi, S.R. and Sutton, J.J.E. (1994) Pharmaceutical composition containing bupropion hydrochloride and a stabilizer. US Pat. 5,358,970.
72. Methocel cellulose ethers technical handbook. http://www.dow.com/PublishedLiterature/dh_03e3/09002f13803e32e6.pdf?filepath=methocel/pdfs/noreg/192–01062.pdf&fromPage=GetDoc.
73. Mitchell, K., Ford, J.L., Armstrong, D.J., Elliott, P.N., Rostron, C. and Hogan, J.E. (1990) The influence of additives on the cloud point, disintegration and dissolution of hydroxypropyl methylcellulose gels and matrix tablets. Int J Pharm, 66, 233–242.
74. Johnson, J.L., Holinej, J. and Williams, M.D. (1993) Influence of ionic strength on matrix integrity and drug release from hydroxypropyl cellulose compacts. Int J Pharm, 90, 151–159.
75. Rajabi-Siahboomi, A.R., Nokhodchi, A. and Rubinstein, M.H. (1998) Compaction behaviour of hydrophilic cellulose ether polymers. Pharm Technol, Tableting and Granulation:1998 yearbook, 32–38.
76. Nokhodchi, A. and Rubinstein, M.H. (2001) An overview of the effects of material and process variables on the compaction and compression properties of hydroxypropylmethyl cellulose and ethylcellulose. STP Pharm Sci, 11, 195–202.
77. Liu, C.-H., Chen, S.-C., Kao, Y.-H., Kao, C.-C., Sokoloski, T.D. and Sheu, M.-T. (1993) Properties of hydroxypropylmethylcellulose granules produced by water spraying. Int J Pharm, 100, 241–248.
78. Dow excipients, foamed binder technology. http://www.dow.com/dowexcipients/applications/foam.htm.
79. Melia, C.D., Rajabi-Siahboomi, A., Hodsdon, A.C., Adler, J. and Mitchell, J.R. (1993) Structure and behavior of hydrophilic matrix sustained release dosage forms. 1. The origin and mechanism of formation of gas bubbles in the hydrated surface layer. Int J Pharm, 100, 263–269.
80. Nokhodchi, A., Ford, J.L., Rowe, P. and Rubinstein, M.H. (1996) The effects of compression rate and force on the compaction properties of different viscosity grades of hydroxypropylmethylcellulose 2208. Int J Pharm, 129, 21–31.
81. Nokhodchi, A., Ford, J.L., Rowe, P.H. and Rubinstein, M.H. (1996) The effect of moisture on the heckel and energy analysis of hydroxypropylmethylcellulose 2208 (HPMC K4M). J Pharm Pharmacol, 48, 1122–1127.
82. Nokhodchi, A., Ford, J.L., Rowe, P.H. and Rubinstein, M.H. (1996) The influence of moisture content on the consolidation properties of hydroxypropylmethylcellulose K4M (HPMC 2208). J Pharm Pharmacol, 48, 1116–1121.
83. Huang, Y., Knanvilkar, K., Moore, A.D. and Hilliard-Lott, M. (2003) Effects of manufacturing variables on *in vitro* dissolution characteristics of extended release tablets formulated with hydroxypropylmethylcellulose. Drug Dev Ind Pharm, 29, 79–88.
84. Reynolds, T.D., Mitchelle, S.A. and Balwinski, K.M. (2002) Investigation of the effect of tablet surface area/volume on drug release from hydroxypropylmethylcellulose controlled release matrix tablets. Drug Dev Ind Pharm, 28, 457–477.
85. Siepmann, J., Kranz, H., Peppas, N.A. and Bodmeier, R. (2000) Calculation of the required size and shape of hydroxypropyl methylcellulose matrices to achieve desired drug release profiles. Int J Pharm, 201, 151–164.

86. Colombo, P., Conte, U., Gazzaniga, A., Maggi, L., Sangalli, M.E., Peppas, N.A. and Manna, A.L. (1990) Drug release modulation by physical restrictions of matrix swelling. Int J Pharm, 63, 43–48.
87. Colombo, P., Catellani, P., Peppas, N.A., Maggi, L. and Conte, U. (1992) Swellling characteristics of hydrophilic matrices for controlled release: new dimensionless number to describe the swelling and release behavior. Int J Pharm, 88, 99–109.
88. Losi, E., Bettini, R., Santi, P., Sonvico, F., Colombo, G., Lofthus, K., Colombo, P. and Peppas, N.A. (2006) Assemblage of novel release modules for the development of adaptable drug delivery systems. J Control Release, 111, 212–218.
89. Levina, M., Wan, P., Jordan, M. Rajabi-Siahboomi, A.R. (2003) The influence of film coatings on performance of hypromellose matrices. http://www.colorcon.com/pharma/mod_rel/methocel/literature/infl_fc_hypromel.pdf.
90. Vuong, H., Levina, M. and Rajabi-Siahboomi, A.R. (2006) The effect of film coating and storage conditions on the performance of metformin HCl 500 mg extended release hypromellose matrices. http://www.colorcon.com/pharma/mod_rel/methocel/literature/fc+stor_perf_metformin.pdf.
91. Tiwari, S.B., Murthy, T.K., Pai, M.R., Mchta, P.R. and Chowdary, P.B. (2003) Controlled release formulation of tramadol hydrochloride using hydrophilic and hydrophobic matrix system. AAPS PharmSciTech, 4, Article 31.
92. Dias, V.D., Gothoskar, A.V., Fegely, K. and Rajabi-Siahboomi, A.R. (2006) Modulation of drug release from hypromellose (HPMC) matrices: suppression of the initial burst effect. http://www.colorcon.com/pharma/mod_rel/methocel/literature/drug_rel_hpmc_burst_suppression.pdf.
93. Takka, S., Rajbhandari, S. and Sakr, A. (2001) Effect of ionic polymers on the release of propranolol hydrochloride from matrix tablets. Eur J Pharm Biopharm, 52, 75–82.
94. Melia, C.D. (1991) Hydrophilic matrix sustained release systems based on polysaccharide carriers. Crit Rev Ther Drug Carrier Syst, 8, 395–421.
95. Bonferoni, M.C., Rossi, S., Ferrari, F., Stavik, E., Pena-Romero, A. and Caramella, C. (2000) Factorial analysis of the influence of dissolution medium on drug release from carrageenan-diltiazem complexes. AAPS PharmSciTech, 1, E15.
96. Sinha, V.R., Singh, A., Singh, S. and Bhinge, J.R. (2007) Compression coated systems for colonic delivery of 5-flurouracil. J Pharm Pharmacol, 59, 359–365.
97. HyperStart® formulation service. http://www.colorcon.com/pharma/mod_rel/methocel/hyperstart_text.html.
98. Fegely, K., Scattergood, L. and Rajabi-Siahboomi, A. (2005) Development of verapamil (240 mg) extended release formulation using Methocel hydrophilic matrices [Application data sheet]. Colorcon Inc., West Point, PA.
99. Isoptin SR, Knoll, verapamil hydrochloride, antihypertensive agent. http://www.rxmed.com/b.main/b2.pharmaceutical/b2.1.monographs/CPS-%20Monographs/CPS-%20(General%20Monographs-%20I)/ISOPTIN%20SR.html.
100. Covera HS (verapamil hydrochloride) extended release tablets, controlled-onset. http://www.pfizer.com/pfizer/download/uspi_covera.pdf.
101. Palmer, F., Levina, M. and Rajabi-Siahboomi, A. (2005) Development of carbamazepine (200 mg) extended release formulation using Methocel hydrophilic matrices [Application data sheet]. Colorcon Inc., West Point, PA.
102. Owen, R.T. (2006) Extended-release carbamazepine for acute bipolar mania: a review. Drugs Today (Barc), 42, 283–289.
103. Palmer, F., Levina, M. and Rajabi-Siahboomi, A.R. (2005) Investigation of a directly compressible metformin HCl 500 mg extended release formulation based on hypromellose. http://www.colorcon.com/pharma/mod_rel/methocel/literature/metformin_500mg.pdf.

Index

Printed in the United States of America.